Cambridge
IGCSE®

Geography

Second Edition

Paul Guinness
Garrett Nagle

HODDER
EDUCATION
AN HACHETTE UK COMPANY

® IGCSE is the registered trademark of Cambridge International Examinations. The questions, example answers, marks awarded and/or comments that appear in this book/CD were written by the authors. In examination the way marks would be awarded to answers like these may be different.

Although every effort has been made to ensure that website addresses are correct at time of going to press, Hodder Education cannot be held responsible for the content of any website mentioned in this book. It is sometimes possible to find a relocated web page by typing in the address of the home page for a website in the URL window of your browser.

Hachette UK's policy is to use papers that are natural, renewable and recyclable products and made from wood grown in sustainable forests. The logging and manufacturing processes are expected to conform to the environmental regulations of the country of origin.

Orders: please contact Bookpoint Ltd, 130 Milton Park, Abingdon, Oxon OX14 4SB. Telephone: (44) 01235 827720. Fax: (44) 01235 400401. Lines are open 9.00–5.00, Monday to Saturday, with a 24-hour message answering service. Visit our website at www.hoddereducation.com

© Paul Guinness and Garrett Nagle 2009

First published in 2009 by

Hodder Education

Carmelite House,

50 Victoria Embankment,

London EC4Y 0DZ

This second edition published 2014

Impression number   4

Year   2016

All rights reserved. Apart from any use permitted under UK copyright law, no part of this publication may be reproduced or transmitted in any form or by any means, electronic or mechanical, including photocopying and recording, or held within any information storage and retrieval system, without permission in writing from the publisher or under licence from the Copyright Licensing Agency Limited. Further details of such licences (for reprographic reproduction) may be obtained from the Copyright Licensing Agency Limited, Saffron House, 6–10 Kirby Street, London EC1N 8TS.

Cover photo © arquiplay77 - Fotolia

Illustrations by Art Construction, Barking Dog Art, Integra Software Services Pvt. Ltd., and Philip Allan updates.

Typeset in 11/13pt ITC Galliard Std and produced by Integra Software Services Pvt. Ltd., Pondicherry, India

Printed and bound in Dubai

A catalogue record for this title is available from the British Library

ISBN 978 1471 807 275

Contents

WORLD MAP

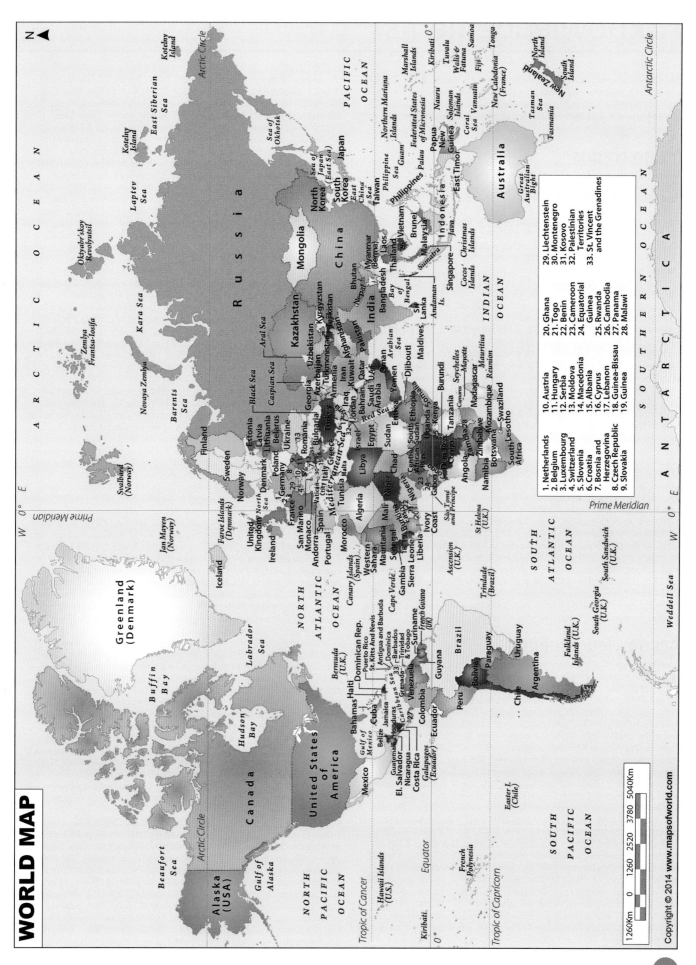

N

1. Netherlands	10. Austria	20. Ghana	29. Liechtenstein
2. Belgium	11. Hungary	21. Togo	30. Montenegro
3. Luxembourg	12. Serbia	22. Benin	31. Kosovo
4. Switzerland	13. Moldova	23. Cameroon	32. Palestinian
5. Slovenia	14. Macedonia	24. Equatorial	Territories
6. Croatia	15. Albania	Guinea	33. St. Vincent
7. Bosnia and	16. Cyprus	25. Rwanda	and the Grenadines
Herzegovina	17. Lebanon	26. Cambodia	
8. Czech Republic	18. Guinea-Bissau	27. Panama	
9. Slovakia	19. Guinea	28. Malawi	

Copyright © 2014 www.mapsofworld.com

| 0 | 1260 | 2520 | 3780 | 5040Km |

1260Km

Acknowledgements

Garrett Nagle would like to dedicate this book to Angela, Rosie, Patrick and Bethany – for their help and patience in the production of the book. Paul Guinness would like to dedicate this book to Mary.
The Publishers would like to thank the following for permission to reproduce copyright material:

Photo credits

Andrew Davis: p.155
Chris Guinness: p.10, p.185 *t*, p.216, p.225, p.234 *tl* and *tr*, p.239
Paul Guinness: p.1, p.2, p.6 *all*, p.8, p.12, p.15, p.17, p.19, p.21, p.22, p.24, p.27, p.31 *all*, p.34, p.161, p.162 *l*, p.166, p.167, p.170 *l&r*, p.171, p.173 *t,c&bl*, p.175, p.177 *all*, p.182, p.183 *all*, p.184, p.185 *b*, p.190 *t*, p.191, p.193, p.196, p.198 *r*, p.200, p.201 *all*, p.202, p.203 *all*, p.207 all, p.208, p.212, p.213 *tr&cr*, p.214, p.219 *all*, p.220, p.222, p.228, p.230, 231 *all*, p.237, p.238, p.240, p.266, p.267, p.278, p.298 *l*
Garrett Nagle: p.35 *all*, p.36 *all*, p.38, p.39, p.40 *all*, p.43, p.50 *all*, p.51 *all*, p.55, p.58, p.60, p.62, p.65, p.66 *all*, p.73, p.74, p.79, p.80, p.81, p.85 *all*, p.86, p.88 *all*, p.89, p.90, p.92, p.93 *all*, p.96, p.104, p.106, p.111 *all*, p.114 *all*, p.118, p.121, p.122 *all*, p.124 *all*, p.127, p.128 *all*, p.129 *all*, p.132, p.136 *all*, p.138 *all*, p.140 *all*, p.143 *all*, p.145 *all*, p.147, p.151 *l&r*, p.152, p.154 *all*, p.156, p.247, p.248, p.249, p.255 *tr, cr&br*, p.257, p.264, p.265 *l*,
p.11 ©Lisa Gogolin – Fotolia; **p.14** © skipaudio/Getty Images/ iStockphoto/Thinkstock; **p.26** © Zuma/Rex Features; **p.30** © Antony SOUTER / Alamy; **p.33** © Hemera Technologies/Getty Images/ Thinkstock; **p.48** *bl* © Laura Wenz; **p.49** © Dan Kitwood/Getty Images; **p.59** ©Trey Campbell/http://www.flickr.com/photos/ treycampbell/4844687638/; **p.61** © Cheong Gy Cheon Museum; **p.68** © David R. Frazier Photolibrary, Inc/Alamy; **p.71** © TonyYao/ Getty Images/iStockphoto/Thinkstock; **p.72** © Peirre Selim/ http://commons.wikimedia.org/wiki/File:Shanghai_-_Pudong_-_ Lujiazui.jpg/http://creativecommons.org/licenses/by/3.0/deed.en (21-Jan-2014); **p.125** © Luis Marden/National Geographic/Getty Images; **p.139** © Matthias Seifert/Reuters/Corbis; **p.144** © Michael Steden/Getty Images/iStockphoto/Thinkstock; **p.153** *l* © Wayne Lynch/All Canada Photos/Corbis; **p.153** *r* © Hagit Berkovich – Fotolia; **p.162** *r* © Rubens Alarcon/Getty Images/Hemera/Thinkstock; **p.173** *br* © Igor Kalamba – Fotolia; **p.181** © Jaguar Land Rover Ltd; **p.187** © Frédéric Soltan/Sygma/Corbis; **p.190** *b* © Antony Njuguna// Reuters/Corbis; **p.206** © AA WorldTravel Library/Alamy; **p.215** © Prill Mediendesign & Fotografie/Getty Images/iStockphoto/Thinkstock; **p.222** © James Randklev/Corbis; **p.224** © Andy Pernick/USBR; **p.234** *br* © James P. Blair/National Geographic/Getty Images; **p.235** © George Esiri//Reuters/Corbis; **p.272** © Chinch Gryniewicz/Ecoscene.
t = top, *b* = bottom, *l* = left, r = right, *c* = centre

Acknowledgements

p.v world map: © 2014 mapsofworld.com **p.2** Figure 2: World population growth by each billion, *World Population Prospects: The 2004 Revision, 2005* (United Nations/Population Reference Bureau,2005/2006), Table 1: World population clock, 2012, *World Population Data Sheet 2012* (United Nations/Population Reference Bureau); **p.3** Figure 3: Population growth in more and less developed countries, 1950–2050, *World Population Data Sheet 2012* (United Nations/Population Reference Bureau); **p.4** Table 2: The world's ten largest countries in terms of population, 2012 and 2050, *World Population Data Sheet 2012* (United Nations/Population Reference Bureau), Table 3: Birth and death rates 2012, *World Population Data Sheet 2012* (United Nations/Population Reference Bureau); **p.5** Figure 6: The demographic transition model, A.Palmer and W. Yates, *Edexcel (A) Advanced Geography* (Philip Allan, 2005); **p.7** Table 4: Countries with the highest and lowest fertility rates, 2012, *World Population Data Sheet 2012* (United Nations/Population Reference Bureau); **p.8** Figure 10: A comparison between female secondary education and total fertility rates (Earth Policy Institute, 2001), reproduced by permission of the publisher, Figure 11: Total fertility rates in industrial, less developed and least developed countries, 1950–2010 (Earth Policy Institute, 2011), reproduced by permission of the publisher, Table 5: Life expectancy at birth, 2012, *World Population Data Sheet 2012* (United Nations/ Population Reference Bureau); **p.9** Figure 14: Projected population change by region, 2005–50, *Population Bulletin*, Vol. 60 No. 4

(Population Reference Bureau, 2005); **p.11** Figure 17: The growth in Kenya's population between 1969 and 2030, *Kenya Population Data Sheet 2011*(United Nations/Population Reference Bureau), Figure 18: Population pyramid for Kenya from *Kenya Population Data Sheet 2011*(United Nations/Population Reference Bureau); **p.12** Figure 20: Russia's population, 1950–2013, *http://en.wikipedia.org/wiki/ Demographics_of_Russia*; **p.13** Figure 21: Optimum population, over-population and under-population, G. Nagle and K. Spencer, *Advanced Geography: Revision Handbook* (Oxford University Press, 1996), reprinted by permission of Oxford University Press; **p.15** Figure 25: Population density map of Australia, *Regional Population Growth, Australia (3218.0)*, Australian Bureau of Statistics, © Commonwealth of Australia; **p.16** Table 7: Comparing Bangladesh and Australia, *World Population Data Sheet 2012* (United Nations/Population Reference Bureau); **p.18** Figure 28: Population growth in France, 2004–13, *www. tradingeconomics.com*; **p.19** Figure 1: Push and pull factors, G. Nagle and K. Spencer, *Advanced Geography: Revision Handbook* (Oxford University Press, 1996), reprinted by permission of Oxford University Press, Figure 2: Types of migration and barriers to migration, G. Nagle and K. Spencer, *Advanced Geography: Revision Handbook* (Oxford University Press, 1996), reprinted by permission of Oxford University Press; **p.20** Figure 3: Refugees and displaced people in the Middle East, *The Sunday Telegraph*, September, 2013; **p.21** Figure 4: International migrant stock by origin and destination, 2010 (United Nations/ Department of Economic and Social Affairs, Population Division, 2012), Figure 5: International migrant stock in the North and South, 1990–2010, (United Nations/Department of Economic and Social Affairs, Population Division, 2012); **p.25** Figure 11: The costs and returns of migration (Pew Research Center), Figure 12: Increase in the Mexican-born population in the USA (Pew Research Center); **p.27** Figure 1: Four population pyramids for 2013 – Niger, Bangladesh, UK, Japan, *CIA World Factbook*; **p.28** Table 1: Population and economic data for the four countries, selected data from *World Population Data Sheet 2012* (United Nations/Population Reference Bureau); **p.29** Figure 2: An annotated population pyramid, M.Harcourt and S.Warren, *Tomorrow's Geography* (Hodder Murray, 2012); **p.31** Figure 1: Dot map showing world population density, P. Guinness and G. Nagle, *AS Geography: Concepts and Cases* (Hodder Murray, 2000); **p.32** Table 1: Variations in world population density, *World Population Data Sheet 2012* (United Nations/Population Reference Bureau); **p.33** Figure 4: Population density of North America, P. Guinness and G.Nagle, *AS Geography: Concepts and Cases* (Hodder Murray, 2000); **p.36** Figure 3: Village Shapes, G. Nagle, *Advanced Geography* (Oxford University Press, 2000), copyright © Garrett Nagle 2000, reprinted by permission of Oxford University Press; **p.38** Figure 6: The relationship between population size and number of services, P. Guinness and G. Nagle, *AS Geography: Concepts and Cases* (Hodder Murray, 2000), Figure 7: Settlement sites in the north-east of the USA, G. Nagle, *Geography Homework Pack for Key Stage 3* (Heinemann, 2000), reproduced by permission of Pearson Education; **p.39** Figure 10: Cloke's model of rural change and accessibility to large urban centres, G. Nagle, *ORG GCSE Geography (Through Diagrams)* (Oxford University Press, 1998), copyright © Garrett Nagle 1998, reprinted by permission of Oxford University Press; **p.41** Figure 15: Key to 1:50000 map of Montego Bay, Jamaica (Government of Jamaica/National Land Department/ Survey Department); **p.42** Figure 14: Map of Montego Bay, Jamaica (Government of Jamaica/National Land Agency/Survey Department); **pp.44** Figure 1: Bid rent theory and urban land use models, G. Nagle, *Thinking Geography* (Hodder Murray, 2000); **p.46** Figure 3: The core frame model, S. Warn, *Managing Change in Human Environments* (Philip Allan Updates, 2001); **p.47** Figure 4: Location of Woodstock and Blikkiesdorp, Cape Town CC-BY-SA-3.0, *http://en.wikipedia.org/wiki/File:2009Blikkiesdorp.JPG*. Released under the GNU Free Documentation License; **p.48** Figure 5: Road network in Woodstock, *Juta's General School Atlas* (Juta Gariep); **pp.54, 55, 56** Figure 12: Land use in New York, *www.nyc.gov/html/ dcp/html/landusefacts.landusefactmaps.html* (New York City/Department of City Planning); **p.57** Figure 13: Land use in Seoul, Young-Han Park et.al., *Atlas of Seoul* (Sung Ji Mun Hwa Co. Ltd, 2000); **p.59** Figure 16: Population change in Detroit, *The Economist*, 27 July, 2013; **p.60** Figure 17: Manufacturing Employment in Detroit, *Federal Reserve economic database, http://acchartfacts.blogspot.co.uk/2010/03/local-charts.html*, © 2009 Chartfacts, www.chartfacts.com; **p.64** Figure 2: Slums in Rio de Janeiro,

GeoFactsheets, 121, reprinted by permission of Curriculum Press, www.curriculum-press.co.uk; **p.69** Figure 7: Transport systems in Rio de Janeiro, *GeoFactsheets, 121*, reprinted by permission of Curriculum Press, www.curriculum-press.co.uk; **p.75** Figure 1: Population by world region 2013 and 2050, Carl Haub and Toshiko Kaneda, *2013 Population Data Sheet* (United Nations/Population Reference Bureau), Figure 2: Main international migration corridors, *http://esa.un.org/unmigration/wallchart2013.htm*, Figure 3: Change in US Population (in thousands) by Age Group, *2012 World Population Data Sheet* (United States Census Bureau/Population Estimates Program); **p.76** Figure 4a: Population density and distribution in Bangladesh, *http://sedac.ciesin.columbia.edu/downloads/maps/grump-v1/grump* (NASA), Figure 4b: Increase in population density in Bangladesh, *http://www.indexmundi.com/facts/bangladesh/population-density*; **p.77** Figure 6: Map HM50CL - Hong Kong Special Administrative Region (Map Publications Centre, Hong Kong/Survey & Mapping Office/Lands Department), © 2011; **p.80** Figure 1: Two types of volcano, *GeoFactsheets, 121*, reprinted by permission of Curriculum Press, www.curriculum-press.co.uk; **p.86** Figure 9: Distribution of plates and tectonic hazards in the Caribbean, *Philip's Certificate Atlas for the Caribbean*, 5th edition (George Philip Maps, 2004); **p.90** Figure 13: Buildings designed for earthquakes, G. Nagle, *Focus Geography: Hazards* (Nelson Thornes, 1998), copyright © Garrett Nagle 1998, reprinted by permission of Oxford University Press; **p.92** Figure 16:1:25,000 map of Soufrière and Plymouth (Directorate of Overseas Surveys/Department for International Development), © Crown copyright; **p.96** Figure 1: Changes in a river downstream, G. Nagle, *AS & A2 Geography for Edexcel B* (Oxford University Press, 2003), copyright © Garrett Nagle 2003, reprinted by permission of Oxford University Press; **p.99** Figure 6: Rates of actual and potential evapotranspiration for South Africa, *https://www.google.com*; **p.101** Figure 9: Groundwater, P. Guinness and G. Nagle, *AS Geography: Concepts and Cases* (Hodder Murray, 2000); **p.102** Figure 10: Types of transport in a river, P Guinness and G. Nagle, *Advanced Geography: Concepts and Cases* (Hodder & Stoughton, 2002); **p.103** Figure 13: Formation of a waterfall, G. Nagle, *Rivers and Water Management* (Hodder Arnold, 2003); **p.105** Figure 16: 1:25000 map of the Niagara Falls area (Government of Canada/Canada Centre for Mapping and Earth Observation/Natural Resources Canada), © Crown copyright; **p.107** Figure 20: Formation of a floodplain and terraces, G. Nagle, *ORG GCSE Geography (Through Diagrams)* (Oxford University Press, 1998), copyright © Garrett Nagle 1998, reprinted by permission of Oxford University Press; **p.109** Figure 23:1:250,000 map extract of the Rhône delta, *IGN map Provence-Alpes-Cote d'Azure 2013*, © IGN France 2012, reproduced by permission of Institut National de L'information Geographique et Forestiere; **p.120** Figure 4: Human activity and longshore drift in West Africa, G. Nagle, *AS & A2 Geography for Edexcel B* (Oxford University Press, 2003) copyright © Garrett Nagle 2003, reprinted by permission of Oxford University Press; **p.121** Figure 5: Features of coastal erosion., G. Nagle, *ORG GCSE Geography (Through Diagrams)* (Oxford University Press, 1998) copyright © Garrett Nagle 1998, reprinted by permission of Oxford University Press; **p.121** Figure 6: The Cape Peninsula, South Africa (Republic of South Africa/Department of Land Affairs/Surveys and Mapping), Figure 7: Formation of a wave-cut platform, G. Nagle, *ORG GCSE Geography (Through Diagrams)* (Oxford University Press, 1998), copyright © Garrett Nagle 1998, reprinted by permission of Oxford University Press; **p.124** Figure 13: The west coast of Antigua, *www.itmb.com*, reproduced by permission of ITMB Publishing; **p.126** Figure 15: Formation of sand dunes, P.Guinness and G. Nagle, *AS Geography: Concepts and Cases* (Hodder Murray, 2000); **p.127** Figure 17: Formation of coral reefs, adapted from S. Warn and C. Roberts, *Coral Reefs: Ecosystem in Crisis?* (Field Studies Council, 2001); **p.133** Figure 24: The path of Typhoon Haiyan, www.channelnewsasia.com; **p.141** Figure 2: Equipment in a weather station, G. Nagle, *Weatherfile GCSE* (Nelson Thornes, 2000), copyright © Garratt Nagle 2000, reprinted by permission of Oxford University Press; **p.145** Figure 9: Cloud types, G. Nagle, *Weatherfile GCSE* (Nelson Thornes, 2000), copyright © Garrett Nagle 2000, reprinted by permission of Oxford University Press; **p.151** Figure 5: Conditions required for the growth of rainforests and hot deserts, G. Nagle, *ORG GCSE Geography (Through Diagrams)* (Oxford University Press, 1998), copyright © Garrett Nagle 1998, reprinted by permission of Oxford University Press; **p.159** Figure 2: extract from 1:55 000 map of Arthur's Pass, New Zealand, *www.newtopo.co.nz*, April, 2013, © NewTopo, reprinted by permission of Geoff Aitken; **p.163** Figure 2: World map showing GNP per capita in 2013, *CIA World Factbook*; **p.163** Table 1: Top 12 and bottom 12 countries in GNP per capita ($), 2013, *CIA World Factbook*; **p.164** Figure 3: The components of the Human Development Index, *http://hdr.undp.org/en/statistics/hdi/*; **p.164** Table 2:

Infant mortality rate by world region, 2012, *World Population Data Sheet, 2012* (United Nations/Population Reference Bureau); **p.166** Figure 6: Stages of Development, Paul Guinness, *Geography for the IB Diploma: Patterns and Change* (Cambridge University Press, 2010), © Cambridge University Press 2010, reprinted with permission; **p.167** Figure 7: Fast and slow development in developing countries, Paul Guinness, *Geography for the IB Diploma: Patterns and Change* (Cambridge University Press, 2010), © Cambridge University Press 2010, reprinted with permission; **p.168** Figure 9: World map showing variations in Gini coefficient 2009, *http://commons.wikimedia.org/wiki/File:Gini_Coefficient_World_CIA_Report_2009-1.png*; **p.169** Figure 10: Simplified model of cumulative causation, Paul Guinness, *Geography for the IB Diploma: Patterns and Change* (Cambridge University Press, 2010), © Cambridge University Press 2010, reprinted with permission, Figure 11: Regional economic divergence and convergence, Paul Guinness, *Geography for the IB Diploma: Patterns and Change* (Cambridge University Press, 2010), © Cambridge University Press 2010, reprinted with permission; **p.171** Figure 15: Southeast Brazil, Paul Guinness, *Geography for the IB Diploma: Patterns and Change* (Cambridge University Press, 2010), © Cambridge University Press 2010, reprinted with permission; **p.172** Figure 16: Core/periphery diagrams, Paul Guinness, *Geography for the IB Diploma: Patterns and Change* (Cambridge University Press, 2010), © Cambridge University Press 2010, reprinted with permission; **p.174** Table 5: Employment structure of a developed country, an NIC and a developing country, *CIA World Factbook*; **p.176** Table 8: The potential advantages and disadvanages of TNCs- Nike to the USA and Vietnam, Paul Guinness, *Geography for the IB Diploma: Global Interactions* (Cambridge University Press, 2011), © Cambridge University Press 2011, reprinted with permission; **p.177** Figure 25: Influences on the globalisation of economic activity, G. Nagle and Paul Guinness, *Cambridge International A and AS Level Geography* (Philip Allan Updates, 2011); **p.178** Figure 27: The global share of GDP, Paul Guinness, *Geography for the IB Diploma: Global Interactions* (Cambridge University Press, 2011), © Cambridge University Press 2011, reprinted with permission; **p.179** Figure 28: Aspects of global urban uniformity, Paul Guinness, *Geography for the IB Diploma: Global Interactions* (Cambridge University Press, 2011), © Cambridge University Press 2011, reprinted with permission; **p.180** Figure 29: The shift of power from nation states, Paul Guinness, *Geography for the IB Diploma: Global Interactions* (Cambridge University Press, 2011), © Cambridge University Press 2011, reprinted with permission, Table 10: The costs and benefits of globalisation to the UK, Paul Guinness, *Geography for the IB Diploma: Global Interactions* (Cambridge University Press, 2011), © Cambridge University Press 2011, reprinted with permission; **p.183** Figure 3: Farming types and levels of development, D.Waugh, *Geography: An Integrated Approach*, 1st Edition (Nelson Thornes, 1990); **p.185** Table 1: Types of irrigation, 'The Water Crisis: A Matter of Life and Death', *Understanding Global Issues*, **p.187** Figure 8: The Lower Ganges valley, D. Waugh, *The New Wider World*, 3rd Edition (Nelson Thornes, 2003), Figure 9: Climate graph for Kolkata, D. Waugh, *The New Wider World*, 3rd Edition (Nelson Thornes, 2003); **p.189** Figure 12: Summary of causes of famine in the Sudan, *GeoFactsheet, 185*, reprinted by permission of Curriculum Press, www.curriculum-press.co.uk; **p.193** Figure 1: Industrial systems diagram, D.Waugh, *The New Wider World*, 3rd Edition (Nelson Thornes, 2003); **p.197** Figure 3: Location of Bangalore (CE Info Systems), © 2008; **p.198** Figure 1: Growth in global tourism, *http://www.coolgeography.co.uk/GCSE/AQA/Tourism/Tourism%20growth/Tourism%20Growth.htm*; **p.199** Figure 3: International tourist arrivals and international tourism receipts, *http://mkt.unwto.org/webform/contact-tourism-trends-and-marketing-strategies-unwto* (UN/World Tourism Organization; Tourism Highlights, 2013 Edition); **p.200** Figure 4: Inbound tourism by purpose of visit, 2012 (UN/World Tourism Organization, UNWTO), Figure 6: The direct and indirect economic impact of the tourist industry, (Global Insight: Tourism Satellite Accounting); Figure 15: Ecotourism in Ecuador's rainforest from *GeoFactsheet,201*, reprinted by permission of Curriculum Press, www.curriculum-press.co.uk; **p.205** Figure 16: Jamaica's National Parks, Jane Dove et.al., *OCR AS Geography* (Heinemann Educational, 2008), reproduced by permission of Pearson Education; **p.207** Figure 1: Changes in world energy consumption by type, 1987–2012, *BP Statistical Review of World Energy*, June 2013; **p.208** Figure 3: Regional energy consumption patterns, 2012, *BP Statistical Review of World Energy*, June 2013; **p.209** Figure 5: Energy consumption per capita, 2012, *BP Statistical Review of World Energy*, June 2013; **p.210** Figure 7: Electricity access deficit, 2010, Emily Badger, *'Where a billion people live without electricity'*, www.theatlanticcities.com; **p.212** Figure 8: Renewable energy consumption by world region, 1992–2012, *BP Statistical Review of World Energy*, June 2013;

p.217 Figure 1: Alternative water supply and management methods, *Alternative Water Supply and Management Methods* (Government of Alberta/Environment and Sustainable Resource Development); **p.218** Figure 3: How cloud seeding works, *Alternative Water Supply and Management Methods* (Government of Alberta/Environment and Sustainable Resource Development); **p.219** Figure 4: Water used for agriculture, industry and domestic for the developed and developing worlds, Paul Guinness, *Geography for the IB Diploma: Patterns and Change* (Cambridge University Press, 2010), © Cambridge University Press 2010, reprinted with permission; **p.220** Figure 8: Physical water scarcity and economic water scarcity worldwide, Paul Guinness, *Geography for the IB Diploma: Patterns and Change* (Cambridge University Press, 2010), © Cambridge University Press 2010, reprinted with permission; **p.222** Table 1: Water investment needs by area, 2005-30, Paul Guinness, *Geography for the IB Diploma: Patterns and Change* (Cambridge University Press, 2010), © Cambridge University Press 2010, reprinted with permission; **p.225** Figure 1: How exposure to pollution can affect human health, P. Byrne et.al., *Edexcel A2 Geography* (Philip Allan Updates); **pp.226-7** Table 2: Major air pollutants, adapted (Environmental Protection Agency); **p.228** Figure 4: Heathrow airport with surrounding noise levels (DEFRA), © Crown Copyright; **p.230** Figure 5: The greenhouse effect, M. Raw, *OCR A2 Geography* (Philip Allan Updates, 2009); **p.232** Figure 9: Causes of land degradation, Paul Guinness, *Geography for the IB Diploma: Patterns and Change* (Cambridge University Press, 2010), © Cambridge University Press 2010, reprinted with permission; **p.233** Figure 10: The causes and process of local soil degradation, Paul Guinness, *Geography for the IB Diploma: Patterns and Change* (Cambridge University Press, 2010), © Cambridge University Press 2010, reprinted with permission; **p.235** Figure 14: The Great Barrier Reef, Australia, *www.great-barrier-reef.biz/Images/100050.gif*; **p.237** Figure 17: The role of ecotourism in rainforest conservation, J. Hill, W. Woodland, R. Hill, *Geography Review*, May, 2007; **p.241** Figure 22: The Pearl river delta, *Financial Times*, 4 February, 2003; **p.242** Emily Badger, *'Where a billion people still live without electricity'*, *http://www.theatlanticcities.com/jobs-and-economy/2013/06/where-billion-people-still-live-without-electricity/5807/*; **p.243** Table 1: Adverse influences on global food production and distribution, Paul Guinness and Brenda Walpole, *Environmental Systems and Societies for the IB Diploma* (Cambridge University Press, 2012), © Cambridge University Press 2012, reprinted with permission; **p.244** Figure 4: Tourism key to development, prosperity and well-being (UN/World Tourism Organization; Tourism Highlights, 2013 Edition); **p.245** Figure 5: Global Nuclear Electricity Production, *http://www.world-nuclear.org/info/Current-and-Future-Generation/Nuclear-Power-in-the-World-Today/*; **p.245** Figure 6: Billions daily affected by water crisis, *http://water.org/water-crisis/one-billion-affected/*; **p.246** Figure 7: Comparing the efficiency of energy conversion in arable and pastoral farming, Paul Guinness and Brenda Walpole, *Environmental Systems and Societies for the IB Diploma* (Cambridge University Press, 2012),

© Cambridge University Press 2012, reprinted with permission, Figure 8: Environmental impact of the increasing demand for meat, Paul Guinness and Brenda Walpole, *Environmental Systems and Societies for the IB Diploma* (Cambridge University Press, 2012), © Cambridge University Press 2012, reprinted with permission; **p.248** Figure 1: Part of the 1:50,000 map of Jamaica (Directorate of Overseas Surveys/Department for International Development), © Crown copyright; **p.250** Figure 4: Part of the 1:25¦000 map of Marmaloda Glacier and Lago di Fadala, Tabacco sheet 06, *www.tabaccoeditrice.com*; **p.251** Figure 5: Part of the 125,000 Ordnance Survey map of northern Montserrat, © Crown Copyright; **p.252** Figure 6: Drawing a cross-section, 1:25,000 section of French IGN Top 75 Tourism et Randonnée map Chaine des Puys Massif du Sancy, *www.ign.fr*, reproduced by permission of Institut National de L'information Geographique et Forestiere; **p.253** Figure 8: Features associated with different stages of a river, G. Nagle and K. Spencer, *Geographical Enquiries* (Stanley Thornes, 1997); **p.254** Figure 9: Part of the West Coast Trail in British Columbia, Canada, *www.itmb.com*, reproduced by permission of ITMB Publishing; **p.255** Figure 10: Extracts from the 1:35,000 map of Antigua, *www.itmb.com*, reproduced by permission of ITMB Publishing; **p.258** Figure 14: Part of the 1:50000 map of Tenerife, Canary Islands, *www.theAA.com/travel*, © KOMPASS-Karten Gmbh; **p.260** Figure: 15:50000 map of St Catharines, Ontario, Canada (Government of Canada/Canada Centre for Mapping and Earth Observation/Natural Resources Canada), © Crown copyright; **p.261** Figure 16: Map of Hong Kong Special Administrative Region (Map Publications Centre, Hong Kong/Survey & Mapping Office/Lands Department), © 2011, Figure 17: Example of the use of pie charts: employment and gross regional domestic product (GRDP) in South Korea, 2000, *Korea Statistical Yearbook 2000*; **p.270** Figure 5: Two questionnaires, one good and one bad, B.Lenon and P.Cleves, *Fieldwork Techniques and Projects in Geography* (Collins Educational, 2001); **p.275** Figure 13: Example of a histogram, B.Lenon and P.Cleves, *Fieldwork Techniques and Projects in Geography* (Collins Educational, 2001); **p.276** Figure 15: Example of a wind rose diagram, B.Lenon and P.Cleves, *Fieldwork Techniques and Projects in Geography* (Collins Educational, 2001); **p.277** Figure 18: An isoline map, *Wideworld*, November 2002 (Philip Allan Updates, 2002); **p.279** Figure 19: Conducting a survey of a sand dune ecosystem, *Wideworld*, September 2002 (Philip Allan Updates, 2002); **p.281** Figure 20: Taking river measurements, David Holmes and Sue Warn, *Fieldwork Investigations: a self-study guide* (Hodder Education, 2003), Figure 21: A river cross-section, David Holmes and Sue Warn, *Fieldwork Investigations: a self-study guide* (Hodder Education, 2003)

Permission for re-use of all © Crown copyright information is granted under the terms of the Open Government Licence (OGL).

Every effort has been made to trace all copyright holders, but if any have been indavertently overlooked the Publishers will be pleased to make the necessary arrangements at the first opportunity.

Introduction

This book has been written to help you while you study for your geography IGCSE. The examples and case studies in the book are from around the world. Geography is about people and places and we hope that you will use your own home area as much as possible to add to the material in this book. We would encourage you also to keep up to date with geographical events – one way is through listening to the news or reading about events in newspapers or on the Internet. Geography is happening every day, everywhere and examiners love to read about new developments – so think about your own geographical location and new geographical events.

This book has been written to follow closely the IGCSE specification. It includes a number of activities to help you succeed with the written assessment and guidance for your coursework. Below are details of the exams and assessment that you will experience. Be prepared – knowing what to expect will help you succeed in your exams. Make sure you also use your teachers' experience – they are an excellent resource waiting to be tapped. Good luck and enjoy your geographical studies.

● Assessment

Scheme of assessment

All candidates will take Paper 1, Paper 2 and either Paper 3 or Paper 4.

Papers 1, 2 and 4 will consist of combined question papers and answer booklets where candidates answer in the spaces provided.

Paper 1 (1 hour 45 minutes) Candidates will be required to answer three questions (3 × 25 marks). They will choose one question out of two on each theme. Questions will be structured with gradients of difficulty, will be resource-based and involve problem solving and free response writing. This paper will mainly be concerned with Assessment Objectives 1, 2, and 3, Knowledge with understanding, Skills and analysis, and Judgement and decision making. 45% of total marks.

Paper 2 (1 hour 30 minutes) (60 marks) Candidates answer all the questions. The paper is based on testing the interpretation and analysis of geographical information, decision making and the application of graphical and other techniques as appropriate. The questions will not require specific information about places but will require the use of a 1:25 000 or 1:50 000 topographical map and will include a full key. 27.5% of total marks.

Either

Paper 3, Coursework (School-based assessment). Teachers set one school-based assignment of up to 2000 words. (60 marks)

Or

Paper 4, Alternative to Coursework (1 hour 30 minutes) (60 marks) Candidates answer two compulsory questions, completing a series of written tasks based on the three themes:

1 Population and Settlement
2 The Natural Environment
3 Economic Development

The questions involve an appreciation of a range of techniques used in fieldwork studies. 27.5% of total marks.

● IGCSE Geography Revision CD Rom

The accompanying CD Rom provides invaluable revision materials and self-testing.

- Definitions of all key terms are provided.
- Topic summaries are provided, enabling quick revision of a topic.
- Multiple choice, mix and match and true or false interactive questions are provided to test yourself on key terms and geographic information.
- Images include selected artwork and photos to help you with your studies and project work.

Paul Guinness
Garrett Nagle

Theme 1 · Population and Settlement

Topics

Population dynamics

People from around the world watching the Olympic Games in London, 2012

Key questions

- How rapidly has the world's population increased?
- What are the reasons for such a rapid increase in the world's population?
- What are the causes of a change in population size?
- What are the reasons for contrasting rates of natural population change?
- What are the causes and consequences of over-population and under-population?
- How effective are population policies in achieving their objectives?

● The rapid increase in the world's population

During most of the early period in which humankind first evolved, global population was very low, reaching perhaps some 125 000 people a million years ago. Ten thousand years ago, when people first began to domesticate animals and cultivate crops, world population was no more than 5 million. Known as the Neolithic Revolution, this period of economic change significantly altered the relationship between people and their environments. But even then the average annual growth rate was less than 0.1 per cent per year.

However, as a result of technological advance the **carrying capacity** of the land improved and population increased. The carrying capacity is the largest population that the resources of a given environment can support. By 3500 BCE, global population reached 30 million and by 2000 years ago, this had risen to about 250 million (Figure 1).

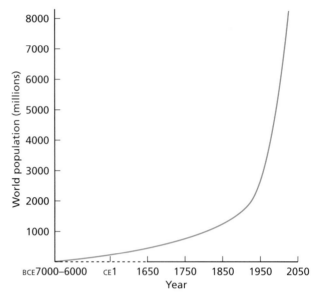

Figure 1 World population growth

Number of years to add each billion (year)

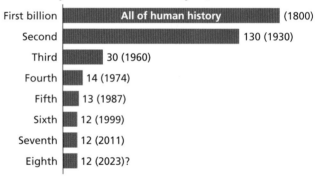

Figure 2 World population growth by each billion

Table 1 World population clock, 2012

Natural increase per …	World	More developed countries	Less developed countries	Less developed countries (excl. China)
Year	84 303 942	1 752 056	82 571 886	76 103 575
Day	230 970	4 745	226 224	208 503
Minute	160	3	157	145

Demographers (people who study human populations) estimate that world population reached 500 million by about 1650. From this time population grew at an increasing rate. By 1800 global population had doubled to reach one billion. Figure 2 shows the time taken for each subsequent billion to be reached, with the global total reaching 7 billion in 2011. It had taken only 12 years for world population to increase from 6 to 7 billion, the same timespan required for the previous billion to be added. It has been estimated that world population will reach 8 billion in 2023.

Table 1 shows population change in 2012, with a global population increase of 84.3 million in that year. This is the result of 140.5 million births and 56.2 million deaths. The bulk of this population increase is in the developing countries. When the number of births exceeds the number of deaths, world population increases. The greater the gap between the number of births and deaths, the greater the population increase. The very rapid growth of the world's population over the last 60 years or so, illustrated by Figures 1 and 2, is the result of the largest ever difference between the number of births and deaths in the world as a whole!

Recent demographic change

Figure 3 shows that both total population and the rate of population growth are much higher in the less developed world than in the more developed world. The fastest rate of growth is taking place in the least developed countries, which is the poorest sub-section of the less developed world. However, only since the Second World War has population growth in the poor countries overtaken that in the rich. The rich countries had their period of high population growth in the nineteenth and early twentieth centuries, while for the less developed countries rapid population growth has occurred since about 1950.

The highest ever global population growth rate was reached in the early to mid 1960s when population growth in the less developed world peaked at 2.4 per cent a year. At this time the term **population explosion** was widely used to describe this rapid population growth. But by the late 1990s the rate of global population growth was down to 1.8 per cent and by 2012 it had reduced further to 1.2 per cent. However, even though the rate of growth has been falling for three decades, **demographic momentum** means that the number of people added each year remains very high. This is because there are so many women in the child-bearing age range.

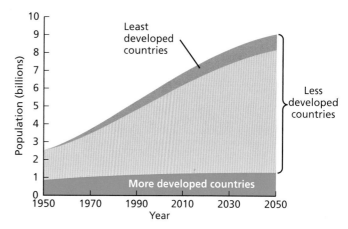

Figure 3 Population growth in more and less developed countries, 1950–2050

The demographic transformation, which took a century to complete in the developed world, has occurred in a single generation in some less developed countries. Fertility has dropped further and faster than most demographers predicted 20 or 30 years ago. Except in Africa, where in around 25 countries families of at least five children are the average and population growth is still over 2.5 per cent per year, birth rates are now declining in virtually every country. According to the Population Reference Bureau: 'Developed countries as a whole will experience little or no population growth in this century, and much of that growth will be from immigration from less developed countries.'

Table 2 shows the ten largest countries in the world in population size in 2012, and their population projections for 2050. In 2012, China and India together accounted for 37 per cent of the world's population. The USA is a long way behind, in third place. While three developed countries were in the top ten in 2012, only one, the USA, is in the forecast for 2050.

Interesting note

The Population Reference Bureau estimates that throughout the history of human population about 108 billion people have lived on Earth. This means that about 6.5 per cent of all people ever born are alive today!

Table 2 The world's ten largest countries in terms of population, 2012 and 2050

2012		2050	
Country	Population (millions)	Country	Population (millions)
China	1350	India	1691
India	1260	China	1311
USA	314	USA	423
Indonesia	241	Nigeria	402
Brazil	194	Pakistan	314
Pakistan	180	Indonesia	309
Nigeria	170	Bangladesh	226
Bangladesh	153	Brazil	213
Russia	143	Congo, Dem. Rep.	194
Japan	128	Ethiopia	166

Activities

1 With the help of Figures 1 and 2, briefly describe the growth of human population over time.
2 Define the term carrying capacity.
3 Comment on the information shown in Table 1.
4 Look at Figure 3. Describe the differences in population growth and projected growth in more developed and less developed countries between 1950 and 2050.
5 Look at Table 2:
   a Show the data for 2012 on an outline map of the world.
   b Briefly describe the changes that are forecast to occur by 2050.

● The causes of a change in population size

The **birth rate** is defined as the number of births per thousand population in a year. If the birth rate of a country is 20/1000 (20 per 1000), this means that on average for every 1000 people in this country 20 births will occur in a year. The **death rate** is the number of deaths per thousand population in a year. If the death rate for the same country is 8/1000, it means that on average for every 1000 people 8 deaths will occur. The difference between the birth rate and the death rate is the **rate of natural change**. If it is positive it is termed **natural increase**. If it is negative it is known as **natural decrease**. In the case given above there is a natural increase of 12/1000 (20/1000 – 8/1000). This is the current rate of

natural increase for the world as a whole – look at the birth and death rates given in Table 3. The rate of natural change may also be shown as a percentage, so in this example 12/1000 is equivalent to 1.2 per cent. Table 3 shows how much birth and death rates vary by world region.

Table 3 Birth and death rates 2012

Region	Birth rate	Death rate
World	20	8
More developed world	11	10
Less developed world	22	8
Africa	36	11
Asia	18	7
Latin America/Caribbean	19	6
North America	13	8
Oceania	18	7
Europe	11	11

Population change in a country is affected by (a) the difference between births and deaths (natural change) and (b) the balance between immigration and emigration (net migration). On Figure 4 the dividing line indicates that the relative contributions of natural change and net migration can vary over time. For most countries natural change is a more important factor in population change than net migration.

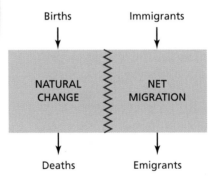

Figure 4 Input-output model of population change

The **immigration rate** is the number of immigrants per thousand population entering a receiving country in a year. The **emigration rate** is the number of emigrants per thousand population leaving a country of origin in a year. The **rate of net migration** is the difference between the rates of immigration and emigration. Figure 5 shows some simple demographic calculations for the imaginary island of Pacifica.

Population at beginning of year: 5000

Population change during the year:
Births: 150   Deaths: 60
Immigrants: 20   Emigrants: 10

Rates of change based on data above
Birth rate: 30/1000   Death rate: 12/1000
Rate of natural change: +18/1000
Immigration rate: 4/1000   Emigration rate: 2/1000
Rate of net migration: +2/1000

Total population at end of the year
= 5100 (natural change of 90 + 10 for net migration)

Figure 5 Pacifica diagram and calculations

Activities

1 Define (a) the birth rate (b) the death rate (c) the rate of natural change.
2 What is net migration?
3 Look at Table 3. Calculate the rate of natural change for each region.
4 Look at Figure 5. Imagine that the population of the island at the beginning of the year was 4000 rather than 5000. Calculate the rates of change for this new starting population figure.

The demographic transition model

The **demographic transition model** helps to explain the causes of a change in population size. Although the populations of no two countries have changed in exactly the same way, some broad generalisations can be made about population growth since the middle of the eighteenth century. These trends are illustrated by the demographic transition model (Figure 6). A model is a simplification of reality, helping us to understand the most important aspects of a process. Demographic transition is the historical shift of birth and death rates from high to low levels in a population.

No country as a whole retains the characteristics of stage 1, which only applies to the most remote societies on Earth, such as isolated tribes in New Guinea and the Amazon basin. All the developed countries of the world are now in stage 4 or stage 5. The poorest of the developing countries

are in stage 2. Most developing countries that have undergone significant social and economic advances are in stage 3, while some of the newly industrialised countries such as South Korea and Taiwan have entered stage 4. Stage 5, natural decrease, is mainly confined to eastern and southern Europe at present.

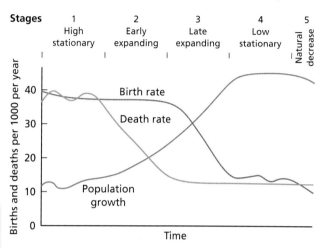

Figure 6 The demographic transition model

- **The High Stationary Stage (stage 1):** The birth rate is high and stable while the death rate is high and fluctuating due to famine, disease and war at particular times. Population growth is very slow and there may be periods of considerable decline. Infant mortality is high and life expectancy low. A high proportion of the population is under the age of 15. Society is pre-industrial with most people living in rural areas, dependent on subsistence agriculture.
- **The Early Expanding Stage (stage 2):** The death rate declines to levels never before experienced. The birth rate remains at its previous level because the **social norms** governing fertility take time to change. As the gap between the two vital rates widens, the rate of natural change increases to a peak at the end of this stage. The infant mortality rate falls and life expectancy increases. The proportion of the population under 15 increases. The main reasons for the decline in the death rate are: better nutrition; improved public health particularly in terms of clean water supply and efficient sewerage systems; and medical advances. Considerable **rural-to-urban migration** occurs during this stage.
- **The Late Expanding Stage (stage 3):** After a period of time social norms adjust to the lower level of mortality and the birth rate begins to

decline. Urbanisation generally slows and the average age increases. Life expectancy continues to increase and infant mortality to decrease. Countries in this stage usually experience lower death rates than nations in the final stage due to their relatively young population structures.

- **The Low Stationary Stage (stage 4):** Both birth and death rates are low. The former is generally slightly higher, fluctuating somewhat due to changing economic conditions. Population growth is slow. Death rates rise slightly as the average age of the population increases. However, life expectancy still improves as **age-specific mortality rates** continue to fall.
- **The Natural Decrease Stage (stage 5):** In a limited but increasing number of countries, mainly European, the birth rate has fallen below the death rate. In the absence of net migration inflows these populations are declining. Examples of **natural decrease** include Germany, Belarus, Bulgaria and Ukraine.

Figure 7 Young people at a popular meeting place in Ulaanbaatar, Mongolia. The country is in Stage 3 of demographic transition.

Contrasts in demographic transition

There are a number of important differences in the way that developing countries have undergone population change compared with the experiences of most developed nations before them. In the developing world:

- birth rates in stages 1 and 2 were generally higher
- the death rate fell much more steeply
- some countries had much larger base populations and thus the impact of high growth in stage 2 and the early part of stage 3 has been much greater

- for those countries in stage 3 the fall in fertility has also been steeper
- the relationship between population change and economic development has been much weaker.

Activities
1 What is a geographical model (such as the model of demographic transition)?
2 Explain the reasons for declining mortality in stage 2.
3 Why does it take some time before fertility follows the fall in mortality?
4 Suggest why the birth rate is lower than the death rate in some countries (stage 5)?
5 How has demographic transition differed in the more developed world and the less developed world?

● Reasons for contrasting rates of population change

Population change is governed by three factors: fertility, mortality and migration. This section examines the influences on fertility and mortality, while migration is covered in more detail in Topic 1.2.

The factors affecting fertility

Figure 8 A school group visiting a 2000-year-old Roman site in Turkey

The most common measure of fertility is the birth rate, but other more detailed measures are used at a more advanced level of study. One of these measures is the **total fertility rate**, which is illustrated in Figures 10 and 11. The total fertility rate is the average number of children a women has during her lifetime. Table 4 shows the countries with the highest and lowest total fertility rates in 2012.

Table 4 Countries with the highest and lowest fertility rates, 2012

Highest	Total fertility rate	Lowest	Total fertility rate
Niger	7.1	Taiwan	1.1
Somalia	6.4	Latvia	1.1
Burundi	6.4	Singapore	1.2
Mali	6.3	Bosnia-Herzegovina	1.2
Angola	6.3	South Korea	1.2
Congo, Dem.Rep.	6.3	Hungary	1.2
Zambia	6.3	Moldova	1.3
Afghanistan	6.2	Poland	1.3
Uganda	6.2	Romania	1.3
Burkina Faso	6.0	Portugal	1.3

The factors affecting fertility can be grouped into four categories:

● **Demographic:** Other population factors, particularly mortality (death) rates, influence fertility. Where infant mortality is high, it is usual for many children to die before reaching adult life. In such societies, parents often have many children to compensate for these expected deaths. The **infant mortality rate** is the number of deaths of children under one year of age per thousand live births per year. In 2012, the infant mortality rate for the world as a whole was 41/1000, ranging from 5/1000 in Europe to 67/1000 in Africa. It is not just coincidence that the continent with the lowest fertility is Europe and the continent with the highest fertility is Africa. The infant mortality rate is generally regarded as a prime indicator of socio-economic progress. Over the world as a whole infant mortality has declined sharply during the last half century.

● **Social/cultural:** In some societies, particularly in Africa, tradition demands high rates of reproduction. Here the opinion of women in the reproductive years may have little influence weighed against intense cultural expectations. Education, especially female literacy, is the key to lower fertility (Figures 9 and 10). With education comes a knowledge of birth control, greater social awareness, more opportunity for employment and a wider choice of action generally. In some countries religion is an important factor. For example, the Muslim and Roman Catholic religions oppose artificial birth control. Most countries that have

population policies have been trying to reduce their fertility by investing in birth control programmes.

● **Economic:** Fertility rates tend to be highest in the world's least developed countries (Figure 11). In many of the least developed countries children are seen as an economic asset because of the work they do, often on very small farms, and also because of the support they are expected to give their parents in old age. In many poor countries there is little or no government support for elderly people. In the developed world the general perception is reversed and the cost of the child dependency years is a major factor in the decision to begin or extend a family. Economic growth allows greater spending on health, housing, nutrition and education, which is important in lowering mortality and in turn reducing fertility. Government statistics published in the UK in 2010 showed that people were getting married five years later than a decade before, with couples in the UK now typically in their mid-30s when getting married. Many other countries have followed this trend. In general an increase in the average age of marriage leads to a fall in the birth rate.

● **Political:** There are many examples in the past century of governments attempting to change the rate of population growth for economic and strategic reasons. During the late 1930s Germany, Italy and Japan all offered inducements and concessions to those with large families. In more recent years Malaysia has adopted a similar policy. However, today, most governments that try to change fertility want to reduce population growth, although some countries such as Russia are concerned about their populations declining because the birth rate has fallen below the death rate.

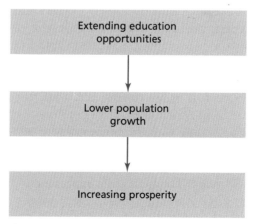

Figure 9 Education and development

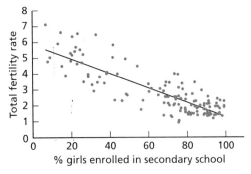

Figure 10 A comparison between female secondary education and total fertility rates

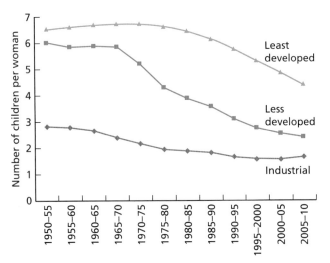

Figure 11 Total fertility rates in industrial, less developed and least developed countries, 1950–2010

Factors affecting mortality

Life expectancy at birth is the average number of years a newborn infant can expect to live under current mortality levels. In 1900 the world average for life expectancy is estimated to have been about 30 years but by 1950–55 it had risen to 46 years. By 1980–85 it had reached a fraction under 60 years and is presently 70 years, with men living to an average of 68 years and women averaging 72 years (Table 5). However, the global average masks significant differences by world region. The highest life expectancy of 79 years is in North America, while the lowest of 58 years is in Africa. Individual countries show an even wider range with the highest figure of 84 years for both genders combined being recorded in San Marino, just ahead of Japan, China and Hong Kong at 83 years. The lowest life expectancy, according to the *2012 World Population Data Sheet*,

was in Sierra Leone at 47 years. Nine other countries had a life expectancy below 50. All are in sub-Saharan Africa apart from Afghanistan.

The twentieth century fall in mortality was particularly marked after the Second World War which had provided a tremendous impetus for research into tropical diseases. Rates of life expectancy at birth have converged significantly between rich and poor countries over the past 50 years in spite of a widening wealth gap. However, it must not be forgotten that the ravages of AIDS in particular have caused recent decreases in life expectancy in some countries.

Figure 12 Graveyard dating from the eighteenth century in the Cotswolds, UK. Inscriptions show that life expectancy in the UK was very low then.

Table 5 Life expectancy at birth, 2012

Region	Both genders	Males	Females
World	70	68	72
More developed world	78	75	81
Less developed world	68	66	70
Africa	58	56	59
Asia	70	68	72
Latin America/Caribbean	74	71	77
North America	79	76	81
Oceania	77	75	79
Europe	77	73	80

The causes of death vary significantly between the developed and developing worlds (Figure 13). Apart from the challenges of the physical environment in many developing countries, a range of social and economic factors contribute to the high rates of infectious diseases. These include:

- poverty
- poor access to healthcare

- antibiotic resistance
- changing human migration patterns
- new infectious agents.

When people live in overcrowded and insanitary conditions, communicable diseases such as tuberculosis and cholera can spread rapidly. Limited access to healthcare and medicines means that otherwise treatable conditions such as malaria and tuberculosis are often fatal to poor people. Poor nutrition and deficient immune systems are also key risk factors for several big killers such as lower respiratory infections, tuberculosis and measles.

What are the main differences between rich and poor countries with respect to causes of death?

Online Q&A
30 April 2012

Q: What are the main differences between rich and poor countries with respect to causes of death?

A: In high-income countries almost 50% of the deaths are among adults 80 and over. The leading causes of death are chronic diseases: cardiovascular disease, chronic obstructive lung disease, cancers, diabetes or dementia. Lung infection remains the only leading infectious cause of death.

In middle-income countries, chronic diseases are the major killers, just as they are in high-income countries. Unlike in high-income countries, however, HIV/AIDS, tuberculosis and road traffic accidents also are leading causes of death.

In low-income countries around 40% of all deaths are among children under the age of 14. Although cardiovascular diseases together represent the leading cause of death in these countries, infectious diseases (above all HIV/AIDS, lower respiratory infection, tuberculosis, diarrhoeal diseases and malaria) together claim more lives. Complications of pregnancy and childbirth together continue to be a leading cause of death, claiming the lives of both infants and mothers.

Figure 13 World Health Organization – What are the main differences between rich and poor countries with respect to causes of death?

Activities

1 Discuss three factors that cause the birth rate to vary from one part of the world to another.
2 Describe the relationship shown in Figure 10 between the total fertility rate and the percentage of girls enrolled in secondary school.
3 Compare the changes in total fertility rates between the three groups of countries shown in Figure 11.
4 Describe and explain the variations in life expectancy at birth shown in Table 5.
5 What are the main differences in the causes of death between countries at different levels of economic development (Figure 13)?

The current demographic divide

Although average population growth has slowed globally, the range of demographic experience has actually widened. Growth rates have remained high in many countries while they have fallen steeply in others. These diverging trends have created a **demographic divide** between countries where population growth remains high and those with very slow-growing, stagnant or declining populations. International migration is now the most unpredictable factor. For example, in the UK the birth rate has risen recently due to high levels of immigration.

Four groups of countries can be recognised in terms of projected population change to 2050 (Figure 14):

- Countries that are projected to decline in population between 2005 and 2050. Less than 15 per cent of the world's population lives in such countries, which include Russia, Germany, Japan and Italy.
- Slow population growth countries, which will increase their populations by 25 per cent at most by 2050. China is the most important country in this group.
- Medium population growth countries, which include the USA, Bangladesh, Brazil and India. For example, the USA is projected to increase its population by 42 per cent between 2005 and 2050.
- High population growth countries, which accounted for only 8 per cent of world population in 2005. Except for a few oil exporting countries, nearly all of the high population growth countries are in the UN's list of least developed countries. Many are in Africa, but the list also includes Afghanistan, Guatemala and Haiti.

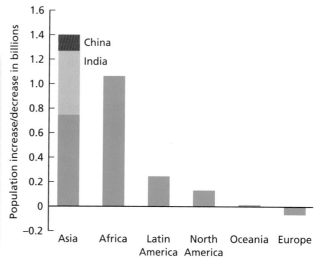

Figure 14 Projected population change by region, 2005–50

Increasing mortality due to HIV/AIDS

Although in general mortality continues to fall around the world, in some countries it is rising. HIV/AIDS is the major reason for such increases in mortality. However, the global battle against AIDS is showing significant signs of success. In 2011, 1.7 million people died from AIDS-related causes worldwide – which was 24 per cent fewer deaths than in 2005. According to UNAIDS in 2012:

- sub-Saharan Africa remains the region most affected, with nearly 1 in every 20 adults living with HIV
- sub-Saharan Africa accounts for 69 per cent of all people living with HIV
- the number of AIDS-related deaths declined by nearly one-third in sub-Saharan Africa between 2005 and 2011
- worldwide, 2.5 million people became newly infected with HIV in 2011
- not all world regions have witnessed a decline in HIV infections – since 2001, the number of people newly infected in the Middle East and North Africa increased by more than 35 per cent.

Figure 15 World AIDS Day is recognised all over the world

The epidemic has been particularly concentrated in southern Africa. The factors responsible for such high rates include:

- poverty and social instability that result in family disruption
- high levels of other sexually transmitted infections
- the low status of women
- sexual violence
- high mobility, which is mainly linked to migratory labour systems
- ineffective leadership during critical periods in the epidemic's spread.

● The impact of HIV/AIDS

- **Labour supply** – the economically active population reduces as more people fall sick and are unable to work. This can have a severe impact on development. In the worst affected countries the epidemic has already reversed many of the development achievements of recent decades. In agriculture, food security is threatened as there are fewer people able to farm and to pass on their skills.
- **Dependency ratio** – those who contract HIV are mainly in the economically active population. An increasing death rate in this age group increases the dependency ratio.
- **Family** – AIDS is impoverishing entire families and many children and old people have to take on the role of carers. Adult deaths, especially of parents, often causes households to be dissolved. The large number

of orphaned children in some areas puts a considerable strain on local communities and on governments in developing countries.
- **Education** – with limited investment in education many young people are still unaware about how to avoid the risk of contracting HIV/AIDS. In addition there are a considerable number of teachers who have AIDS and are too ill to work. UNICEF has stressed how the loss of a significant number of teachers is a serious blow to the future development of low-income countries.
- **Poverty** – there is a vicious cycle between AIDS and poverty. AIDS prevents development and increases the impact of poverty. Poverty worsens the AIDS situation due to economic burdens such as debt repayments and drug/medical costs.
- **Infant and child mortality** – mortality rates increase as AIDS can be passed from mother to child.

Activities

1 With reference to Figure 14, explain what you understand by the term 'demographic divide'.
2 What are the factors responsible for such high prevalence rates of HIV/AIDS in southern Africa?
3 How does HIV/AIDS affect communities where prevalence rates are high?

Case study: Kenya – a country with a high rate of population growth

Figure 16 Overcrowding in Kibera – Nairobi's and Africa's largest slum

Kenya has a high rate of population growth (Figure 17). Between 1969 and 2009 the country's population increased more than threefold from 10.9 million to 38.6 million. This has been due to several factors:

● A high number of births per woman: in 2009 the Kenyan average was 4.6 children per woman. This compares with the current global average of 2.4.
● Falling death rates, particularly in infant mortality: in 2012, the Population Reference Bureau estimated that infant mortality in Kenya had fallen to 47/1000. This compares with the average for Africa as a whole of 67/1000.
● A steady and significant increase in life expectancy: in 2012 life expectancy was 62 years in Kenya compared with 58 for Africa as a whole.

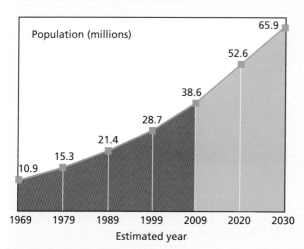

Figure 17 The growth in Kenya's population between 1969 and 2030

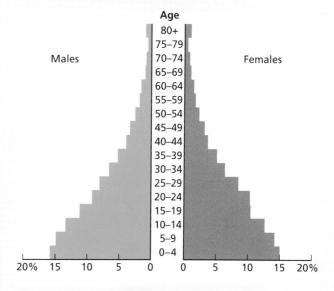

Figure 18 Population pyramid for Kenya

Although Kenya's total fertility rate is falling, in line with most other African countries, even if it drops to the forecast of 3.7 children per woman by 2030 the population will still grow to 65.9 million (Figure 17). This will be almost six times the population in 1969. Such rapid population increase puts heavy pressure on a country's resources, particularly food, water, housing, health and education.
    An analysis of family planning in Kenya showed that:

● women with more education have fewer children
● fewer than one-half of births are attended by a skilled provider (doctor, nurse, midwife, etc.)
● the poorest women have the highest unmet need for family planning
● many adolescents have sex before age 15
● birth spacing of at least two years has a big impact on child health and well-being.

Kenya has a very high youth dependency ratio, with over 40 per cent of the population under 15. Figure 18 shows that Kenya has a classic population pyramid for a country with a high population growth rate. A rapidly growing population results in a lower amount of land per capita available to farmers and their children. This is a major issue as about 70 per cent of the country's population live in rural areas. A recent survey showed that 67 per cent of farmers thought that the size of their land holding was not sufficient for their children. Young people who cannot find work on the land often migrate to urban areas. Such rural-to-urban migration is a significant phenomenon in Africa. However, although the economy is growing, youth unemployment is a considerable problem as the rate of population increase is greater than the rate of job creation. The World Bank estimates that 50 per cent of Kenya's population will live in urban areas by 2033.

Although the poverty rate fell from 47 per cent in 2005 to 38 per cent in 2012, Kenya remains amongst the most unequal countries in Africa. While clear progress has been made in health, education, infrastructure and other aspects of society, a significant proportion of the population continue to live in fragile conditions with sub-standard access to water, sanitation and energy. The situation is particularly difficult in the north and north-east.

Case study analysis

1 Use the atlas map at the beginning of the book to describe the location of Kenya.
2 Using Figure 17, produce a table to show population increase in Kenya between 1969 and 2030.
3 Study Figure 18 and describe the population structure of Kenya. Refer to the populations: (a) under 15 (b) 15–64 (c) 65 and over.
4 Briefly explain the reasons for Kenya's population structure.
5 What are the problems associated with high population growth in Kenya?

Case study: Population decline in Russia

Figure 19 The trans-Siberian railway. Many communities in Asiatic Russia have declined in population

In 2012, Russia's birth rate was 13/1000 while the country's death rate was 14/1000, giving a rate of natural change of –0.1 per cent. Such natural decrease is common in eastern Europe. Russia's population reached its highest level of almost 148.7 million in 1991 (Figure 20 and Table 6), just before the break-up of the Soviet Union. Since then it has been mainly in decline, although very slight increases have been registered in some recent years. In 2012 the Russian population stood at 143 million. With a population density of just over 8 km², this is one of the most sparsely populated countries in the world.

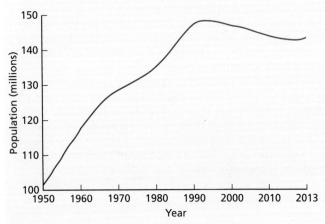

Figure 20 Russia's population, 1950–2013

Table 6 Russia's population, 1991–2050

Year	Population (millions)
1991	148.7
2012	143.2
2025 (estimate)	140.8
2050 (estimate)	127.8

The decline in Russia's population has been due to a combination of economic and social factors. Population decline or very slow growth has been due to:

● low birth rates
● high death rates, particularly among men
● emigration.

The change in recent decades from a communist centrally planned economy to a market economy has resulted in some people being much better off, while many other people struggle to make a reasonable living. Inequality has increased considerably in Russia, with unemployment and poverty being major concerns for many people. The cost of raising children is perceived to be high when both parents need to work to make ends meet. These circumstances have had a big impact on decisions to start a family and on decisions to extend a family.

Education standards for women in Russia are high and thus women in general have the decisive say in decisions about family size. The use of contraception is high with 80 per cent of married women aged 15–49 using various methods of contraception.

The difference in life expectancy between men and women in Russia is considerable. In 2012 life expectancy for women was 75 years, but only 63 for men. This extremely low rate for men in a European country has been attributed to very high intakes of alcohol, the high incidence of smoking, pollution, poverty and the ravages of HIV/AIDS and other diseases. The high male death rate has resulted in there being 10.7 million more women than men in Russia.

Population decline has had its greatest impact in rural areas with 8500 villages said to have been abandoned since 2002. The cold northern regions of Russia have experienced the highest levels of **depopulation**. Such are the concerns of many Russians about the future that a sociological survey in June 2011 found that one-fifth of the Russian population are potential emigrants!

In 2008 Russia began honouring families with four or more children with a Paternal Glory medal. The government has urged Russians to have more children, sometimes suggesting that it is a matter of public duty.

Interesting note

In 2012 worldwide, fifteen countries had birth rates lower than their death rates, thus registering a natural decrease in population. Latvia and Serbia had the highest rates of natural decrease at –0.5% (–5/1000).

Over-population and under-population

The idea of **optimum population** has been mainly understood in an economic sense (Figure 21). At first, an increasing population allows for a fuller exploitation of a country's resource base, causing living standards to rise. However, beyond a certain level rising numbers place increasing pressure on resources and living standards begin to decline. The highest average living standards mark the optimum population. Before that population is reached, the country or region can be said to be **under-populated**. As the population rises beyond the optimum, the country or region can be said to be **over-populated**. In terms of the planet as a whole, there are many indications that human population is pushing up against the limits of the Earth's resources. For example:

- One-quarter of the world's children have protein-energy malnutrition.
- The long-term trend for grain production per person is falling.

Case study analysis

1 Use the atlas map at the beginning of the book to describe the location of Russia.
2 Describe the changes in Russia's population shown in Figure 20 and Table 6.
3 Discuss the reasons for population decline in Russia.

- About 40 per cent of agricultural land is moderately degraded and 9 per cent is highly degraded.
- Water scarcity already affects every continent and 4 of every 10 people in the world.
- A quarter of all fish stocks are overharvested.
- There are concerns that global peak oil production will come as early as the next decade.

Where an individual country is placed in terms of its relationship between population and resources is a matter of opinion. There may be a big difference in the views of people living in the same country. Such views can change over time, particularly with economic cycles.

The Netherlands and the UK are two of the most densely populated countries in Europe. Not everyone in these countries thinks they are over-populated, but it does seem that an increasing number of people are of this opinion. In the UK, an organisation called the Optimum Population Trust states that 30 million is the optimum population for the country. At present the population of the UK is about 63 million. Signs of population pressure in both the UK and the Netherlands include:

- intense competition for land
- heavy traffic congestion
- high house prices
- high environmental impact of economic activity
- pressure on water resources.

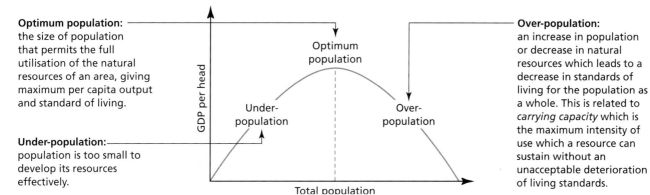

Optimum population: the size of population that permits the full utilisation of the natural resources of an area, giving maximum per capita output and standard of living.

Under-population: population is too small to develop its resources effectively.

Over-population: an increase in population or decrease in natural resources which leads to a decrease in standards of living for the population as a whole. This is related to *carrying capacity* which is the maximum intensity of use which a resource can sustain without an unacceptable deterioration of living standards.

Figure 21 Optimum population, over-population and under-population

Two of the most sparsely populated developed countries in the world are Australia and Canada. Throughout the history of both countries the general view has been that they would benefit from higher populations. Thus Australia and Canada have welcomed significant numbers of immigrants. However, in recent years, with an uncertain economic climate, both countries have been much more selective in terms of immigration. Although both countries are very large in size, they have large areas of inhospitable landscape.

In the developing world, China and Bangladesh are countries that many would view as over-populated. The 'one child' policy confirms the Chinese government view. Bangladesh has one of the highest population densities in the world and struggles to provide for many in its population. In contrast, Malaysia has been concerned that it was under-populated. In 1982, when that country's population was below 15 million, the government announced that the country should aim for an ultimate population of 70 million. A range of benefits were put in place to encourage people to have larger families.

Case study: Bangladesh – an over-populated country?

'Spiraling population strains Bangladesh's sustainability.'
OneWorld South Asia, April, 2009

If small, largely urbanised countries such as Singapore and small island states such as the Maldives are not considered, then Bangladesh (Figure 22) has the highest population density in the world. At 1062 people per km² its population density is about twenty times the global average. It compares with 383 per km² in India and 141 per km² in China – two countries that many people associate with over-population. Over-population cannot be judged by population density alone, but it is a useful starting point when considering the relationship between population and resources in a country.

Figure 23 Urban overcrowding in Bangladesh

Bangladesh is a relatively small and resource-poor country with a land area of 147 000 km². This compares with 244 000 km² in the UK. Yet the population of Bangladesh is 153 million compared with 63 million in the UK.

The lack of natural resources is a major factor in over-population in Bangladesh, as is rapid population growth. The current rate of natural increase in Bangladesh is 1.6 per cent. When Bangladesh became an independent country in 1971 its population was about 75 million, just under half of the total today! The Population Reference Bureau estimates that the population of Bangladesh will rise to 183 million by 2025 and 226 million by 2050. There is already intense competition for the available resources in Bangladesh. How will the country cope in the future when it is already experiencing large-scale poverty and so many other problems?

Extremely high population pressure and the deprivation associated with it is characteristic of both rural and urban areas. Four-fifths of the population live in rural areas. The very small amount of cultivable land per person has resulted in a very high level of rural poverty. Most families have to survive on extremely small plots of land. This means that there is not enough work available for many people. About 40 per cent of the population is underemployed, working a limited number of hours a

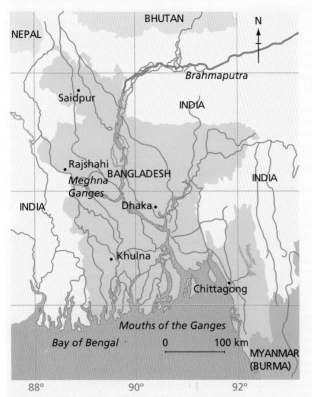

Figure 22 Map of Bangladesh

week at low wages. **Underemployment** is not just confined to the countryside, but affects urban areas too.

The regular threat of cyclones and flooding makes this problem much worse. Eighty per cent of the country is situated on the floodplains of the Ganges, Brahmaputra, Meghna and those of several other minor rivers. Much of the country is close to sea level and about 40 per cent is regularly flooded during the monsoon season. Major floods can cause considerable loss of life and destroy vital infrastructure, often setting back development many years. Where possible, people move to higher land, increasing the already overcrowded nature of such areas. Major floods increase the level of rural-to-urban migration, with the majority of migrants heading for the capital city Dhaka. Other urban areas such as Chittagong, Khulna and Rajshahi are also growing in population at very rapid rates. Around 8000 hectares of cultivable land is lost every year due to urbanisation, industrialisation and the expansion of infrastructure. This is potential food production that Bangladesh can ill afford to lose.

Living conditions in Dhaka and the other main urban areas are in a continuous state of deterioration. Many people lack basic amenities such as electricity and clean drinking water. Dhaka has become one of the most crowded cities in the world with a population density of 43 000 per km². This rapidly growing megacity was recently ranked as the least habitable city among 140 cities surveyed by the Economist Intelligence Unit. The 2012 Human Development Index, which ranks all the countries of the world according to their quality of life, placed Bangladesh 146th in the world.

Land is being lost to rising sea levels, a process associated with global warming. The United Nations Intergovernmental Panel on Climate Change has predicted that as sea levels rise, by 2050 about 35 million people from Bangladesh will cross the border into India in search of more secure living conditions.

Poor governance and corruption have undoubtedly hindered development in Bangladesh. However, national and international efforts to improve the lives of the population have registered progress. For example, the World Bank noted in 2013 that the number of people in poverty in Bangladesh had fallen from 63 million in 2000 to 47 million in 2010. The question is: can Bangladesh continue to reduce poverty in the future?

Case study analysis

1 Describe the geographical location of Bangladesh.
2 How does the physical geography of the country make life difficult for its people?
3 What evidence would you produce to support the statement that 'Bangladesh is an over-populated country'?

Case study: Australia – an under-populated country?

Figure 24 Sydney, Australia

Australia is generally regarded as an example of an under-populated country although there are some experts who would disagree because much of the interior of the country is so inhospitable. In the world's sixth largest country in land area, most of the population is concentrated in two widely separated coastal regions – the south-east and east, and the south-west (Figure 25) – and within these regions there is considerable concentration in urban areas.

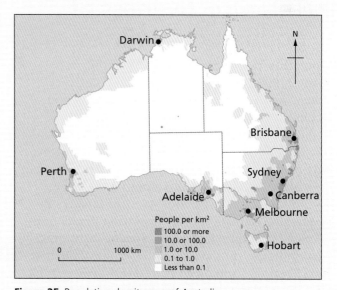

Figure 25 Population density map of Australia

This country of continental size had a population of only 22 million people in 2012 (Table 7). The population of Australia is forecast to rise to 26 million in 2025 and 33 million by 2050. The current population density is only 3 per km², one of the lowest figures in the world.

Table 7 Comparing Bangladesh and Australia

Indicator	Bangladesh	Australia
Population 2012	153 million	22 million
Population forecast 2025	183 million	26 million
Population density 2012	1062 km²	3 km²
Rate of natural change 2012	1.6%	0.7%
Net migration rate 2012	–3/1000	+8/1000
Infant mortality rate	43/1000	4/1000
Life expectancy	69 years	82 years
% urban	25%	82%
GNI PPP per capita ($)	1810	36 910

Australia is a resource-rich nation, exporting raw materials in demand on the global market all over the world. The country's major resources include coal, iron ore, copper, gold, natural gas and uranium. Australia also has great potential for renewable energy, particularly in terms of wind and solar power. Such an abundant resource base has attracted a high level of foreign direct investment. Countries that need to import large amounts of natural resources, such as China, Japan and Korea, have been major investors in Australia. The country has a well-developed infrastructure and a relatively highly skilled population which enjoys a generally high income. Australia exudes an image of an affluent outdoor lifestyle that attracts potential migrants from many different countries. Australia was ranked second in the world (after Norway) according to the 2012 Human Development Index.

Net migration is a good measure of how attractive a country is to people from other countries. While Bangladesh has negative net migration (Table 7), Australia has one of the highest positive net migration figures in the world. Australia's extremely high Gross National Income per capita is not just a major attraction to potential international migrants, it is also a useful statement of the opportunities available in the country and the relationship between population and resources.

Another useful indicator of the population/resources relationship is unemployment. In 2012, Australia's unemployment rate was 5.2 per cent, a low figure by global standards. Unlike Bangladesh, underemployment is not a significant problem. Australia scores highly for virtually all measures of the quality of life, including health and education. Although Australia's population is highly concentrated in certain areas, there are undoubtedly more genuine opportunities for population increase here than in most other parts of the world.

Case study analysis

1 Describe the location of Australia.
2 Use Figure 25 to briefly comment on the population distribution of Australia.
3 What evidence would you produce to support the statement that 'Australia is an under-populated country'?

⬤ The effectiveness of population policies

Population policy encompasses all of the measures taken by a government aimed at influencing population size, growth, distribution, or composition. Such policies may promote large families (**pro-natalist policies**) or immigration to increase population size, or encourage fewer births (**anti-natalist policies**) to reduce population growth. A population policy may also aim to modify the distribution of the population over the country by encouraging migration or by displacing populations.

A significant number of governments have officially stated positions on the level of the national birth rate. However, forming an opinion on demographic issues is one thing, but establishing a policy to do something about it is much further along the line. Thus not all nations stating an opinion on population have gone as far as establishing a formal policy.

Most countries that have tried to control fertility have sought to curtail it. In 1952 India became the first developing country to introduce a policy designed to reduce fertility and to aid development with a government backed family planning programme. Because India's family planning programme was perceived to be working, it was not long before many other developing nations followed India's policy of government investment to reduce fertility. In India the birth rate fell from 45/1000 in 1951–61 to 41/1000 in 1961–71. By 1987 it was down to 33/1000, falling further to 29/1000 in 1995. By 2012 it had dropped to 22/1000.

Anti-natalist policy in China

China, with a population in excess of 1.3 billion, operates the world's most severe family planning programme. Although it is the fourth largest country in the world in land area, 25 per cent of China is infertile desert or mountain and only 10 per cent of the total area can be used for arable farming. Most of the best land is in the east and south, reflected in the extremely high population densities found in these regions. Thus the balance between population and resources has been a major cause of concern for much of the latter part of the twentieth century and the early part of the present century.

In the aftermath of the communist revolution in 1949, population growth was encouraged for economic, military and strategic reasons. Sterilisation and abortion were banned and families received a benefit payment for every child. However, by 1954 China's population had reached 600 million and the government was now worried about the pressure on food supplies and other resources. Consequently, the country's first birth control programme was introduced in 1956. This was to prove short-lived, for in 1958 the 'Great Leap Forward' began. Its objective was rapid industrialisation and modernisation. The government was now concerned that progress might be hindered by labour shortages and so births were again encouraged. But by 1962 the government had changed its mind, heavily influenced by a catastrophic famine due in large part to the relative neglect of agriculture during the pursuit of industrialisation. An estimated 20 million people died during the famine. A new phase of birth control ensued in 1964 but just as the new programme was beginning to have some effect, a new social upheaval, the Cultural Revolution, got underway. This period, during which the birth rate peaked at 45/1000, lasted from 1966 to 1971.

With order restored, a third family planning campaign was launched in the early 1970s with the slogan 'Late, sparse, few'. However, towards the end of the decade the government felt that its impact might falter and in 1979 the controversial 'one child' policy was imposed. The Chinese demographer Liu Zeng calculated that China's optimum population was 700 million, and he looked for this figure to be achieved by 2080.

Figure 26 shows changes in the birth and death rates in China since 1950, and estimates to 2050. The impact of the one child policy is very clear to see. Some organisations, including the UN Fund for Population Activities, have praised China's policy on birth control. Many others see it as a fundamental violation of civil liberties because it has placed such extreme pressure on couples to obey the policy. In July 2009, newspapers in the UK and elsewhere reported that dozens of babies had been taken from parents who had breached China's policy, and sold for adoption abroad.

The one child policy has been most effective in urban areas where the traditional bias of couples wanting a son is less. However, the story is different in rural areas where the strong desire for a male heir remains the norm. In most provincial rural areas,

government policy has now relaxed so that couples can now have two children without penalties.

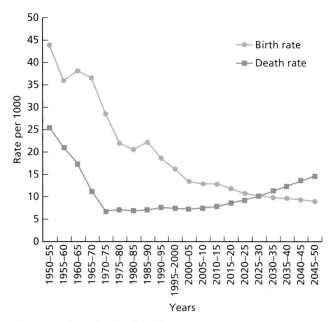

Figure 26 China's birth and death rates, 1950–2050

Figure 27 Beijing – crowds at the Forbidden City

Although the one child policy has been very effective in reducing China's birth rate, it has caused other problems:

- The policy has had a considerable impact on the gender ratio which at birth in China is currently 119 boys to 100 girls. This compares with the natural rate of 106 to 100. This is already causing social problems which are likely to multiply in the future. Selective abortion after pre-natal screening is a major cause of the wide gap between the actual rate and the natural rate. But even if a female child is born, her lifespan may be sharply curtailed by infanticide or deliberate neglect.
- A paper published in 2008 estimated that China had 32 million more men aged under 20 than

women. The imbalance is greatest in rural areas because women are 'marrying out' into cities. In recent years, reference has been made to the 'Four-Two-One' problem whereby one adult child is left with having to provide support for his or her two parents and four grandparents.

- China's low birth rate, 12/1000 in 2012, has contributed to the country's ageing population which has now become a major concern for the government.

In recent years there has been increasing debate within China about the one child policy. An article in one UK newspaper in August 2013 was entitled 'China to terminate its one child policy as ageing crisis looms'. It is likely that there will be much debate about this issue in the coming years.

Pro-natalist policy in France

A relatively small but growing number of countries now see their fertility as being too low. Such countries are concerned about:

- the socioeconomic implications of population ageing
- the decrease in the supply of labour
- the long-term prospect of population decline.

France's relatively high fertility level (in European terms) can be partly explained by its long-term active family policy, adopted in the 1980s to accommodate the entry of women into the labour force. The policy seems to have created especially positive attitudes towards two- and three-child families in France. France has taken steps to encourage fertility on a number of occasions over the last 70 years. In 1939 the government passed the 'Code de la Famille' which:

- offered financial incentives to mothers who stayed at home to look after children
- subsidised holidays
- banned the sale of contraceptives (this stopped in 1967).

More recent measures to encourage couples to have more children include:

- longer maternity and paternity leave: maternity leave, on near full pay, ranges from 20 weeks for the first child to 40 or more for the third child
- higher child benefits

- improved tax allowances for larger families until the youngest child reaches 18
- pension scheme for mothers/housewives
- 30 per cent reduction on all public transport for three-child families
- child-oriented policies, for example provision of crèches and day nurseries – state-supported daycare centres and nursery schools are available for infants starting at the age of three months, with parents paying a sliding scale according to income
- preferential treatment in the allocation of government housing.

Overall, France is trying to reduce the economic cost to parents of having children. The country is close to the replacement level of 2.1 children per woman. The 2012 Population Data Sheet put France's total fertility rate at 2.0. The only country in Europe with a higher rate was Ireland at 2.1. In 2010, when the population of France rose by 0.53 per cent (Figure 28), there were 802 000 births and 540 000 deaths. During the same year the net migration figure was +75 000.

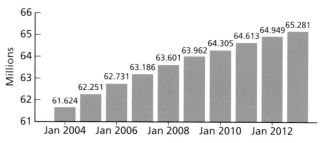

Figure 28 Population growth in France, 2004–13

Although the average age of French mothers at childbirth is still rising, it is still less than in many other European countries. Within France, the highest level of fertility is among the immigrant population, and even for those born in France the average is 1.8 babies. French economists argue that although higher fertility means more expenditure on childcare facilities and education, in the longer term it gives the country a more sustainable age structure.

French politicians have talked about demography as a 'source of vitality' for the country. Some French commentators also argue that there is a better work/life balance in France compared with many other European countries.

1.2 Migration

Chinatown, San Francisco – a major Chinese community in an American city

Key questions
- What is migration?
- What are the causes of migration?
- What are the impacts of migration on areas of both origin and destination?

The nature of and reasons for population migration

Migration is the movement of people across a specified boundary, national or international, to establish a new permanent place of residence. The UN defines 'permanent' as a change of residence lasting more than one year. Migration has been a major process in shaping the world as it is today. Its impact has been economic, social, cultural, political and environmental. Few people now go through life without changing residence several times. How many people in your class have lived in (a) different parts of your country (b) other countries?

Push and pull factors

Figure 1 shows the main **push and pull factors** relating to migration. Push factors are negative conditions at the point or origin which encourage or force people to move. For example, a high level of unemployment is a major push factor in a region or a country. In contrast, pull factors are positive conditions at the point of destination which encourage people to migrate. An important pull factor is often much higher wages in another country or region. The nature of push and pull factors varies from country to country (and from person to person) and changes over time.

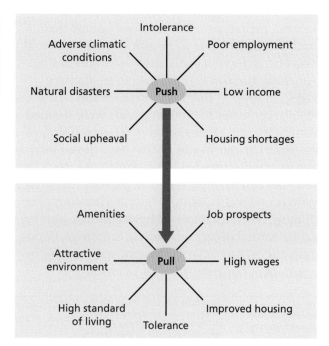

Figure 1 Push and pull factors

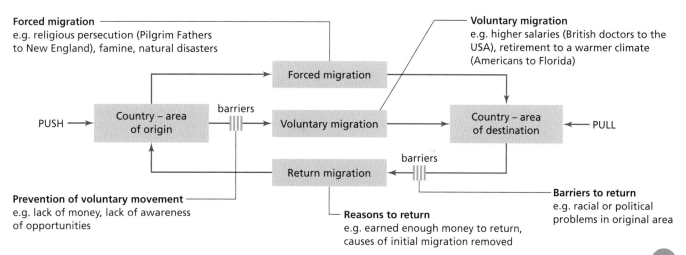

Forced migration
e.g. religious persecution (Pilgrim Fathers to New England), famine, natural disasters

Voluntary migration
e.g. higher salaries (British doctors to the USA), retirement to a warmer climate (Americans to Florida)

Prevention of voluntary movement
e.g. lack of money, lack of awareness of opportunities

Reasons to return
e.g. earned enough money to return, causes of initial migration removed

Barriers to return
e.g. racial or political problems in original area

Figure 2 Types of migration and barriers to migration

Voluntary and involuntary migrations

Figure 2 shows the main types of migration and the barriers that potential migrants can face. Today, immigration laws present the greatest obstacle to most potential international migrants whereas in the past the physical dangers encountered on the journey often presented the greatest difficulty. The cost of migration was also generally higher in the past in real terms (taking account of inflation) than it is today. Most countries now attempt to manage immigration carefully, being most eager to attract people whose skills are in demand.

The big distinction is between **voluntary migration** and **involuntary (forced) migration**. In voluntary migration the individual has a free choice about whether to migrate or not. In involuntary migrations, people are made to move against their will and this may be due to human or environmental factors. The abduction and transport of Africans to the Americas as slaves was the largest involuntary migration in history. In the seventeenth and eighteenth centuries 15 million people were shipped across the Atlantic Ocean as slaves. The expulsion of Asians from Uganda in the 1970s when the country was under the dictatorship of Idi Amin, and the forcible movement of people from parts of the former Yugoslavia under the policy of 'ethnic cleansing', are much more recent examples. Migrations may also be forced by natural disasters (volcanic eruptions, floods, drought and so on) or by environmental catastrophe such as nuclear contamination in Chernobyl.

In the latter part of the twentieth century and the beginning of the twenty-first century, some of the world's most violent and protracted conflicts have been in the developing world, particularly in Africa, the Middle East and Asia. These troubles have led to numerous population movements on a significant scale. Not all have crossed international frontiers to merit the term **refugee** movements. Instead many are **internally displaced people**. The current conflict in Syria has produced large numbers of both refugees and internally displaced people (Figure 3), as has conflict in Iraq, Afghanistan and the former Yugoslavia. Major natural disasters such as the Pakistan floods of 2010 and the Haiti earthquake in the same year created large numbers of internally displaced people.

The United Nations High Commission for Refugees (UNHCR) put the number of forcibly displaced people worldwide at 42.5 million at the end of 2012. This included 15.4 million refugees, the remainder being internally displaced people.

Migration trends

Figure 4 shows international migrant stock by origin and destination. The terms 'North' and 'South' refer to the developed world and the developing world respectively. The diagram shows that 213 million people live outside the country of their birth – more than ever before. This is about 3 per cent of the world's population. Foreign-born populations are rising in both developed and developing countries (Figure 5). The number of international migrants doubled in the 25 years to 2010.

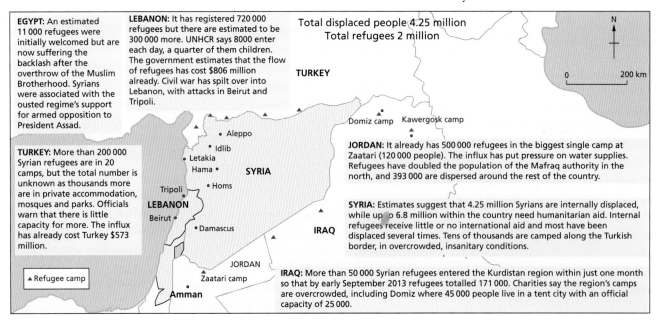

Figure 3 Syria: refugees and internally displaced people, September 2013

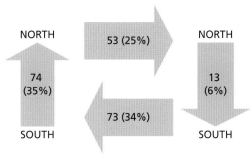

Figures in millions and percentages

Figure 4 International migrant stock by origin and destination, 2010

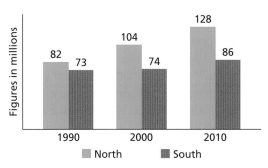

Figure 5 International migrant stock in the North and South, 1990–2010

Recent migration data show that:

- With the growth in the importance of labour-related migration and international student mobility, migration has become increasingly temporary and circular in nature. The international mobility of highly skilled workers increased substantially in the 1990s.
- The spatial impact of migration has spread, with an increasing number of countries affected either as points of origin or destination. While many traditional migration streams remained strong, significant new streams have developed.
- The proportion of female migrants has steadily increased (now almost 50 per cent of all migrants). For some countries of origin, women now make up the majority of contract workers (for example the Philippines, Sri Lanka, Thailand and Indonesia).
- The great majority of international migrants move from developing to developed countries. However, there are also strong migration links between some developing countries, in particular between low and middle income countries (Figure 4).
- Developed countries have reinforced controls, in part in response to security issues, but also to

combat illegal immigration and networks that deal in trafficking and exploitation of human beings.

Globalisation in all its aspects has led to an increased awareness of opportunities in other countries. With advances in transportation and communication and a reduction in the real cost of both, the world's population has never had a higher level of potential mobility. Also, in various ways, economic and social development has made people more mobile and created the conditions for emigration. Many developing countries are looking to developed countries to adopt a more favourable attitude to international migration, arguing that it brings benefits to both developed and developing countries.

Figure 6 Southall, the centre of London's Indian community

Internal population movements

Population movement within countries is at a much higher level than movements between countries. In both developed and developing countries significant movements of people take place from poorer regions to richer regions as people seek employment and higher standards of living. In developing countries, much of this migration is in fact from rural to urban areas. Developed countries had their period of high rural-to-urban migration in the nineteenth century and the early part of the twentieth century. The developing countries have been undergoing high rural-to-urban migration since about 1950, resulting in the very rapid growth of urban areas such as Cairo, Nairobi, São Paulo and Dhaka. The largest rural-to-urban migration in history is now taking place in China where more than 150 million people have moved from the countryside to the rapidly expanding urban/industrial areas to satisfy the demand for workers in China's factories.

In Brazil, there has been a large migration from the poor Northeast region to the more affluent southeast. Within the northeast, movement from rural areas is greatest in the Sertão, the dry interior which suffers intensely from unreliable rainfall. However, poor living standards and a general lack of opportunity in the cities of the northeast have also been a powerful incentive to move. Explaining the attraction of urban areas in the southeast demands more than the 'bright lights' scenario that is still sometimes used. The Todaro model presents a more realistic explanation. According to this model, migrants are all too well aware that they may not find employment by moving to, say, São Paulo. However, they calculate that the probability of employment, and other factors that are important to the quality of life of the individual and the family, is greater in the preferred destination than at their point of origin.

Depopulation and counterurbanisation

In developed countries two significant trends can be identified concerning the redistribution of population since the late eighteenth century. The first, urbanisation, lasted until about 1970, while the second, counterurbanisation, has been dominant since that time.

The process of urbanisation had a considerable impact on many rural areas where depopulation occurred because of it. Depopulation is an absolute decline in the population of an area, usually due to a high level of out-migration. It is generally the most isolated rural areas that are affected. Figure 7 shows how the process of **rural depopulation** can occur.

Counterurbanisation is the process of population decentralisation as people move from large urban areas to smaller urban settlements and rural areas. The objective is usually to seek a better quality of life by getting away from the problems of large cities. This process has resulted in a renaissance in the demographic fortunes of rural areas and is often referred to as the 'population turnaround'. There has been considerable debate as to whether this trend will be long-term or relatively short-lived.

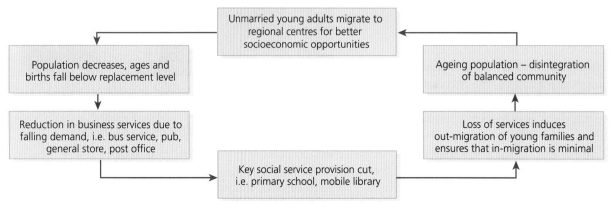

Figure 7 Model of rural depopulation

Figure 8 Rural depopulation in northern Spain

Interesting note

In 2012, the country hosting the largest number of refugees was Pakistan, with 1.6 million refugees. Afghanistan was the biggest source country, a position it has held for 32 years. One out of every four refugees worldwide is Afghan, with 95 per cent located in Pakistan or Iran.

Activities

1 Define migration.
2 What is the difference between voluntary and involuntary migration?
3 Discuss three significant push factors in migration.
4 Suggest how the barriers to migration have changed over time.
5 State the difference between a refugee and an internally displaced person.
6 Summarise the data presented in Figures 4 and 5.
7 Briefly describe two types of internal population movements.

● The impacts of migration

'The overall economic gains from international migration for sending countries, receiving countries, and the migrants themselves are substantial.'

The World Bank, 2012

Migration has played a major role in shaping the global cultural map. The process of migration is essentially a series of exchanges between places. In some parts of the world international migration is a long-established process. The truly cosmopolitan nature of major cities such as London, New York and Paris is clear evidence of this. In other parts of the world there is little evidence of cultural diversity because international migration has never been a significant phenomenon in the recent past.

Figure 9 is an attempt to summarise some of the possible impacts of international migration. Many of these factors are also relevant to internal migration. Because migration can be such an emotive issue you may not agree with all of these statements, and you may consider that some important factors have been omitted.

Impact on countries of origin

Remittances are often seen as the most positive impact on the country of origin. Remittances are money sent back by migrants to their families in their home community. They are a major economic factor in developing countries. Remittances to developing countries are estimated to have totalled just over $400 billion in 2012. Global remittance flows, including those to high-income countries, were an estimated $529 billion in 2012. Remittances exceed considerably the amount of official aid received by developing countries.

The impact of international migration		
Impact on countries of origin	**Impact on countries of destination**	**Impact on migrants themselves**
Positive		
• Remittances are a major source of income in some countries. • Emigration can ease the levels of unemployment and underemployment. • Reduces pressure on health and education services and on housing. • Return migrants can bring new skills, ideas and money into a community.	• Increase in the pool of available labour may reduce the cost of labour to businesses and help reduce inflation. • Migrants may bring important skills to their destination. • Increasing cultural diversity can enrich receiving communities. • An influx of young migrants can reduce the rate of population ageing.	• Wages are higher than in the country of origin. • There is a wider choice of job opportunities. • A greater opportunity to develop new skills. • They have the ability to support family members in the country of origin through remittances. • Some migrants have the opportunity to learn a new language.
Negative		
• Loss of young adult workers who may have vital skills, e.g. doctors, nurses, teachers, engineers (the 'brain-drain' effect). • An ageing population in communities with a large outflow of (young) migrants. • Agricultural output may suffer if the labour force falls below a certain level. • Migrants returning on a temporary or permanent basis may question traditional values, causing divisions in the community.	• Migrants may be perceived as taking jobs from people in the long-established population. • Increased pressure on housing stock and on services such as health and education. • A significant change in the ethnic balance of a country or region may cause tension. • A larger population can have a negative impact on the environment.	• The financial cost of migration can be high. • Migration means separation from family and friends in the country of origin. • There may be problems settling into a new culture (assimilation). • Migrants can be exploited by unscrupulous employers. • Some migrations, particularly those that are illegal, can involve hazardous journeys.

Figure 9 Matrix showing the impact of migration

Remittances can account for over 20 per cent of annual GDP in some developing countries such as Tajikistan, Liberia, Lesotho and Nepal. They have been described as 'globalisation bottom up'. Migration advocates stress that these revenue flows:

- help alleviate poverty
- spur investment and create a multiplier effect
- cushion the impact of global recession when private capital flows decrease.

The major sources of remittances are the USA, and countries in western Europe and the Persian Gulf. In 2012, the top recipients of remittances were: India ($69 billion), China ($60 billion), the Philippines ($24 billion) and Mexico ($23 billion). Other large recipients included Nigeria, Egypt, Pakistan, Bangladesh, Vietnam and Lebanon.

Figure 10 MoneyGram sign – remittances are an important element of international migration

Other possible advantages of emigration include reducing population pressure on resources such as food and water, and lowering levels of unemployment and underemployment which are major problems in many poor countries. Emigration can also reduce pressure on the housing stock and on key services such as health and education which are heavy areas of expenditure for low-income countries. The links that migrants maintain with their home communities can also help to develop new skills and technologies, particularly when migrants return home permanently and establish new businesses.

In terms of disadvantages, the loss of workers with important skills – the so-called brain-drain

effect – is of concern in many countries of origin. In extreme cases there may not be enough people to continue to farm effectively in some communities. Population structure can be adversely affected if migration is very gender-selective, and the loss of many young people can advance the ageing of the population.

Impact on countries of destination

Any increase in the labour force is generally welcomed by businesses, particularly if migrants have skills that are in short supply. For example, in the UK the National Health Service (NHS) relies heavily on foreign nurses and doctors. Greater competition in the labour force tends to limit wage rises which helps to keep inflation low. Low inflation is an important factor in economic stability. Many people value an increase in cultural diversity which can enrich communities. An influx of young migrants can help to reduce the rate of population ageing and lower the dependency ratio. This has a positive financial impact on countries.

The negative impact of immigration is more contentious. A significant influx of migrants can put pressure on the available housing stock, causing overcrowding and pushing up prices. Similarly, it will increase demand on health, education and other services to varying degrees. Trade unions often voice concerns over migration levels if they feel that their members are losing out on employment prospects because of an increase in the level of competition for work. In areas where there has been a significant change in the ethnic balance, tension between ethnic groups may increase.

Each receiving country has its own sources of immigrants. This is the result of historical, economic and geographical relationships. Earlier generations of migrants form networks that help new arrivals to overcome legal and other obstacles.

Impact on migrants themselves

The prospect of higher wages, a wider choice of job opportunities and the chance to develop new skills are all important pull factors to potential migrants. For many this will enable them to provide financial support to their families in the country of origin. Other benefits may also become apparent, such as developing language skills.

On the debit side, the financial cost of migration may be significant, but may be shared by family members in anticipation of the remittances to follow. For many migrants, by far the greatest cost is separation from family and friends in their home country. The 'culture shock' of settling in a new country may be lessened if migrants find housing and employment in an established migrant community in their new country. The risk of exploitation is a real concern for many migrants. Receiving countries vary in how effectively they challenge such practices.

The costs and benefits of internal migration

Figure 11 provides a useful framework for understanding the costs and returns from migration between rural and urban areas. It highlights the main factors that determine how rural areas are affected by migration – namely the two-way transfers of labour, money, skills and attitudes. However, while all the linkages seem fairly obvious, none is easy to quantify. Therefore, apart from some very clear-cut cases, it is often difficult to decide which is greater: the cost or the benefits of migration.

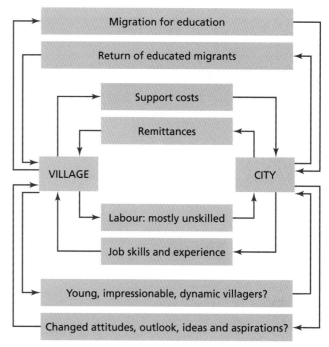

Figure 11 The costs and returns of migration

Climate change

It is predicted that climate change will force mass migrations in the future. In 2009 the International Organization for Migration estimated that worsening tropical storms, desert droughts and rising sea levels will displace 200 million people by 2050.

Case study: International migration from Mexico to the USA

One of the largest labour migrations in the world has been from Mexico to the United States, a rare example where a developed country borders a developing country. This migration has largely been the result of:

- much higher average incomes in the USA
- lower unemployment rates in the USA
- the faster growth of the labour force in Mexico, with significantly higher population growth in Mexico than in the USA
- the overall quality of life: on virtually every aspect of the quality of life conditions are better in the USA than in Mexico.

Most migration has taken place in the last three decades. Although previous surges occurred in the 1920s and 1950s when the American government allowed the recruitment of Mexican workers as guest workers, persistent **mass migration** between the two countries did not take hold until the late twentieth century.

There is a very strong concentration of the US Mexican population in the four states along the Mexican border: California, Arizona, New Mexico and Texas. The main reasons for this spatial distribution are:

- proximity to the border
- the location of demand for immigrant farm workers
- urban areas where the Mexican community is long-established.

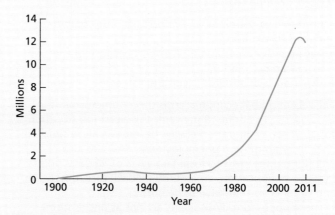

Figure 12 Increase in the Mexican-born population in the USA

Figure 13 Fencing along the US border with Mexico, and the US Border Patrol

Figure 12 shows the increase in the Mexican-born population in the USA. About one in ten Mexican citizens (12 million) live in the USA, half of them illegally. This is the largest immigrant community in the world. However, in the last year or so there is evidence that this migration has gone into reverse (Figure 12) because:

- tougher economic conditions in the USA have made migration less attractive
- the US Border Patrol has made illegal immigration much more difficult to achieve.

In the USA the Federation for American Immigration Reform (FAIR) has opposed large-scale immigration from Mexico, arguing that it:

- undermines the employment opportunities of low-skilled US workers
- has negative environmental effects because of the increased population
- threatens established US cultural values.

The recent global economic crisis saw unemployment in the USA rise to about 10 per cent, the worst job situation for 25 years. Immigration always becomes a more sensitive issue in times of high unemployment. FAIR has also highlighted the costs to local taxpayers of illegal workers in terms of education, emergency medical care, detention, and other costs that have to be borne.

Those opposed to FAIR see its actions as uncharitable and arguably racist. Such individuals and groups highlight the advantages that Mexican and other migrant groups have brought to the country.

The impact on Mexico

Sustained large-scale labour migration has had a range of impacts on Mexico, some of them clear and others debatable. Significant impacts include:

- the high value of remittances, which totalled over $22 billion in 2011 – this is the world's biggest flow of remittances and as a national source of income for Mexico is only exceeded by its oil exports
- reduced unemployment pressure as migrants tend to leave areas where unemployment is particularly high
- lower pressure on housing stock and public services
- changes in population structure with emigration of young adults, particularly males
- loss of skilled and enterprising people
- migrants returning to Mexico with changed values and attitudes.

It remains to be seen whether the recent migration reversal is temporary or permanent. Much depends on the relative fortunes of the US and Mexican economies.

Case study analysis

1. What are the main reasons for such a high level of international migration from Mexico to the USA?
2. Describe the change in the Mexican-born population of the USA shown in Figure 12.
3. Suggest why immigration from Mexico is a controversial issue in the USA.
4. Briefly discuss the impact of emigration on Mexico.

Population structure

A 60th birthday party with three generations present

Key questions

- How does population structure vary between countries at different levels of economic development?
- What are the implications of different types of population structure?

● Variations in population structure

The structure of a population is the result of the processes of fertility, mortality and migration. The most studied aspects of **population structure** are age and sex (gender). Other aspects of population structure that can also be studied include race, language, religion, and social/occupational group.

Population pyramids

Age and sex structure can be illustrated by the use of **population pyramids** (Figure 1). Pyramids can be used to portray either absolute or relative data. Absolute data show the figures in thousands or millions while relative data shows the numbers involved in percentages. Each bar represents a five year age-group. The male population is represented to the left of the vertical axis with females to the right.

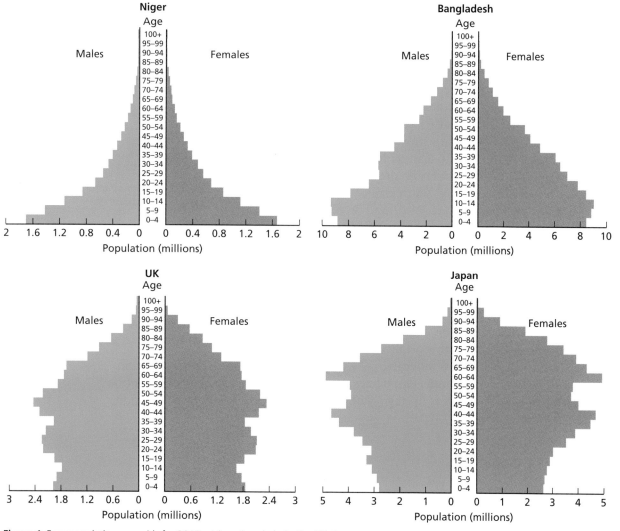

Figure 1 Four population pyramids for 2013 – Niger, Bangladesh, the UK, Japan

Demographic transition and changing population structure

Population pyramids change significantly in shape as a country progresses through demographic transition and economic development:

- The wide base of Niger's pyramid reflects extremely high fertility. The birth rate in Niger is 46/1000, one of the highest in the world. The marked decrease in width of each successive bar indicates relatively high mortality and limited life expectancy. The death rate at 11/1000 is high, particularly considering how young the population is. The infant mortality rate is a very high at 81/1000. Life expectancy in Niger is 58 years, 52 per cent of the population is under 15, with only 3 per cent aged 65 or over. Niger is in stage 2 of demographic transition.

- The base of the pyramid for Bangladesh is narrower than that of Niger, reflecting a considerable fall in fertility after decades of government-promoted birth control programmes. The fact that the 0–4 and 5–9 bars are narrower than the bar immediately above is evidence of recent falls in fertility. The birth rate is currently 23/1000. Falling mortality and lengthening life expectancy is reflected in the relatively wide bars in the teenage and young adult age groups. The death rate at 6/1000 is almost half that of Niger. The infant mortality rate is 43/1000. Life expectancy in Bangladesh is 69 years, 31 per cent of the population are under 15, while 5 per cent are aged 65 or over. Bangladesh is an example of a country in stage 3 of demographic transition.

- In the pyramid for the UK, much lower fertility still is illustrated by narrowing of the base. The birth rate in the UK is only 13/1000. The relatively uniform width of the bars for the working-age population indicates a significantly higher life expectancy than for Bangladesh. The death rate in the UK is 9/1000, with an infant mortality rate of 4.3/1000. Life expectancy is 80 years, 18 per cent of the population are under 15, while 17 per cent are 65 or over. The UK is in stage 4 of demographic transition.

- The final pyramid, for Japan, has a distinctly inverted base reflecting the lowest fertility of all four countries. The birth rate is 9/1000. The width of the rest of the pyramid is a consequence of the highest life expectancy of all four countries. The death rate is 10/1000 with infant mortality at 2.3/1000. Life expectancy is 83 years. Japan has only 13 per cent of its population under 15, with 24 per cent aged 65 or over. With the birth rate lower than the death rate, Japan is experiencing natural decrease and thus the country is in stage 5 of demographic transition.

Table 1 compares the basic demographic rates for the four countries with Gross National Income (GNI) per capita (per person). The data are given at Purchasing Power Parity (PPP), which takes account of the different costs of living in the four countries. Not all of the figures for natural increase match the figures for birth rate and death rate due to rounding. The difference in GNI per capita between Niger and Bangladesh is consistent with the general difference between countries in stages 2 and 3 of demographic transition, although many countries in stage 3 have much higher GNI per capita figures than Bangladesh. The much higher GNI figures for the UK and Japan are also to be expected. There is no significant difference in GNI per capita between countries in stages 4 and 5 of demographic transition.

Table 1 Population and economic data for the four countries, 2012

Country	Birth rate (per1000)	Death rate (per1000)	Rate of natural increase (%)	GNI PPP per capita ($)
Niger	46	11	3.5	720
Bangladesh	23	6	1.6	1810
UK	13	9	0.4	35840
Japan	9	10	−0.2	34610

Figure 2 provides some useful tips for understanding population pyramids. A good starting point is to divide the pyramid into three sections:

- the young dependent population
- the economically active population
- the elderly dependent population.

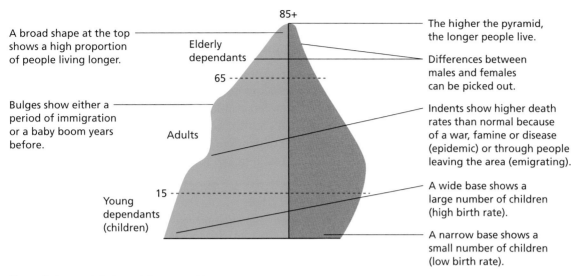

Figure 2 An annotated population pyramid

> ## Interesting note
> At 52 per cent, Niger has the highest percentage of population under 15 in the world. Japan and Monaco have 24 per cent of their populations aged 65 and over, again the highest in the world.

Population structure: differences within countries

In countries where there is strong rural-to-urban migration, the population structures of the areas affected can be markedly different. These differences show up clearly on population pyramids. Out-migration from rural areas is age-selective, with single young adults and young adults with children dominating this process. Thus the bars for these age groups in rural areas affected by out-migration will indicate fewer people than expected in these age groups.

In contrast, the population pyramids for urban areas attracting migrants will show age-selective in-migration, with substantially more people in these age groups than expected. Such migrations may also be sex-selective. If this is the case it should be apparent on the population pyramids.

The implications of different types of population structure

The fact that Niger and Bangladesh have large numbers of young people in their populations has implications for both countries. For example, these young people have to be housed, fed, educated and looked after in terms of health. All this costs money, and governments have to allocate resources to cater for these needs. Conversely, countries such as the UK and Japan have large numbers of older people in their populations. Older people have different needs which governments have to provide for.

The dependency ratio

Dependants are people who are too young or too old to work. The **dependency ratio** is the relationship between the working or economically active population and the non-working population. The formula for calculating the dependency ratio is as follows:

$$\text{The dependency ratio} = \frac{\% \text{ population aged } 0\text{–}14 + \% \text{ population aged } 65 \text{ and over}}{\% \text{ population aged } 15\text{–}64} \times 100$$

A dependency ratio of 60 means that for every 100 people in the economically active population there are 60 people dependent on them. The dependency ratio in developed countries is usually between 50 and 75. In contrast, developing countries typically have higher ratios which may reach over 100. In developing countries, children form the great majority of the dependent population. In contrast, in developed countries there is much more of a balance between young and old dependants.

The dependency ratio is important because the economically active population will in general contribute more to the economy in terms of income tax, sales taxes and the taxes on the profits made by businesses. In contrast, the dependent population tend to be bigger recipients of government funding, particularly for education, healthcare and public pensions. An increase in the dependency ratio can cause significant financial problems for governments if it does not have the financial reserves to cope with such a change.

The dependency ratio is an internationally agreed measure. Partly because of this it is a very crude indicator. For example:

- In developed countries, few people leave education before the age of 18 and a significant number will go on to university and not get a job before the age of 21. In addition, while some people will retire before the age of 65, others will go on working beyond this age. Also a significant number of people in the economically active age group, such as parents staying at home to look after children, do not work for various reasons. The number of people in this situation can vary considerably from one country to another.
- In developing countries a significant proportion of children are working full- or part-time before the

age of 15. In some developing countries there is very high unemployment and underemployment within the economically active age group.

However, despite its limitations the dependency ratio does allow reasonable comparisons to be made between countries. It is also useful to see how individual countries change over time. Once an analysis using the dependency ratio has been made, more detailed research can look into any apparent anomalies.

Activities

1 Who are dependants?
2 Define the term dependency ratio.
3 Identify one limitation of the dependency ratio.
4 Calculate the dependency ratios for Niger, Bangladesh, the UK and Japan from the information given earlier in this Topic.

Case study: The Gambia – a country with a high dependent population

Figure 3 A school in the Gambia

The Gambia, in West Africa, is a small country with a young population which has placed big demands on the resources of the country. The population of 1.8 million is forecast to grow to 4 million by 2050. 95 per cent of the country's population are Muslim and until recently religious leaders were against the use of contraception. In addition, cultural tradition meant that women had little influence on family size. Children were viewed as an economic asset because of their help with crop production and tending animals. One in three children aged 10–14 are working. The country suffered from high infant and maternal mortality. In 2012 the infant mortality rate was 70/1000. With 44 per cent of the population classed as young dependants and only 2 per cent as elderly dependants, the dependency ratio is 85. This means that for every 100 people in the economically active population in the Gambia there are 85 people dependent on them.

The World Health Organization has stressed the link between rapid population growth and poverty for the

Gambia and other countries. Many parents in the Gambia struggle to provide basic housing for their families. There is huge overcrowding and a lack of sanitation, with many children sharing the same bed. Rates of unemployment and underemployment are high, and wages are low with parents struggling to provide even the basics for large families.

The government has insufficient financial resources for education and health. Because there are not enough schools, many of them operate a two-shift system, with one group of pupils attending in the morning and a different group attending in the afternoon. The shortage of teachers means that some are working 12 hours a day. General facilities are poor and sanitation facilities are inadequate. School books are in very short supply.

Another sign of population pressure is the large number of trees being chopped down for firewood. As a result desertification is increasing at a rapid rate. 'Forest educators' are working in rural areas in particular in an attempt to improve this situation.

In recent years the government has introduced a family planning campaign which has been accepted by religious leaders. It has been working with a non-governmental organisation (NGO) called Futures, to deliver contraceptives and family planning advice to rural areas. The scheme has been subsidised by the World Health Organization. To some extent there has also been a change in male attitudes to family size and contraception. This has been very important to the success of the campaign.

Case study analysis

1 Describe the location of the Gambia.
2 What is the dependency ratio in the Gambia and why is it so high?
3 What are the problems of such a high rate of dependency in the Gambia?

Population density and distribution

The coast of Antarctica – the continent that has no permanent residents

Figure 2 Low population density on the west coast of Ireland

Key question

● What are the factors that influence the density and distribution of population?

Population density is the average number of people per square kilometre (km²) in a country or region. **Population distribution** is the way that the population is spread out over a given area, from a small region to the Earth as a whole. Figure 1 shows the global distribution of population using a dot map. Areas with a high population density are said to be **densely populated**. Regions with a low population density are **sparsely populated**.

Figure 3 High population density in Cairo, Egypt

Table 1 shows the density of population by world region. The huge overall contrast between the more developed and less developed worlds is very clear. The average density in the less developed world is more than two and a half times that of the more developed

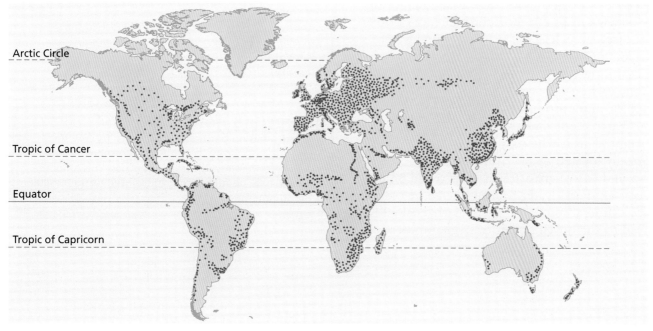

Figure 1 Dot map showing areas of high population density

world. North America (16 per km²) and Oceania (4 per km²) have the lowest population densities of all the world regions. However, the overall difference between the developed and developing worlds is largely accounted for by the extremely high figure for Asia (134 per km²). Population density is increasing most in regions and countries that have the fastest rates of population growth.

Table 1 Variations in world population density, 2012

Region	Population density (people per km²)
World	52
More developed world	27
Less developed world	70
Africa	35
Asia	134
Latin America/Caribbean	29
North America	16
Europe	32
Oceania	4

The average density figure for each region masks considerable variations. The most uniform distributions of population occur where there is little variation in the physical and human environments. Steep contrasts in these environments are sharply reflected in population density. People have always avoided hostile environments if a reasonable choice has been available. Look at an atlas map of the world illustrating population density. Now look at world maps of relief, temperature, precipitation and vegetation. Note the low densities associated with high altitudes, polar regions, deserts and rainforests. More detailed maps can show the influence of other physical factors such as soil fertility, natural water supply and mineral resources.

Areas of low soil fertility have been avoided from the earliest times of settlement as people looked for more productive areas in which to settle. Water supply has always been vitally important. This is why so many settlements are historically located by (a) rivers (b) lakes (c) springs and (d) where artesian wells could be dug to access aquifers (water-bearing rocks) below the surface. Mineral resources, particularly coalfields, have led to the development of large numbers of

settlements in many countries. Although mining may eventually cease when the resource runs out, the investment in infrastructure (housing, railways, roads etc.) over time usually means that the settlement will continue. The Ruhr coalfield in western Germany, once the most productive in Europe, now has only a few working mines remaining. However, it is still one of the most densely populated regions of Europe. But mining settlements in very hostile environments such as Alaska, Siberia and the Sahara desert may be abandoned when mining stops.

The more advanced a country is, the more important the elements of human infrastructure become in influencing population density and distribution. While a combination of physical factors will have decided the initial location of the major urban areas, once such towns or cities reach a certain size, economies of scale and the process of cumulative causation ensure further growth. As a country advances, the importance of agriculture decreases and employment relies more and more on the secondary and tertiary sectors of the economy which are largely urban based. The lines of communication and infrastructure between major urban centres provide opportunities for further urban and industrial location.

Activities

1 Define (a) population density (b) population distribution.
2 To what extent does population density vary by world region (Table 1)?
3 Briefly discuss the factors that influence low population density.
4 What are the factors encouraging high population density?

Interesting note

Antarctica has no permanent residents. Its very small temporary population is made up of researchers and scientists from many different countries. This amounts to about 1000 people during the winter, rising to 5000 in the summer months.

Case study: Sparsely and densely populated areas in North America

North America has a low population density compared with most other parts of the world. The USA has an average of 33 persons per km², while Canada has only 3 per km². In both countries population is highly concentrated in some areas while large expanses of land elsewhere are very sparsely settled (Figure 4). Very few people live in the cold, dry and mountainous regions.

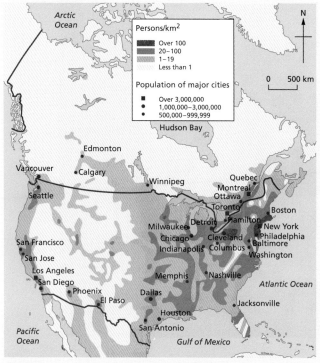

Figure 4 Population density of North America

The Canadian northlands: a sparsely populated region

Figure 5 A remote community in the Canadian northlands

The Canadian northlands comprise that part of Canada lying north of 55°N. Figure 4 shows that virtually the whole area has a population density of less than one person per km².

The influence of low temperature is very clear in the north and largely explains why 75 per cent of Canadians live within 160 km of the main border with the USA. Winters are cold with most of the region having a mean January temperature below –20 °C. Summers are short, becoming increasingly shorter further north. The climate in much of the region is beyond the limits of agriculture, which is a key factor in explaining the very low rural population density.

Much of the northlands are affected by permafrost. Here the ground is permanently frozen to a depth of about 300 metres. In summer the top metre or so thaws out, resulting in a marshy, waterlogged landscape. Life is extremely difficult in the permafrost environment of the northlands and, apart from the Inuit and other native groups, the few people living there are mainly involved in the exploitation of raw materials and in maintaining defence installations, although the role of tourism is expanding.

The great distances separating the generally small communities in the northlands and the severe environmental conditions in this vast region have created substantial economic, engineering and maintenance difficulties for transportation development. Immense areas of the northlands are lacking in surface communications. Not one of the railway lines extending into the northlands crosses the Arctic Circle! The road system is also very sparse, the most important elements of which are the Alaska, Mackenzie and Dempster highways. The northern limit of the Mackenzie Highway is Yellowknife on the northern shore of the Great Slave Lake. The town, which has a population of over 19 000, was founded in 1935 after the discovery of rich deposits of gold. Yellowknife is the capital and largest settlement in the Northwest Territories.

The use of water transport is dictated by location and season, with many water transportation routes frozen over for much of the year. For many communities, air transport is the only link they have to the outside world.

Case study analysis

1 With the help of an atlas, draw a sketch map of the Canadian northlands. Show the main physical and human features of the region.
2 What is the average population density of this region?
3 Why is this the most sparsely populated region in North America?

The north-east of the USA: a densely populated region

Figure 6 Chicago, a major city in north-eastern USA

In the USA the greatest concentration of population is in the north-east, the first area of substantial European settlement. By the end of the nineteenth century it had become the greatest manufacturing region in the world and in the USA it became known as 'the manufacturing belt'. The region stretches inland from Boston and Washington to Chicago and St Louis. By the 1960s the very highly urbanised area between Boston and Washington had reached the level of a **megalopolis**. This is the term used to describe an area where many conurbations exist in relatively close proximity. The region is sometimes referred to as 'Boswash' after the main cities at its northern and southern extremities. Apart from Boston and Washington the other main cities in this region are New York, Philadelphia and Baltimore.

New York is classed as a 'global city' because it is one of the world's three great financial cities along with Tokyo and London. With a population of 8.4 million, New York is the most densely populated city in the USA. The population of New York alone is much greater than the entire Canadian northlands. The population of the larger Metropolitan Area of New York is 18.9 million.

The region also includes many smaller urban areas. Much of the area has an average density over 100 per km². Population densities are of course much higher in the main urban areas. The rural parts of the region are generally fertile and intensively farmed. The climate and soils at this latitude are conducive to agriculture, unlike much of the Canadian northlands. Many people living in the rural communities commute to work in the towns and cities. The region has the most highly developed transport networks in North America. Although other parts of the country are growing at a faster rate, the intense concentration of job opportunities in the north-east will ensure that it remains the most densely populated part of the continent in the foreseeable future.

Case study analysis

1 With the aid of an atlas, draw a sketch map of the north-east region to show its main physical and human features.
2 What is the population density of this region?
3 Why is this the most densely populated region in North America?

1.5 Settlements and service provision

Dispersed settlement, Arabba, Italy

Key questions

- What are the main patterns of settlement?
- What are the factors that influence the sites, growth and functions of settlements?
- What are settlement hierarchies and how do they affect people?

Most of us live in settlements, and most of us take them for granted. And yet there is a huge variety of settlements, and they are changing rapidly. For example, some settlements in rural areas differ greatly from those in urban areas, although the distinction between them is becoming less clear. In developing countries large cities are growing at the expense of rural areas, despite a recent movement out of some very large cities or 'supercities'. Population change, technological developments and changing lifestyles are having a tremendous impact on settlement geography.

In this section we look at the size, development and function of rural and urban settlements. We begin with rural settlements and examine their pattern, site and situation, function and hierarchy. We examine the characteristics of land use and describe the problems of urban areas in the developed and developing worlds, and consider possible solutions to these problems. We also look at the impacts on the environment as a result of urbanisation, and examine possible solutions to reduce these impacts.

● Rural settlement

A settlement is defined as a place in which people live and where they carry out a variety of activities, such as residence, trade, agriculture and manufacturing. Most rural settlements are hamlets and villages, although not all are. The study of rural settlement includes:

- pattern
- form (or shape)
- site and situation
- function and hierarchy
- change.

Pattern

A **dispersed settlement** pattern is one in which individual houses and farms are widely scattered throughout the countryside (Figure 1). It occurs when farms or houses are set among their fields or spread out along roads, rather than concentrated on one point. They are common in sparsely populated areas, such as the Australian outback and the Sahel region of Africa, and in recently settled areas, such as after the creation of the Dutch Polders. The enclosure of large areas of common grazing land into smaller fields separated by hedges, led to a dispersed settlement pattern. This happened because it became more convenient to build farmhouses out in the fields of the newly established farms. Similarly, the break-up of large estates (particularly in England during the sixteenth and seventeenth century) also led to a dispersed settlement pattern. In areas where the physical geography is quite extreme (too hot or cold, wet or dry) there is likely to be a low population density, and a poor transport network, which discourages settlement.

Figure 1 Dispersed settlement – Dingle Peninsula, west coast of Ireland

Nucleated settlements are those in which houses and other buildings are tightly clustered around a central feature such as a church, village green or crossroads (Figure 2). Very few houses are found in the surrounding fields. Such nucleated settlements are usually termed **hamlets** or **villages** according to their size and/or **function**. A number of factors favour nucleation:

- joint and cooperative working of the land – people live in nearby settlements
- defence, for example hilltop locations, sites within a meander or within walled cities such as Jericho
- shortage of water causing people to locate in areas close to springs
- swampy conditions which force settlements to locate on dry ground
- near important junctions and crossroads as these favour trade and communications
- in some countries, the government has encouraged people to live in nucleated settlements, such as the Ujaama scheme in Tanzania, the kibbutzim in Israel and the communes in China.

A linear pattern occurs when settlements are found along a geographical feature, for example, along a river valley, a major transport route (see Figure 7 on page 38 for the potential site of linear settlements).

Figure 2 Nucleated settlement, Mgwali, Eastern Cape, South Africa

Village form

Village **form** refers to shape (Figure 3). In a **linear settlement**, houses are spread out along a road or a river. This suggests the importance of trade and transport during the growth of the village. Linear villages are also found where poor drainage prohibits growth in a certain direction. In the rainforests of Sarawak (Malaysia), many of the longhouses are generally spread alongside rivers (Figure 4).

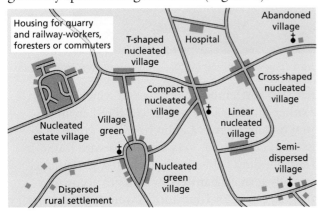

Figure 3 Village shapes

Figure 4 A Malaysian longhouse

Cruciform settlements occur at the intersection of roads and usually consist of lines of buildings radiating out from the crossroads. The exact shape depends on the position of the roads and the amount of infilling that has since taken place. By contrast, a **green village** consists of dwellings and other buildings, such as a church, clustered around a small village green or common, or other open space. In South Africa ring villages are formed where the houses, called kraals, are built around an open area.

Site and situation

The **site** of a settlement is the actual land on which a settlement is built, whereas the **situation** or position is the relationship between a particular settlement and its surrounding area. In the past geographers have emphasised the importance of physical conditions on the pattern of settlement, land tenure and the type of agriculture practised. Increasingly, social and economic factors are important, especially in explaining recent changes in rural settlements.

Early settlers took into account advantages and disadvantages of alternative sites for agriculture and housing. These included:

- availability of water – necessary for drinking, cooking, washing, as a source of food supply, and for transport
- freedom from flooding – but close to the flooded areas as river deposits form fertile soils
- level sites to build on – but they are less easy to defend
- local timber for construction and fuel
- **aspect**, for example sunny south-facing slopes (in the northern hemisphere) as these are warmer than north-facing slopes and are therefore better for crop growth
- proximity to rich soils for cultivation and lush pasture for grazing
- the potential for trade and commerce, such as close to bridges or weirs, near confluence sites, at heads of estuaries, points of navigation and upland gaps.

A **dry point site** is an elevated site in an area of otherwise poor natural drainage. It includes small hills (knolls) and islands. Gravel terraces along major rivers are well favoured. Water supply and fertile alluvial soils as well as the use of the valley as a line of communication are all positive advantages.

A **wet point site** is a site with reliable supply of water from springs or wells in an otherwise dry area. Spring line villages at the foot of the chalk and limestone ridges are good examples. **Spring line settlements** occur when there is a line of sites where water is available.

Some hilltop villages suggest that the site was chosen to avoid flooding in a marshy area as well as for defence. Villages at important river crossings are excellent centres of communications.

Settlement hierarchy

The term **hierarchy** means 'order'. Settlements are often ordered in terms of their size. Dispersed, individual households are at the base of the rural settlement hierarchy. At the next level are hamlets (Figure 5). A hamlet is a very small settlement, consisting of a small number of houses or farms, with very few services. The trade generated by the population, which is often less than 100 people, will only support **low-order services** such as a general store, a small post office or a pub. By contrast, a village is much larger in population. Hence it can support a wider range of services, including school, church or chapel, community centre and a small range of shops (Table 1). Higher up the hierarchy are towns and cities offering many more services and different types of service. As Table 1 shows, there are more settlements lower down the hierarchy – the higher up you go, the fewer the number of each type of settlement. Thus, for example, there are far fewer cities in a country than there are villages.

Rural settlements offer certain functions and services. Only basic or low-order functions are found in the smaller hamlets whereas the same functions and services are found in larger settlements (villages and market towns) together with more specialised ones – **high-order functions**. The market towns draw custom from the surrounding villages and hamlets as well as serving their own population.

The maximum distance that a person is prepared to travel to buy a good is known as the **range** of a good. Low-order goods have a small range whereas high-order goods have a large range. The number of people needed to support a good or service is known

as the **threshold** population. Low-order goods may only need a small number of people (for example 1000) to support a small shop, whereas a large department store might require 50 000 people in order for it to survive and make a profit.

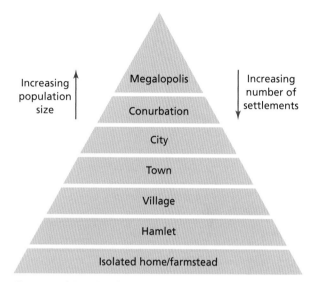

Figure 5 A hierarchy of settlement

Table 1 A simple rural hierarchy

Hamlet	Village	Small market town
General store	General store	General store
Post office	Post office	Post office
	Butcher	Butcher
	Garage	Garage
	Grocer	Grocer
	Hardware store	Hardware store
	Primary school	Primary school, baker, butcher, bike shop, chemist, electrical/TV/radio shop, furniture store, hairdresser, local government offices, restaurant, shoe shop, solicitor, supermarket, undertaker

The area that a settlement serves is known as its **sphere of influence**. Hamlets and villages generally have low spheres of influence whereas larger towns and cities have a large sphere of influence. The definition of hamlet, village and town is not always very clear-cut and these terms represent features that are part of a sliding-scale (continuum) rather than distinct categories.

In general, as population size in settlements increases the number and range of services increases (Figure 6). However, there are exceptions. Some small settlements, notably those with a tourist-related function, may be small in size but have many services. In contrast, some **dormitory** (**commuter**) **settlements** may be quite large but offer few functions or services other than a residential one. In these settlements, people live (reside) in the village but work and shop elsewhere.

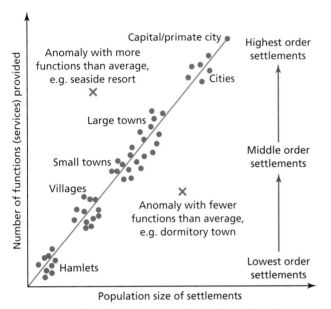

Figure 6 The relationship between population size and number of services

Factors affecting the size, growth and function of settlements

A number of factors affect settlement size, growth and function. In extreme environments settlements are generally small. This is because the environment is too harsh to provide much food. Areas that are too hot or too cold, wet or dry usually have small, isolated settlements. In contrast, settlements have managed to grow in areas where food production is favoured. If there is more food produced than the farmers need, then non-farming services can be supported. In the early days these included builders, craftsmen, teachers, traders, administrators and so on. Thus, settlements in the more favoured areas had greater potential for growth, and for a wider range of services and functions.

Some environments naturally favoured growth and hence a large size. In the north-east of the USA, settlements on the lowland coastal plain were able to farm and trade (Figure 7). Those that had links inland as well, such as New York, were doubly favoured (Figure 8).

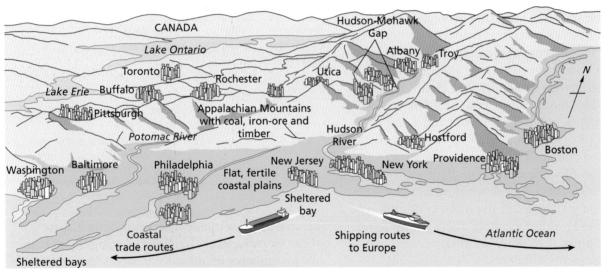

Figure 7 Settlement sites in the north-east of the USA

Figure 8 New York developed because of its excellent trading position – inland as well as overseas

Trade and communications have always been important. Cairo grew as a result of being located at the meeting point of the African, Asian and European trade routes. It also benefited from having a royal family, being the government centre, and having a university and all kinds of linked trades and industries such as food and drink, and textiles. Similarly, Paris grew because of its excellent location on the Seine. Not only could the river be crossed at this point, it could also be used for trade (Figure 9).

Other centres had good raw materials. In South Africa, the gold deposits near Johannesburg, and the diamonds at Kimberley and Bloemfontein, caused these settlements to grow as important mining and industrial areas.

Functions change over time. Many settlements that were formerly fishing villages have become important tourist resorts. The Spanish costas are a good example. Many Caribbean settlements, such as Soufrière in St Lucia, have evolved into important tourist destinations. In the developed world, many rural settlements have now become dormitory settlements – this is related to good **accessibility** to nearby urban centres (Figure 10). Increasingly, many rural settlements in the developed world are also becoming centres of industry, as new science parks locate in areas such as Silicon Valley in California, formerly an agricultural region. South Korea has industrialised and urbanised over the last fifty years or so, and the rural population had declined to just 17 per cent by 2011.

Interesting note
Baniachong in Bangladesh claims to be the world's largest village. The area covers about 75 km² and contains around 70 000 people!

Figure 9 The Seine was a vital factor in enabling the growth of Paris into a city of international importance

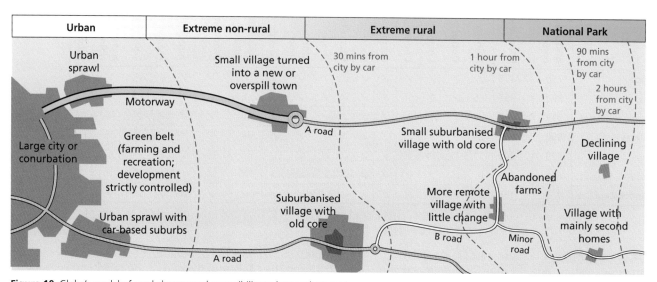

Figure 10 Cloke's model of rural change and accessibility to large urban centres

Other centres have become important due to political factors. New capital cities such as Brasilia, Canberra and Ottawa have developed central administrative roles. Other planned cities, such as Putrajaya in Malaysia and Incheon in South Korea, have become centres of high-tech industry.

Case study: Population size and number of services in Lozère

Figure 11 Lozère environment

Table 2 Population change in St-André-Capcèze, 1800–2005

Year	Population
1800	437
1821	455
1841	479
1861	427
1881	383
1901	316
1921	222
1931	190
1962	148
1982	104
1999	145
2006	174

Table 3 Population change at Lozère

Year	Population	Year	Population
1801	130 000	1921	108 000
1821	135 000	1941	94 000
1841	142 000	1961	82 000
1861	138 000	1981	74 000
1881	144 000	2001	76 000
1901	130 000	2011	73 000 (estimate)

Table 4 Services in Lozère

Settlement	Altitude in metres	Population	Railway	Doctor	Chemist	Dentist	Restaurant	Hotel	Post office	Shops	Mobile shop	Cinema	Swimming pool	Swimming (river, lake)	Tennis	Fishing	Canoeing	Horse riding	Skiing
Mende	750	12 378	✓	✓	✓	✓	✓	✓	✓	✓	✓	✓	✓	✓	✓	✓	✓	✓	25
Badaroux	800	897	0.5	✓	6	6	✓	✓	✓	✓	✓	6	6	✓	4	✓	6	✓	12
Bagnois-les-Bains	913	229	6	✓	✓	20	✓	✓	✓	✓	✓	✓	20	✓	✓	✓	✓	16	14
Cubières	900	197	25	9	9	25	✓	✓	9	✓	✓	25	25	20	9	✓	25	25	9
Altier	725	209	11	11	11	11	✓	✓	✓	✓	✓	11	11	11	11	✓	11	16	25
Villefort	605	639	✓	✓	✓	✓	✓	✓	✓	✓	✓	✓	✓	✓	✓	✓	✓	✓	✓
St-André-Capcèze	450	168	✓	✓	✓	✓	✓	✓	✓	✓	✓	✓	✓	✓	✓	✓	✓	✓	14

Key Services available to tourists and residents in settlement: ✓
Numbers show distance in km to nearest service, i.e. 25 = 25 km distant

Lozère is a department in south-east France. It is a mountainous region, and the main economic activities are farming and tourism (Figure 11). However, due to the mountainous relief and poor-quality soil, farming is mainly cattle rearing. Surprisingly, the region has a very low rate of unemployment. This is due to a long history of out-migration of young people in search of work.

Table 2 shows how the population of St-André-Capcèze fell between the 1860s and the end of the twentieth century. However, in recent years the population has increased slightly. This is due to improved communications and easier travel – but the population is an ageing one. Tourism offers some employment, but the jobs are seasonal, part-time, unskilled and often quite poorly paid.

Case study analysis

1 Describe the landscape of Lozère as shown in Figures 11 a and b. Suggest the economic opportunities and difficulties that these landscape produce for their inhabitants.

2 a Draw a line graph to show the change in population in Lozère between 1801 and 2011 (Table 3).
  b Describe the changes in population in the graph you have drawn.
  c Suggest reasons for the changes in population between:
    i 1801 and 1881
    ii 1881 and 1981
    iii 1981 and 2011.

d Compare the population changes in the Lozère department with those in St-André-Capcèze (Table 2).

3 Table 4 shows data for services in seven settlements in Lozère.
  a Choose a suitable method to plot population size against the number of services.
  b Describe the relationship between population size and the number of services for the region.
  c Identify one exception to the pattern and suggest how, and why, it does not fit the pattern.
  d Suggest a hierarchy of settlements based on the information provided.

Activities

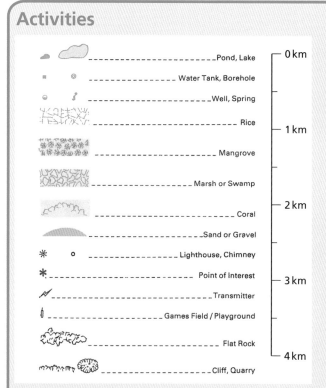

Figure 12 Key to 1:50 000 map of Montego Bay

Study Figure 13, a 1:50 000 map of Montego Bay, Jamaica. Use the key (Figure 12) to help you.

1 What is the grid square reference of (a) the hospital in Montego Bay and (b) the factory on Torboy (Bogue Islands)?

2 What is the grid square reference for (a) the hotel (H) at Doctors Cave and (b) the Fairfield Estate?

3 What is the length of the longest runway at Sangster International Airport?

4 a How far is it, 'as the crow flies' (in a straight line), from the hotel (H) on Bogue Islands to the main buildings at Sangster International Airport?
  b How far is it, by road, from the hotel (H) on Bogue Islands to the main buildings at Sangster International Airport?

5 In which direction is Gordons Crossing from the settlement of Montego Bay?

6 Describe the site of Montego Bay. Suggest why the area grew into an important tourist destination.

7 What types of settlement are found at Pitfour Pen (5598) and Wales Pond (5396)?

8 Suggest reasons for the lack of settlements in grid squares 5497 and 5199.

9 Suggest reasons for the growth of settlements at Bogue (5198) and Granville (5599).

10 Find an example of (a) dispersed settlement and (b) nucleated settlement on the map. Suggest why each type of settlement has that pattern in the area where it is found.

11 Using the map extract, work out a settlement hierarchy for the area. Name and locate an example of (a) an area of isolated, individual buildings, (b) a village, (c) a minor town, (d) a town and (e) a large town.

Figure 13 1:50 000 map of Montego Bay, Jamaica

1.6 Urban settlements

Downtown Seoul

Key questions
- What are the characteristics of urban land use?
- How does urban land use vary between countries at different levels of development?
- What is the effect of change in land use and rapid urban growth?

● Urban land use

The growth of cities in the nineteenth and early twentieth centuries produced a form of city that was easily recognisable by its **urban land use**. It included a central commercial area, a surrounding industrial zone with densely packed housing, and outer zones of suburban expansion and development. Geographers have spent a lot of time modelling these cities to explain 'how they work'.

Every model is a simplification. No city will 'fit\' these models perfectly, but there are parts of every model that can be applied to most cities in the developed world. All models are useful because they focus our attention on one or two key factors.

Interesting note
Hong Kong is the only city in the world with more completed skyscrapers than New York City.

Land value (bid rent)

The value of land varies with different land uses. For example, it varies for retail, office and residential land uses. Retail land uses are attracted to more expensive central areas. Land at the centre of a city is the most expensive for two main reasons: it is the most accessible land to public transport, and there is only a small amount available. Land prices generally decrease away from the central area, although there are secondary peaks at the intersections of main roads and ring roads. Change in levels of accessibility, due to private transport as opposed to public transport, explains why areas on the edge of town are often now more accessible than inner areas.

Burgess's concentric model (1925)

This is the basic model (Figure 1b). Burgess assumed that new migrants to a city moved into inner city areas where housing was cheapest and it was closest to the sources of employment. Over time residents move out of the inner city area as they become wealthier. In his model, housing quality and social class increase with distance from the city centre. Land in the centre is dominated by commerce as this sector is best able to afford the high land prices, and requires highly accessible sites. In the early twentieth century, public transport made the central city the most accessible part of town. Beyond the centre is a manufacturing zone that also includes high-density, low-quality housing to accommodate the workers. As the city grows and the **central business district (CBD)** expands, the concentric rings of land use are pushed further out. The area of immediate change next to the expanding CBD is known as the *zone in transition* (usually from residential to commercial).

Hoyt's sector model (1939)

Homer Hoyt's **sector model** emphasised the importance of transport routes and the incompatibility of certain land uses (Figure 1c). Sectors develop along important routeways, while certain land uses, such as high-class residential and manufacturing industry, deter each other and are separated by buffer zones or physical features.

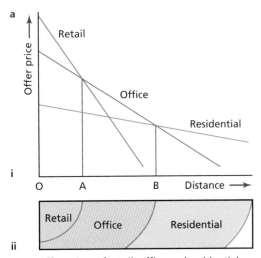

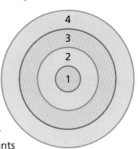

b Concentric zone model (Burgess, 1925)

- model based on Chicago in the 1920s
- the city is growing spatially due to immigration and natural increase
- the area around the CBD has the lowest status and highest density housing
- residents move outwards with increasing social class and their homes are taken by new migrants

Key to diagrams b and c

1 CBD (central business district)
2 Zone in transition/light manufacturing
3 Low-class residential

4 Medium-class residential
5 High-class residential
6 Heavy manufacturing

a i
Offer prices of retail, office and residential uses with distance from the city centre:
i section across the urban value surface
ii plan of the urban value surface

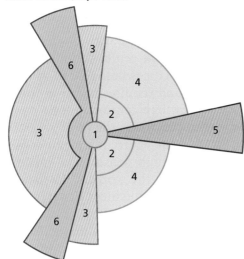

c Sector model (Hoyt, 1939)

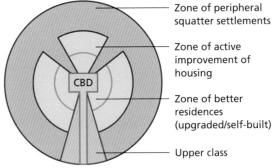

d Latin American city structure

Zone of peripheral squatter settlements

Zone of active improvement of housing

Zone of better residences (upgraded/self-built)

Upper class

Figure 1 Bid rent theory and urban land use models

Urban land use in developing countries

There are a number of models of cities in developing countries. One of the most common is the model of a Latin American city (Figure 1d). The CBD has developed around the colonial core, and there is a commercial avenue extending from it. This has become the spine of a sector containing open areas and parks, and homes for the upper- and middle-income classes. These areas have good-quality streets, schools and public services. Further out are the more recent **suburbs**, with more haphazard housing and fewer services. More recent squatter housing is found at the edge of the city. Older and more established squatter housing is found along some sectors that extend in towards the city centre. Conditions in these areas near the city centre are better than in the more recent areas at the edge. In addition, those living in the central areas are closer to centres of employment and are more likely to find work. Industrial areas are found scattered along major transport routes, with the latest developments at the edge.

Land use zoning in developing countries

A number of models describe and explain the development of cities in developing countries. These include several key points:

- The rich generally live close to the city centre whereas the very poor are more likely to be found on the periphery.
- Better-quality land is occupied by the wealthy.
- Segregation by wealth, race and ethnicity is evident.
- Manufacturing is scattered throughout the city.

Activities

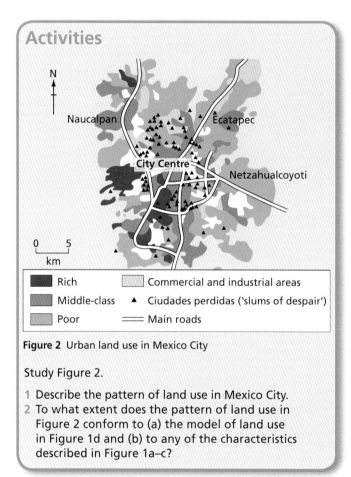

Figure 2 Urban land use in Mexico City

Study Figure 2.

1 Describe the pattern of land use in Mexico City.
2 To what extent does the pattern of land use in Figure 2 conform to (a) the model of land use in Figure 1d and (b) to any of the characteristics described in Figure 1a–c?

● Internal structure of towns and cities

The central business district

The central business district (CBD) is the commercial and economic core of the city, the area that is most accessible to public transport and the location with the highest land values. It has a number of characteristic features.

- Multistorey development – high land values force buildings to grow upwards, so the total floor space of the CBD is much greater than the ground space.
- Concentration of retailing – high levels of accessibility attract shops with high range and threshold characteristics, such as department stores in the most central areas, while specialist shops are found in less accessible areas.
- Concentration of public transport – there is a convergence of bus routes on the CBD.
- Concentration of offices – centrality favours office development.
- Vertical zoning – shops occupy the lower floors for better accessibility, while offices occupy upper floors.
- Functional grouping – similar shops and similar functions tend to locate together (increasing their thresholds).
- Low residential population – high bid rents can only be met by luxury apartments.
- Highest pedestrian flows – due to the attractions of a variety of commercial outlets and service facilities.
- Traffic restrictions are greatest in the CBD – pedestrianisation has reduced access for cars since the 1960s.
- The CBD changes over time – there is an assimilation zone (the direction in which the CBD is expanding) and there is a discard zone (the direction from which it is moving away).

There are, however, many problems in the CBD, such as a lack of space, the high cost of land, congestion, pollution, a lack of sites, planning restrictions, and strict government control.

The core–frame concept

The core-frame concept suggests that the CBD can be divided into two – an inner **core** where most of the department stores and specialist shops are found, and an outer **frame** where coach and train stations, offices and warehouses may be located (Figure 3). The CBD core and the frame are closely connected and the CBD core may advance into the frame, just as the frame may advance into the core as parts of the CBD become run down.

Core and frame elements of the CBD

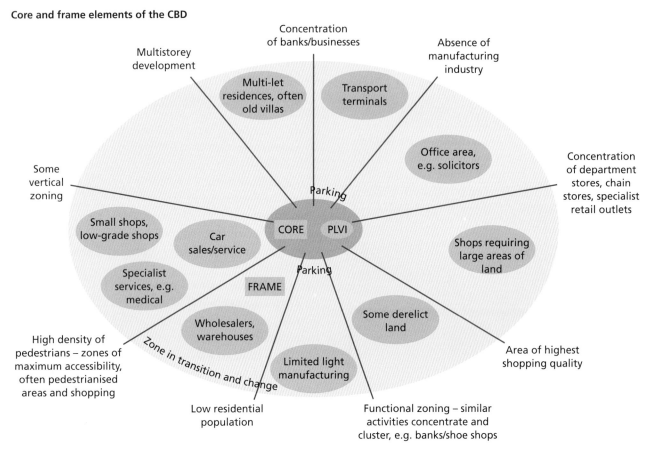

Multistorey development

Concentration of banks/businesses

Absence of manufacturing industry

Concentration of department stores, chain stores, specialist retail outlets

Some vertical zoning

Multi-let residences, often old villas

Transport terminals

Office area, e.g. solicitors

Parking

CORE · PLVI

Small shops, low-grade shops

Car sales/service

FRAME

Parking

Shops requiring large areas of land

Specialist services, e.g. medical

Some derelict land

Area of highest shopping quality

High density of pedestrians – zones of maximum accessibility, often pedestrianised areas and shopping

Zone in transition and change

Wholesalers, warehouses

Limited light manufacturing

Low residential population

Functional zoning – similar activities concentrate and cluster, e.g. banks/shoe shops

PLVI = peak land value intersection: the highest rated, busiest, most accessible part of a CBD

Factors influencing CBD decline

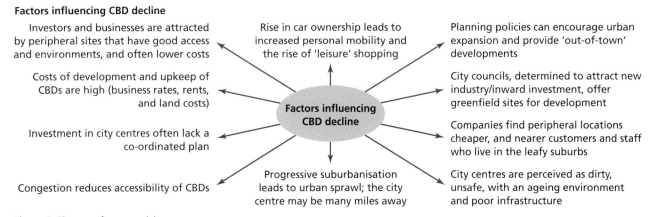

Investors and businesses are attracted by peripheral sites that have good access and environments, and often lower costs

Rise in car ownership leads to increased personal mobility and the rise of 'leisure' shopping

Planning policies can encourage urban expansion and provide 'out-of-town' developments

Costs of development and upkeep of CBDs are high (business rates, rents, and land costs)

City councils, determined to attract new industry/inward investment, offer greenfield sites for development

Factors influencing CBD decline

Investment in city centres often lack a co-ordinated plan

Companies find peripheral locations cheaper, and nearer customers and staff who live in the leafy suburbs

Congestion reduces accessibility of CBDs

Progressive suburbanisation leads to urban sprawl; the city centre may be many miles away

City centres are perceived as dirty, unsafe, with an ageing environment and poor infrastructure

Figure 3 The core–frame model

Residential zones

In most developed countries, as a general rule, residential densities decrease with distance from the CBD. This is due to a number of reasons:

- Historically, more central areas developed first and supported high population densities.
- There is greater availability of land with increased distance from the CBD.

- Improvements in transport and technology allow people to live further away from their place of work in lower-density areas.

However, this pattern can be disrupted by:

- low densities in the CBD, as residential land use cannot compete with commercial land use to meet the high bid rents

- the location of high-rise peripheral estates, increasing densities at the margins of the urban area
- 'green belt' restrictions which artificially raise population densities in the suburbs.

Population densities tend to change over time, with peak densities decreasing and average densities increasing.

The pattern of population density declining with distance can be observed in most cities, but this pattern also changes over time. After a period of expansion, city centres start to decline following suburbanisation. This is sometimes followed by a repopulation of the inner city if the centre is redeveloped.

Case study: Gentrification and relocation in Cape Town, South Africa

In South Africa, the phenomenon of **gentrification** is commonly associated with the resurrection of downtown Johannesburg and the rebirth of Woodstock in Cape Town. Both areas share a common denominator for gentrification: a growing middle class with disposable incomes and a taste for all things 'designer'.

Site

Woodstock is an inner-city suburb of Cape Town, South Africa. It is located between the docks of Table Bay and the lower slopes of Devil's Peak, about 1 km east of the centre of Cape Town (Figure 4). Woodstock covers an area of less than 5 km² and has a population of over 11 500. This gives it a population density of approximately 2300 per km².

History

In the middle of the nineteenth century, notably after the arrival of the railway line, Woodstock became a fashionable seaside suburb with a beach that stretched as far as the Castle of Good Hope. However, in the 1870s and 1880s Woodstock grew rapidly due to ease of access to the harbour, improved transport, and increased industrialisation. The first glass manufactured in South Africa was made at the Woodstock Glass Factory in 1879.

By the 1950s Woodstock had ceased to be a seaside resort. However, many people began to move into Woodstock during the 1970s and 1980s, creating the foundation for the **urban renewal** that was to start in the late 1990s.

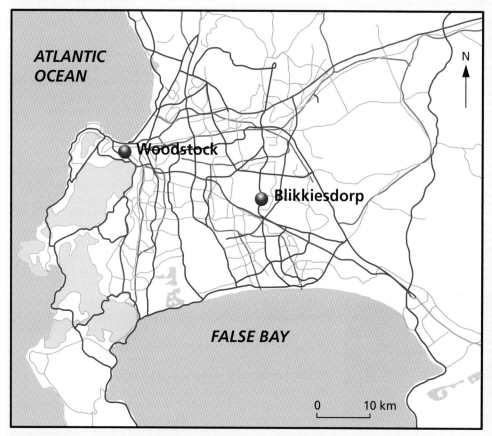

Figure 4 Location of Woodstock and Blikkiesdorp, Cape Town

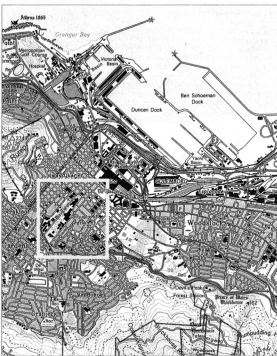

Figure 5 Road network in Woodstock

Urban renewal

Many of the lower parts of Woodstock became run down in the second half of the twentieth century, and litter, crime and drugs became serious issues (compare this with some of the issues in Detroit – see page 60). But young professional workers took advantage of affordable Victorian semi-detached homes, many of these being renovated and restored. Fashionable restaurants, ICT and other businesses and offices have sprung up in converted warehouses, abandoned buildings and even in the former Castle brewery (Figure 6).

Figure 6 Services in Woodstock

Contrasting views of gentrification

There are different views on the regeneration of this part of Cape Town. Some claim that any upgrading or regeneration will be labelled 'gentrification', with negative connotations of less wealthy local people being forced out by profit-hungry developers. However, another view is that these early developers are out-of-work young professionals who, seeing an opportunity, through hard work make the best of their creativity and contacts. Neighbourhoods with the right mix of potential and urban decay attract students, young professionals and artists in search of cheap accommodation. Once they have made the area acceptably safe and more economically viable, the area can become the target for property developers to enter the market. This causes the rents and the price of land to rise, forcing the poorer residents to cash in and sell their properties. Cape Town's 'design community' has recently made Woodstock its home. Here, artists and young designers still manage to retain a foothold, though they may have to settle for smaller studios and higher rents.

Often, the problem with gentrification is as much about the people who are displaced as it is about those who manage to stay on but feel increasingly alienated by the changing image of their neighbourhood. For those who stay on, the problem of traffic congestion and car security issues increase. But there is much more at stake for many local residents than the loss of parking.

According to the Anti-Eviction Campaign, some families have not been able to afford the escalating rates of Woodstock and other central areas, leading to their eviction and removal to such locations as Blikkiesdorp.

Relocating the urban poor: Blikkiesdorp, Cape Town

Blikkiesdorp, which is Afrikaans for 'Tin Can Town', was given its name by residents because of the row-upon-row of tin-like one-room structures throughout the settlement (Figure 7). It was built in 2007 and comprises approximately 1600 one-room structures. Its population today is estimated at around 15 000.

Figure 7 Living conditions in Blikkiesdorp

The official name of Blikkiesdorp is 'Symphony Way Temporary Relocation Area'. It is a relocation camp made up of corrugated iron shacks. The structures have walls and roofs made of thin tin and zinc sheets. Washing, sanitation and water facilities are shared between four of these structures.

Conditions and criticisms

Blikkiesdorp is regarded as unsafe and it has a high crime rate and substandard living conditions. Extremes of weather mean that it can be extremely hot or cold, and sometimes very windy, blowing sand through the shacks. Residents have been reported to be suffering from depression.

Case study analysis
1 Compare the site and situation of Woodstock and Blikkiesdorp as shown in Figure 4.
2 Describe the pattern of roads as seen in Figure 5.
3 Identify and comment on the types of service shown in Figure 6.
4 Describe the living conditions as suggested by Figure 7.

Industrial areas

There are a number of industrial zones in most cities in developed countries. These include:

- traditional inner city areas close to railways and/or canals
- areas that require access to water, for example here industries such as imports and exports are located close to docks
- areas where there is good access and good availability of land
- edge-of-town/**greenfield** suburban sites close to airports.

Open spaces

In general the amount of open space increases towards the edge of town. This is because the value of land is lower towards the edge, and there is more land available. Nevertheless, there are important areas of open space in many urban areas. Central Park in New York is a good example. In the centre, any areas of open land tend to be small. Many of the open spaces are related to areas that

are next to rivers or formerly belonged to wealthy landowners.

Transport routes

Most city centres are characterised by small, congested roads. As the roads were built when the cities were still small, they were quite small. Now, as private transport is the main form of transport, the volume of traffic for the roads is too great. In contrast, towards the edge of town there are larger motorways and ring roads. These take advantage of the space available. Natural routeways, such as river valleys, are important for the orientation of roads. However, given that many cities are in lowland areas, constraints of the natural environment are generally not great.

The rural-urban fringe

The **rural-urban fringe** is the area at the edge of a city where it meets the countryside. There are many pressures on the rural-urban fringe. These include:

- **urban sprawl**
- more housing
- industrial growth

- recreational pressures for golf courses and sports stadia
- transport
- agricultural developments.

The nature of the pressure depends on the type of urban fringe (Table 1). For example, an area of growth such as Barra de Tijuca outside Rio de Janeiro can be contrasted with an area of decline such as some parts of Detroit.

Table 1 Issues in the urban fringe

Land use	Positive aspects	Negative aspects
Agriculture	Some areas have well-managed farms and smallholdings.	Farms often suffer litter, trespass and vandalism; some land is left derelict in the hope that planning permission for development will be granted.
Development	Some developments are well-sited and landscaped, such as business and science parks.	Some developments, such as out-of-town shopping areas, cause serious pollution. Many businesses are unregulated, e.g. scrap metal/caravan storage.
Urban services	Some, such as reservoirs or cemeteries, may be attractive.	Mineral workings, sewage works, landfill sites and Sunday markets (car boot sales) can be unattractive and polluting.
Transport infrastructure	Some cycleways improve access and promote new development.	Motorways destroy countryside, especially near junctions.
Recreation and sport	Country parks, sports fields and golf courses can lead to conservation.	Stock car racing and scrambling erode ecosystems and create localised litter and pollution.
Landscape and nature conservation	Conservation areas may be included at the edge of the city.	There may be degraded land, e.g. land ruined by fly-tipping; many conservation areas are under threat.

The growth of out-of-town shopping centres

Shopping in many more developed countries has changed from an industry dominated by small firms to one being led by large companies. The retailing revolution has focused on superstores, **hypermarkets** and **out-of-town** shopping centres (Figure 8). These are located on 'greenfield' suburban sites with good accessibility and plenty of space for parking and future expansion. The increasing use of out-of-town shopping centres, and the trend for less frequent shopping, has led to the closure of many small shops that relied on regular sales of daily items.

The changes in retailing have been brought about by:

- suburbanisation of more affluent households
- technological change, for example, more families own a deep-freezer
- economic change, with higher standards of living, especially including car ownership
- traffic congestion and inflated land prices in city centres
- social changes, such as more working women.

The initial out-of-town developments came in the late 1960s and early 1970s. Now more than 20 per cent of shopping expenditure in developed countries takes place in out-of-town stores.

Figure 8 An out-of-town shopping centre in Bandar Seri Begawan, Brunei

Activities

Figure 9 An out-of-town shopping centre near Oxford, UK

Study Figure 9.

1 What are the advantages of this site for the supermarket?
2 Which population groups benefit from out-of-town developments such as the one shown in the photograph? Give reasons for your choice.
3 Describe the land uses shown in Figure 10a–d, which show contrasting areas in Bandar Seri Begawan, Brunei. Identify each of the main land uses and their likely location in the city.

a

c

b

d

Figure 10 Land use in Bandar Seri Begawan, Brunei

Table 2 The advantages and disadvantages of out-of-town shopping centres

Advantages	Disadvantages
There is plenty of free parking.	They destroy large amounts of undeveloped greenfield sites.
There is lots of space so shops are not cramped.	They destroy valuable habitats.
They are new developments so are usually quite attractive.	They lead to pollution and environmental problems at the edge of town.
They are easily accessible by car.	An increase in impermeable surfaces (shops, car parks, roads etc.) may lead to an increase in flooding and a decrease in water quality.
Being large means the shops can sell large volumes of goods and often at slightly lower prices.	They only help those with cars (or those lucky enough to live on the route of a courtesy bus) – people who do not benefit might include the elderly, those without a car, those who cannot drive.
Individual shops are larger so can offer a greater range of goods than smaller shops.	Successful out-of-town developments may take trade away from city centres and lead to a decline in sales in the CBD.
Being on the edge of town means the land price is lower so the cost of development is kept down.	Small businesses and family firms may not be able to compete with the large multinational companies that dominate out-of-town developments – there may be a loss of the 'personal touch'.
Developments on the edge of town reduce the environmental pressures and problems in city centres.	They cause congestion in out-of-town areas.
Many new jobs may be created both in the short term (construction industry) and in the long term (retail industry and linked industries such as transport, warehousing, storage, catering etc.).	Many of the jobs created are unskilled.

Edge towns – a new form of development on the rural-urban fringe

Case study: Barra da Tijuca, an 'edge city' in Brazil

Barra da Tijuca is a recent 'edge city' – an exclusive edge-of-town development – and is now a key Olympic site. Land in southern Rio de Janeiro was already fully developed by the early 1960s. At that time Lucio Costa, a French-born urban planner, produced a plan to develop the area known as Jacarepagua, with Barra da Tijuca as its centre.

Barra da Tijuca holds many attractions for Rio's expanding rich middle class, with a pleasant environment, mountain views, forests, lagoons and 20 km of beaches – but it is still only 30 minutes away by motorway from the increasingly congested and polluted city centre. The area expanded from 2500 inhabitants in 1960 to 98 000 in 1991, growing by 139 per cent during the 1980s. Barra is the most recent example of the decentralisation of Rio's wealthy population.

By contrast, the lowlands of Jacarepagua are an area of flat scrub, marshland, lagoons and coastal spits, but it had the potential for enlarging the urban area by a further 122 km². Its neighbourhoods include Anil, Gardenia Azul and Cidade de Deus, and the area was developed from the 1960s onwards. However, it still suffers from a lack of public transport and services, difficult access, overcrowding and poor maintenance.

The 2016 Olympic Park will be built on the lowlands of Jacarepagua to the west of Barra. Improving sanitation systems to reverse long-term pollution of the surrounding lagoons and canals was the first stage. Power lines are to be laid underground. The future site of the athletes' village for Rio 2016, 'Ilha Pura', will become a new neighbourhood after the games. Thirty-one apartment blocks are planned – larger than London's athletes' village – and families will be able to move into these after the event.

However, the existing Jacarepagua race track and a fishing community are still located in the 1.2 million square metres that will become the focus of the games, including an athletics track, swimming and cycling arena, a media centre and a luxury hotel.

Case study analysis

1 Describe the situation of Barra da Tijuca.
2 Comment on population changes in Barra since 1961.
3 In what ways might the 2016 Olympic Games affect Barra and its surrounding area?

Case study: Urban sprawl in Sydney

Urban sprawl is the unplanned and uncontrolled physical expansion of an urban area into the surrounding countryside. It is closely linked to the process of suburbanisation.

Over 75 per cent of Australia's population live in large and/or rapidly growing urban areas along the coastal belt. Sydney accounts for 20 per cent of the national total population and is expected to grow to over 5 million by 2016. Much of Sydney's growth has been associated with urban sprawl. For example, centres at the edge of Sydney, such as Camden, Wyong and Penrith, have been growing more rapidly than long-established suburbs within Sydney. In addition remote rural areas have declined as younger populations have moved to large urban centres in search of employment and entertainment.

The Sydney–Canberra corridor

The Sydney–Canberra corridor extends over about 300 km and contains around 200 000 people. When the populations of Sydney and Canberra are combined, this accounts for 25 per cent of Australia's total population.

The corridor is long but generally sparsely populated. Many sections are used for farming and forestry. Most of the pressures for growth are in the areas closest to Sydney and Canberra. Commuting belts stretch to Wingecarribee south of Sydney, and to Yass and Goulburn north of Canberra. There is increasing decentralisation of population out of both Sydney and Canberra, because people want to enjoy a small-town lifestyle. At the same time, decentralisation of employment to centres such as Parramatta, Liverpool and Campbelltown have reinforced movements into those parts of the corridor. Improvements in road and rail transport have facilitated the growth of these centres at the rural-urban fringe.

There has also been a change in the work patterns in parts of the corridor. Increasingly, ICT is enabling more people to work from home and is reducing the centralisation of Sydney and Canberra.

Change in the corridor

The highest population growth rates are in Sydney and Canberra but settlements close to these centres have also grown rapidly, whereas those furthest away have had the slowest growth. For example, Wollondilly, south

of Sydney, has changed from a farming and mining centre to one based on retailing, services, wholesale and manufacturing. In contrast, Goulburn, further from Sydney, has stagnated owing to the decline in the wool and other pastoral industries, and the closure of the railway workshops. The opening of the Hume Freeway in 1992, which by-passed the town, led to a large-scale decrease in through traffic, and the loss to the town of around US$15–38 million each year.

Queanbeyan, close to Canberra, has grown rapidly as a regional state government office centre and as an industrial centre but over 60 per cent of Queanbeyan's workforce commutes to Canberra.

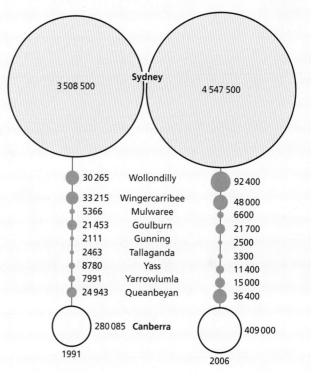

Figure 11 Growth and population distribution in the Sydney–Canberra corridor, 1991 and 2006

Case study analysis

1 Define the term 'urban sprawl'.
2 In what ways is the Sydney–Canberra corridor changing?
3 How has technology allowed the corridor to change?

Case study: Land use in New York

New York City's land area covers about 825 km². The distribution of commercial land use is dominated by two areas, Midtown Manhattan and Downtown (Lower) Manhattan (Figure 12). Lower Manhattan is the centre for finance and banking, containing Wall Street and the Stock Exchange. In contrast, midtown Manhattan has the main shops (Fifth Avenue), theatres (Broadway), hotels and landmark buildings such as the Rockefeller Centre and the Empire Sate Building. Commercial areas occupy less than 4 per cent of the city's land, but they use space intensively. Most of the city's 3.6 million jobs are in commercial areas, ranging from the office towers of Manhattan and the regional business districts of downtown Brooklyn, to the local shopping corridors throughout the city.

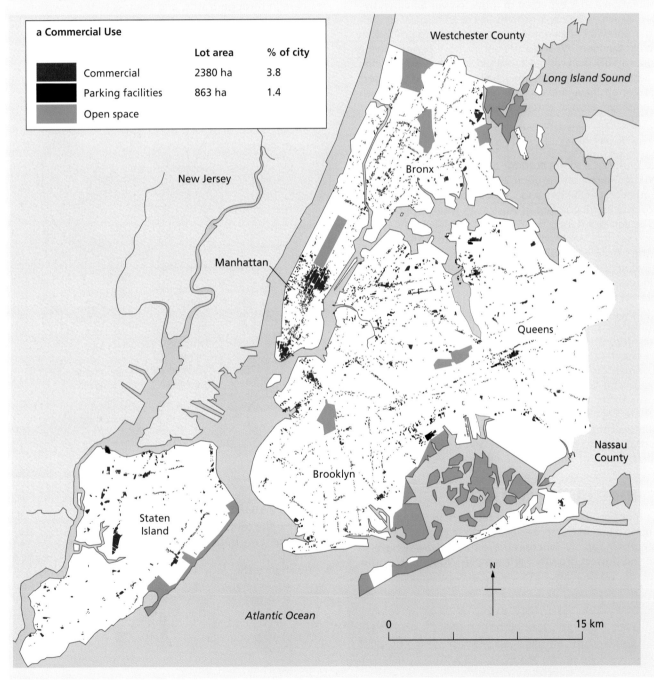

a Commercial Use		
	Lot area	% of city
Commercial	2380 ha	3.8
Parking facilities	863 ha	1.4
Open space		

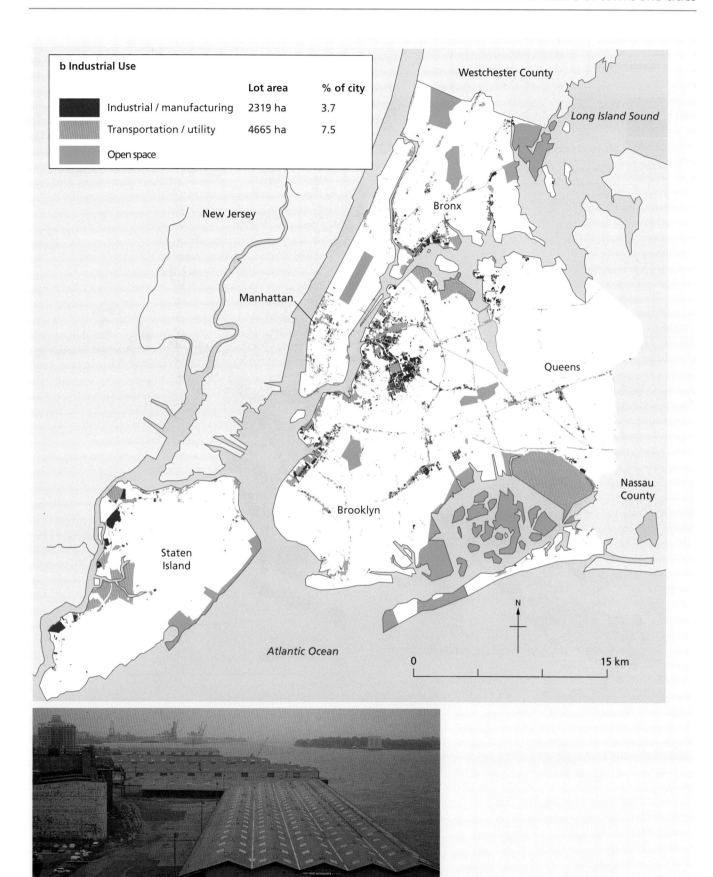

b Industrial Use

	Lot area	% of city
Industrial / manufacturing	2319 ha	3.7
Transportation / utility	4665 ha	7.5
Open space		

Westchester County

Long Island Sound

New Jersey

Bronx

Manhattan

Queens

Nassau County

Brooklyn

Staten Island

Atlantic Ocean

N

0   15 km

BROOKLYNWORKS

Warehousing and industrial land use, Brooklyn, New York

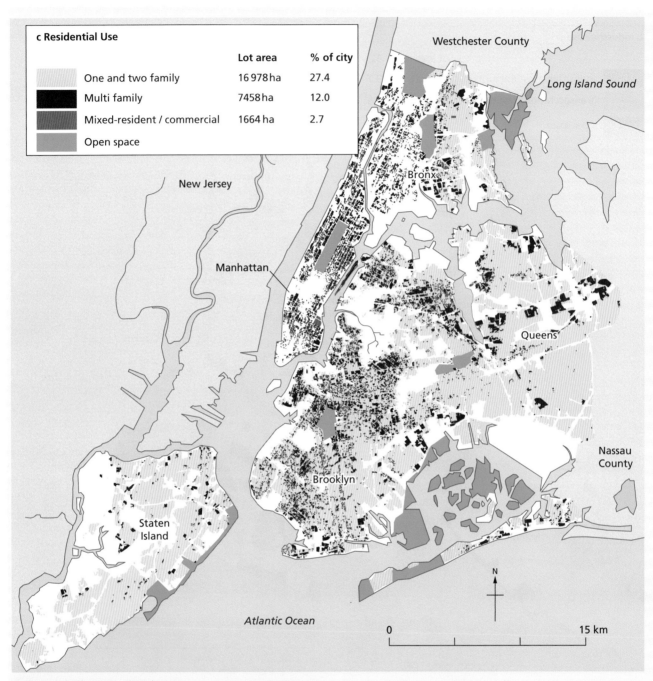

c Residential Use	Lot area	% of city
One and two family	16 978 ha	27.4
Multi family	7458 ha	12.0
Mixed-resident / commercial	1664 ha	2.7
Open space		

Figure 12 Land use in New York

Industrial uses, warehouses and factories occupy 4 per cent of the city's total lot area. They are found primarily in the South Bronx, along either side of Newtown Creek in Brooklyn and Queens, and along the western shores of Brooklyn and Staten Island. Riverfront locations are very important.

Low-density residences, the largest use of city land, are found mostly in Staten Island, eastern Queens, southern Brooklyn, and north-west and eastern Bronx. In contrast, medium- to high-density residential buildings (three or more dwelling units) contain more than two-thirds of the city's housing units but occupy 12 per cent of the city's

total lot area. The highest-density residences are found mainly in Manhattan, and four- to twelve-story apartment houses are common in many parts of the Bronx, Brooklyn and Queens.

Public facilities and institutions – including schools, hospitals and nursing homes, museums and performance centres, places of worship, police stations and fire houses, courts and detention centres – are spread throughout the city and occupy 7 per cent of the city's land.

Approximately 25 per cent of the city's lot area is occupied by public parks, playgrounds and nature preserves, cemeteries, amusement areas, beaches, stadiums and golf courses. Approximately 8 per cent of the city's land is classified as vacant. Staten Island has the most vacant land with more than 2145 hectares, Manhattan the least with less than 160.

Table 3 Land use (%) in New York City by borough

	1	2	3	4	5	6	7	8	9	10	11	12
Bronx	18.1	15.5	2.7	4.3	3.8	2.8	11.6	31.1	2.0	4.3	3.8	100
Brooklyn	22.7	16.0	3.4	3.1	4.9	4.2	6.0	33.9	1.8	3.1	0.9	100
Manhattan	1.3	23.9	12.2	10.2	2.4	6.6	11.7	25.1	1.7	3.0	1.9	100
Queens	36.2	10.6	1.5	3.2	3.7	11.8	4.5	19.7	1.3	5.2	2.3	100
Staten Island	33.6	3.1	0.5	3.4	2.9	7.8	9.6	20.7	0.5	17.5	0.5	100
New York City	27.4	12.0	2.7	3.8	3.7	7.5	7.3	25.4	1.4	6.9	1.8	100

Key

1 Low-density residential areas
2 High-density residential
3 High-density apartments/commercial
4 Commercial/office
5 Industrial/manufacturing
6 Transport/utility

7 Public facilities and institutions
8 Open space
9 Parking facilities
10 Vacant land
11 Miscellaneous
12 Total

Case study analysis

Refer to Figure 12 and Table 3.

1 Describe the distribution of commercial land in New York City.
2 Comment on the distribution of industrial land.

3 Compare the distribution of low-density residential areas with that of high-density residential areas. How do these compare with the distribution of apartments?
4 Compare and contrast the main land uses in Manhattan with those in Queens.

Activities

Table 4 Urban land use in Seoul (%)

Urbanised area	
Residential area	19.5
Commercial and business area	5.9
Mixed residential and business area	12.8
Industrial area	1.0
Public facilities	5.4
Transport	10.9
Urban infrastructure facilities	1.1
Construction sites and wasteland	2.0
Total	58.6
Forest and open space	
Grassland	0
Farmland	0
Rivers and wetland	8.1
Forest	30.9
Unsurveyable	2.4
Total	41.4

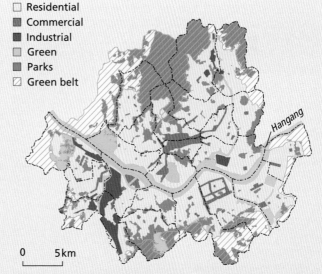

☐ Residential
▨ Commercial
■ Industrial
▢ Green
▦ Parks
▨ Green belt

Figure 13 Land use in Seoul

1 a Using the data in Table 4, draw a pie chart to show land uses in Seoul.
  b Comment on the results you have drawn.
2 Refer to Figure 13.
  a Describe the distribution of industrial land use on the map.
  b Describe, and suggest reasons for, the distribution of open space (green, parks and green belt).
  c Describe the distribution of commercial land use.

● Problems associated with urban growth

A number of problems are associated with the growth of urban areas. These include:

- congestion in the CBD
- very high land prices in the city centre
- overcrowding
- housing shortages
- traffic congestion
- unemployment
- racial conflict
- urban decay and dereliction
- **deprivation**
- pollution of air and water.

These problems are found in most large cities worldwide. Cities in developing countries have the added problems of shanty housing and squatter settlements.

Case study: New York and Detroit

Urban problems in New York

New York's population is declining and changing:

- The total population has fallen by over 10 per cent in the last decade.
- The white population has fallen from 87 per cent to 65 per cent since 1950.
- The middle-class population has fallen by 2 million in the past 20 years.
- The number of elderly residents has risen by 21 per cent since 1950.
- New York has one of the largest black populations in the country, making up 9.1 per cent of the population.
- The population of the South Bronx has declined by nearly 50 per cent.
- Almost 1 million New Yorkers depend on welfare payments.

The inner areas are declining both in terms of population and employment. Much of the housing and industry is dated (Figure 14). By contrast, the suburban periphery, has expanded in terms of population and employment. Only 20 per cent of the workers in Westchester County (a New York suburb and one of the wealthiest counties in the USA) now commute into the city for their work.

Up to 25 per cent of its citizens now live in poverty. A good example of an **inner city** area is the South Bronx. There, the average income is 40 per cent that of the country as a whole, and one-third of residents are on welfare support.

New York's problems arise from:

- changes in the composition of the labour force
- the high living standards which the US economy guarantees for most citizens
- the social strains set up by the continuing existence of an undereducated, unskilled, underpaid and underprivileged minority
- the massive outward movement of the middle class, spurred on by good roads and increased living standards
- the counter movement of lower-income families into the inner city.

Figure 14 Old tenement housing in New York

Urban decline in Detroit

Detroit was once the USA's fourth largest city. Indeed, in 1960 it had the highest per capita income in the USA. Now up to 104 km² of the 360 km² inner city have been reclaimed by nature. Up to 40 000 buildings and parcels of land are vacant (Figure 15). Property prices have fallen by 80 per cent or more. In 2013 there was a three-bedroomed house on Albany Street for sale at $1!

Figure 15 Detroit – a forsaken city

Detroit is the largest US city to declare bankruptcy. Its long-term debts are estimated at over $18 billion, or $27 000 for every resident.

Between 1900 and 1950 Detroit prospered. General Motors (GM), Ford and Chrysler made most of the cars sold in the USA. Detroit's population increased from about 300 000 in 1900 to 1.8 million in 1950, but fell to just 700 000 in 2013 (Figure 16).

Population change in Detroit (million)

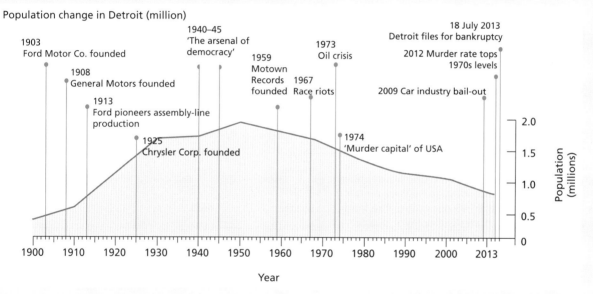

Figure 16 Population change in Detroit

Many of the people are poor and relatively poorly educated – over 80 per cent have no more than a High School diploma. Delivering services to sparsely populated neighbourhoods would be difficult even if the city could afford it.

The causes of Detroit's troubles are complex:

- fading car sales and therefore less tax revenue from the city's large firms
- a shrinking population – many of the richer people have moved away from the city
- very high pension and social welfare costs – the city has an ageing population.

Detroit has paid the price for being too dependent on a single industry – the motor car. It attracted many black workers from the American south to work in its factories, and inequalities in working conditions and living conditions led to race riots in 1943 and 1967. Many white people abandoned the city during the 'white flight' of the 1950s, 1960s and 1970s.

Only 30 per cent of the jobs available in the city are taken by Detroit residents. Over 60 per cent of Detroit's population who work, do so outside of the city. Unemployment had reached 30 per cent by 2013. Over 33 per cent of Detroit's population and nearly 50 per cent of its children live below the poverty line. Nearly 50 per cent of Detroit adults are 'functionally illiterate' and 29 schools closed down in 2009 alone. Detroit's population is now 81.6 per cent Afro-American.

According to a report in *The Economist*, law and order has completely broken down in the inner city, and drugs and prostitution are commonplace. Detroit's murder rate is at a 40-year high. Of the city's 85 000 street lights, half are usually out of service because thieves have stripped them for copper. Only one-third of its ambulances are in working order.

However, there is some hope. Urban farms are appearing. Young people – especially artists and musicians – are moving into Detroit to make use of the abandoned and affordable urban spaces. Policies set in place, such as low rents, good universities and tax breaks, are attempting to attract businesses back to the city.

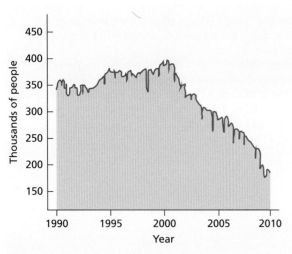

Figure 17 Change in manufacturing employment in Detroit

Case study analysis

1   a   Suggest reasons for Detroit's growth between 1900 and 1950.

    b   Suggest why Detroit's population declined after 1950.

    c   Suggest the likely impacts of a falling population size.

2   Describe the trend in manufacturing employment in Detroit between 1990 and 2010 (Figure 17). Suggest the likely impacts of the changes you have described.

Case study: Problems in Seoul

Housing shortage

Seoul's population grew from 2.5 million in 1960 to over 10 million in 1990 and over 20 million in the Greater Seoul region by 2005. In-migration and the trend towards nuclear families (two generations) rather than the extended family (three generations in the one house) have created a major housing shortage, despite massive building programmes (Figure 18). Less than 45 per cent of the land around Seoul is available for urban development due to the steep terrain and surrounding mountains.

The type of housing is changing too. The typical one-storey one-family house with an inner courtyard is being replaced by high-rise apartment blocks. Such flats have increased from 4 per cent of housing in 1970 to 35 per cent in 1990 and 50 per cent in 2005. Until recently most of the housing was to the north of the river, but a number of satellite towns have been built to the south of the river. This has evened out population density, which on average is over 16 000 people per km² (Table 5).

Figure 18 High-rise buildings in Seoul

Table 5 Population densities in selected cities

	Population/km2
Seoul	16 364
Tokyo	13 092
Beijing	4 810
Singapore	4 773
London	4 671
Paris	8 084
New York	9 721
Los Angeles	3 037

Traffic congestion

Like New York, Seoul experiences massive traffic congestion. In 1975 Korea manufactured fewer than 20 000 cars. By 1994 there were over 2 million cars registered in the Seoul area. Despite improvements to the motorway network, the increase in the population of Seoul and the number of cars in the area means that congestion has increased. In addition, many of the roads in central Seoul are relatively small and unable to handle the large volumes of traffic.

Pollution

As Seoul has grown the amount of air and water pollution has increased. A good example is the Cheong Gye Cheon River in central Seoul (Figure 19a). It had become heavily polluted with lead, chromium and manganese and was a health risk. Restoration of the river has been a central part of the regeneration of central Seoul (Figures 19b and 19c). Previously up to 87 per cent of the city's sewage flowed untreated into the Hangang River. Now Seoul has the capacity to treat up to 3 million tonnes of sewage each day.

a

b

c

Figure 19 The Cheong Gye Cheon River in central Seoul
  a In the 1950s
  b Natural section following restoration
  c Cultural and social section following restoration

Activities

1 Suggest several different reasons why there is poverty in New York.
2 Why is air pollution a problem in large cities?
3 Describe the conditions in Cheong Gye Cheong as shown in Figure 19.

1.7 Urbanisation

Glenmore, South Africa

Key questions
- What is urbanisation?
- What are the reasons for rapid urban growth?
- What are the impacts of urban growth on both urban and rural areas?
- What are the strategies to reduce the negative impacts of urban growth?

Urbanisation is an increase in the percentage of a population living in urban areas. It is one of the most significant geographical phenomena of the twentieth century. Urbanisation takes place when the urban population is growing more rapidly than the population as a whole. It is caused by a number of interrelated factors including:

- migration to urban areas
- higher birth rates in urban areas due to the youthful age structure
- higher death rates in rural areas due to diseases, unreliable food supply, famine, decreased standard of living in rural areas, poor water, hygiene and medication
- rural areas being reclassified as urban areas (this would normally accompany the above factors).

More than two-thirds of the world's urban population is now in Africa, Asia, Latin America and the Caribbean. The population in urban areas in developing countries will grow from 1.9 billion in 2000 to 3.9 billion by 2030. In developed countries, however, the urban population is expected to increase very slowly, from 0.9 billion in 2000 to 1 billion in 2030.

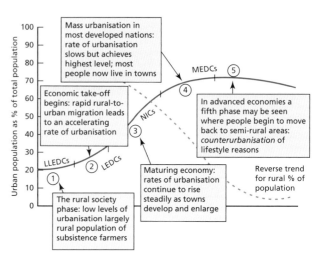

Figure 1 The process of urbanisation

The overall population growth rate for the world for that period is 1 per cent, while the growth rate for urban areas is nearly double, at 1.8 per cent. At that rate, the world's urban population will double in 38 years. Growth will be even more rapid in the urban areas of developing countries, averaging 2.3 per cent per year, with a doubling time of 30 years.

The urbanisation process (Figure 1) in developed countries has stabilised, with about 75 per cent of the population living in urban areas. Latin America and the Caribbean were 50 per cent urbanised by 1960 but here urbanisation is now about 75 per cent. Africa is still predominantly rural, with only 37.3 per cent living in urban areas in 1999, but with a growth rate of 4.87 per cent, the continent of Africa has the fastest rate of urbanisation. By 2030, Asia and Africa will each have more urban dwellers than any other major area of the world. Two aspects of this rapid growth have been the increase in the number of large cities and the historically unprecedented size of the largest cities.

A **megacity** is a city with over 10 million people. In 1950 there was only one megacity: New York City. In 2000 there were 19 megacities, 424 cities with a population of 1 to 10 million, and 433 cities in the 0.5 to 1 million category. By 2015, it is expected that there will be 23 cities with a population over 10 million and of these, 19 will be in developing countries (Table 1).

Table 1 The world's megacities, 1950–2015

	1950		1975		2000		2015	
1	New York	12.3	Tokyo	19.8	Tokyo	26.4	Tokyo	26.4
2			New York	15.9	Mexico City	18.1	Mumbai	26.1
3			Shanghai	11.4	Mumbai	18.1	Lagos	23.2
4			Mexico City	11.2	São Paulo	17.8	Dhaka	21.1
5			São Paulo	10.0	New York	16.6	São Paulo	20.4
6					Lagos	13.4	Karachi	19.2
7					Los Angeles	13.1	Mexico City	19.2
8					Kolkata	12.9	New York	17.4
9					Shanghai	12.9	Jakarta	17.3
10					Buenos Aires	12.6	Kolkata	17.3
11					Dhaka	12.3	Shanghai	17.0
12					Karachi	11.8	Delhi	16.8
13					Delhi	11.7	Manila	14.8
14					Jakarta	11.0	Los Angeles	14.1
15					Osaka	11.0	Buenos Aires	14.1
16					Metro Manila	10.9	Cairo	13.8
17					Beijing	10.8	Istanbul	12.5
18					Rio de Janeiro	10.6	Beijing	12.3
19					Cairo	10.6	Rio de Janeiro	11.9
20							Osaka	11.0
21							Tianjin	10.7
22							Hyderabad	10.5
23							Bangkok	10.1

Most of the world's megacities had slower population growth rates during the 1980s and 1990s, and most of the larger cities are significantly smaller than had been expected. For instance, Mexico City had around 18 million people in 2000 – not the 31 million predicted in 1980. Kolkata in India had fewer than 13 million inhabitants in 2000, not the 40–50 million people predicted in the 1970s.

Several factors help explain this:

- In many cities in the developing world, slow economic growth (or economic decline) attracted fewer people.
- The capacity of cities outside the very large metropolitan centres to attract a significant proportion of new investment was limited.
- Lower rates of natural increase have occurred, as fertility rates have come down.

However, there were some large cities whose population growth rates remained high from the 1980s through to the 2010s, for example Dhaka in Bangladesh and many cities in India and China. Strong economic performance by such cities is the most important factor in explaining this. This growth attracts many young migrants, for whom birth rates are higher in urban areas, and death rates are lower there too.

Given the association between economic growth and urbanisation, a steady increase in the level of urbanisation in low-income nations is only likely to take place if they also have a steadily growing economy.

● Urban problems in developing countries

Table 2 Slum population by region, 2001

Major area/region	Total population (millions)	Urban population (millions)	Urban population (% of total population)	Estimated slum population (thousands)	Estimated slum population (% of urban population)
World	6134	2923	47.7	923 986	31.6
Developed regions	1194	902	75.5	54 068	6.0
Europe	726	534	73.6	33 062	6.2
Other	467	367	78.6	21 006	5.7
Developing regions	4940	2022	40.9	869 918	43.0
Northern Africa	146	76	52.0	21 355	28.2
Sub-Saharan Africa	667	231	34.6	166 208	71.9
Latin America and the Caribbean (LAC)	527	399	75.8	127 567	31.9
Eastern Asia	1364	533	39.1	193 824	36.4
South-central Asia	1507	452	30.0	262 354	58.8
South-eastern Asia	530	203	38.3	56 781	28.0
Western Asia	192	125	64.9	41 331	33.1
Oceania	8	2	26.7	499	24.1
Least developed countries	685	179	26.2	140 114	78.2
Landlocked developing countries	275	84	30.4	47 303	56.5
Small island developing states	52	30	57.9	7 321	24.4

Interesting note

Some of the world's largest slums could be millionaire cities! Neza-Chalco-Itza in Mexico City has a population of 4 million people. Orangi Town in Karachi has an estimated population of 1.5 million, and Dharavi in Mumbai has a population of over 1 million residents. Khayelitsha in Cape Town had 400 000 in 2005, and Kibera in Nairobi has a population of between 200 000 and 1 million!

Case study: Urban problems in Rio de Janeiro

Rio de Janeiro is a city of contrasts, with a huge gap between rich and poor.

Rio's urban problems and their causes

Millions of people have migrated from rural areas and other urban areas in Brazil to Rio de Janeiro in search of work and a better standard of living. The problem is that there are neither enough jobs nor houses for everyone. Therefore, many migrants are forced to make their own homes (on land they do not own) and get whatever casual work they can, often in the informal sector.

Rio de Janeiro has many **slum** communities or *comunidade*. The natural increase in population is much higher in the most recent favelas (squatter settlements, largely in the outer suburbs and rural–urban fringe). As spontaneous settlements are forced to develop on available land, most of the sites have been used in the central and inner urban areas. Many favelas were moved to outer suburban areas. In inner urban areas, newly established favelas were frequently forced to develop on steep hillsides, where landslides are a threat. About 17 per cent of Rio's population are favela-dwellers. They occupy just 6.3 per cent of Rio's land area (Figure 2).

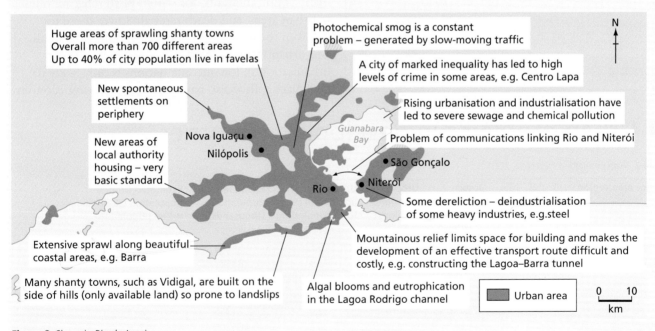

Figure 2 Slums in Rio de Janeiro

There are four main types of slum in Rio:
- squatter settlements or favelas – dense invasions of land with self-built housing on land lacking in infrastructure
- illegal subdivisions of land and/or housing
- invasions – irregular occupations of land still in the process of becoming fully established, generally found in

'risky' areas such as around and beneath viaducts, under electricity lines, on the edge of railways or in public streets and squares
- corticos – old decaying housing that has been rented out without any legal basis. These are mainly located in the central areas and the port area.

Population growth is very rapid in some slums. For example, Rio das Pedras, located in a flooded swamp area, grew to 18 000 within its first two years.

Recently there has been **decentralisation** of activities from the CBD towards the edge of the city. This has been mirrored by the movement of high-income classes to the coastal expansion areas to the east of the city.

The inequality in wealth in Rio is staggering: the richest 1 per cent of the population earns 12 per cent of the income and the poorest 50 per cent earns just 13 per cent of the city's income. The southern zone of the city is the richest area.

Rio's period of extremely rapid growth (1960–80) has slowed but growth is still at the rate of over a million a decade. Public services such as education and health have become inadequate as a result of rapid urban growth. The steep mountains that surround the narrow, flat coastal strips of land have affected the physical growth of Rio. The mountainous relief limits the space for building and makes the development of an effective transport network more difficult. The few existing transport routes have to be used by everyone, which leads to traffic congestion. The mountains surrounding the city trap photochemical smog created by exhaust fumes, resulting in poor air quality.

Raw sewage has been draining straight into the bay, with population growth and industrial growth in Rio worsening the problems this causes. Today there is no marine life left in much of the bay, commercial fishing has decreased by 90 per cent in the last 20 years, and swimming from beaches in the interior of the bay is not advisable. The Lagoa Rodrigo de Freitas is the lake inland from Ipanema and Leblon. Here, in February 2000 the release of raw sewage led to algal blooms and eutrophication which resulted in the death of 132 tonnes of fish.

Case study analysis

1 What proportion of Rio's population live in slums?
2 How much of Rio's land is occupied by slums?
3 Describe the different types of slums in Rio.
4 Comment on the inequality of wealth in Rio.
5 Suggest why there is a housing shortage in Rio.

Activities

1 a Choose an appropriate method to show the global distribution of people living in slums (Table 2).
  b Where is the frequency of slum dwellers highest? Where is it lowest?
  c Comment on your conclusions.
2 Describe one environmental problem that is the result of rapid urban growth.
3 Explain why natural hazards may have a major impact in areas of slum housing.

● Solutions to urban problems

The housing crisis in developing countries

Provision of enough quality housing is another a major problem in developing countries. There are at least four aspects to the management of housing stock:

- quality of housing – with proper water, sanitation, electricity and space (Figure 3)
- quantity of housing – having enough units to meet demand
- availability and affordability of housing
- housing tenure (ownership or rental).

Figure 3 Poor-quality housing in Oisins, Barbados

There are a variety of possible solutions to the housing problem:

- government support for low-income self-built housing
- subsidies for home building
- flexible loans to help **shanty town** dwellers
- slum upgrading in central areas
- improved private and public rental housing
- support for the informal sector/small businesses operating at home
- site and service schemes
- encouragement of community schemes
- construction of health and educational services.

Housing the urban poor

Governments of developing countries would not be able to solve their housing problems even if they were to try. The best that they could be expected to do in an environment of general poverty is to improve living conditions. They should try to:

- reduce the number of people living at densities of more than 1.5 individuals in each room
- increase access to electricity and drinking water
- improve sanitation
- prevent families from moving into areas that are physically unsafe
- encourage households to improve the quality of their accommodation.

A sensible approach is to destroy slums as seldom as possible, on the grounds that every displaced family needs to be rehoused, and removing families is often disastrous. Governments should also avoid building formal housing for the very poor. Sensible governments will attempt to upgrade inadequate accommodation by providing infrastructure and services of an appropriate standard.

There are no easy solutions to housing problems in developing countries because poor housing is merely one manifestation of general poverty. Decent shelter can never be provided while there is widespread poverty.

Urban agriculture

The phrase 'urban agriculture' initially sounds like a contradiction in terms. However, the phenomenon has grown in significance in cities in the developing world over the past 20 years. Evidence suggests that in some cities, urban agriculture may already occupy up to 35 per cent of the land area, may employ up to 36 per cent of the population, and may supply up to 50 per cent of urban fresh vegetable needs.

Table 3 Urban agriculture issues

Advantages	Concerns
Vital or useful supplement to food procurement strategies	Conflict over water supply, particularly in arid or semi-arid areas
Various environmental benefits	Health concerns, particularly from use of contaminated wastes
Employment creation for the jobless	Conflicting urban land issues
Provides a survival strategy for low-income urban residents	Focus on the urban cultivation activities rather than in relation to broader urban management issues
Urban agriculture makes use of urban wastes	Urban agriculture can benefit only the wealthier city dwellers in some cases

Food produced locally in urban areas may have several added benefits. First, it employs a proportion of the city's population. In Addis Ababa dwellers can save between 10 and 20 per cent of their income through urban cultivation. It also diversifies the sources of food, resulting in a more secure supply (Table 3).

New cities

For more developed countries there are more options. At one end of the scale are new towns and cities, such as Brasilia in Brazil, Canberra in Australia, and Gongju-Yongi in South Korea. Gongju-Yongi is a 42 billion-dollar scheme to reduce the importance of Seoul as Korea's capital by 2020 (Figure 4). The relocation is necessary to ease chronic overcrowding in Seoul, redistribution of the state's wealth, and to reduce the danger of a military attack from North Korea. Previous developments have concentrated huge amounts of money, power and up to half of Korea's population in Seoul. Construction began in 2007. Another impressive scheme is the Malaysian new town of Putrajaya.

Figure 4 Gongju-Yongi – the proposed new capital for South Korea: planning board (top) and the scene in 2007 (bottom)

Managing urban problems

Managing transport in cities

Table 4 Traffic problems in cities

Developed world	Developing world
• Increased number of motor vehicles	• Private car ownership is lower
• Increased dependence on cars as public transport declines	• Less dependence on the car, but growing
• Major concentration of economic activities in CBDs	• Many cars are poorly maintained and are heavy polluters
• Inadequate provision of roads and parking	• Growing centralisation and development of CBDs increases traffic in urban areas
• Frequent roadworks	• Heavy reliance on affordable public transport
• Roads overwhelmed by sheer volume of traffic	• Journeys are shorter but getting longer
• Urban sprawl results in low-density built-up areas, and increasingly long journeys to work	• Rapid growth has led to enormous urban sprawl and longer journeys
• Development of out-of-town retail and employment leads to cross-city commuting	• Out-of-town developments are beginning as economic development occurs, e.g. Bogota, Colombia

Table 5 Attempts to manage the transport issue

'Carrots'	'Sticks'
• Park-and-ride schemes – parking at the terminal for a major bus or train route, e.g. Oxford (UK), Brisbane (Australia)	• High car parking charges in city centres, e.g. Copenhagen (Denmark), London, Oxford
• Subsidised public transport systems, e.g. Oxford, Zurich (Switzerland), Brisbane	• Restricted city-centre parking, e.g. Copenhagen, Cambridge (UK)
• Modern electronic bus systems with consumer information on frequency, e.g. Brisbane, Curitiba (Brazil); rapid transit systems – supertrams on dedicated tracks, e.g. Zurich, or underground trains, e.g. Newcastle, Cairo (Egypt)	• Road tolls and road pricing: congestion charges, e.g. Durham (UK), Bergen (Norway) and central London, so that people have to pay to drive through congested areas of the city centre
• Providing bus lanes to speed up buses, e.g. Oxford, London	

Sustainable development in Curitiba

Curitiba, a city in south-west Brazil, is an excellent model for sustainable urban development (Figure 5). It has experienced rapid population growth, from 300 000 in 1950 to over 2.1 million in 1990, but has managed to avoid all the problems normally associated with it. This success is largely due to innovative planning:

- Public transport is preferred over private cars.
- The environment is used rather than changed.
- Cheap, low-technology solutions are used rather than high-technology ones.
- Development occurs through the participation of citizens (bottom-up development) rather than top-down development (centralised planning).

Figure 5 Location of Curitiba in Brazil

Table 6 Sustainable solutions to flooding

Problems (1950s/60s)	Solutions (late 1960s onwards)
Many streams had been covered to form underground canals which restricted water flow.	Natural drainage was preserved – these natural flood plains are used as parks.
Houses and other buildings had been built too close to rivers.	Certain low-lying areas are off-limits.
New buildings were built on poorly drained land on the periphery of the city.	Parks have been extensively planted with trees; existing buildings have been converted into new sports and leisure facilities.
An increase in roads and concrete surfaces accelerated runoff.	Bus routes and cycle paths integrate the parks into the urban life of the city.

Transport

Transport in Curitiba is highly integrated. The road network and public transport system have structural axes. These allow the city to expand but keep shops, workplaces and homes closely linked. There are five main axes of the three parallel roadways:

- express routes – a central road with two express bus lanes
- direct routes
- local roads.

Curitiba's mass transport system is based on the bus. Interdistrict and feeder bus routes complement the express bus lanes along the structural axes. Everything is geared towards the speed of the journey and convenience of passengers:

- a single fare allows transfer from express routes to interdistrict and local buses
- extra wide doors allow passengers to crowd on quickly
- double- and triple-length buses allow for rush hour loads.

The rationale for the bus system was economic as well as sustainability. A subway would have cost up to $80 million per km whereas the express busways were only $200 000 per km. The bus companies are paid by the kilometre of road they serve, not the number of passengers. This ensures that all areas of the city are served.

Rio de Janeiro

Housing

Areas of spontaneous housing in Rio used to be bulldozed without warning. However, the authorities were unable to offer enough alternative housing with the result that the favelas grew again. The authorities have now allowed these areas to become permanent.

The Favela Bairro Project (Favela Neighbourhood Project) began in Rio in 1994. It aimed to recognise the favelas as neighbourhoods of the city in their own right and to provide the inhabitants with essential services. Approximately 120 medium-sized favelas (those with 500–2500 households) were chosen. The primary phase of the project addressed the built environment, aiming to provide:

- paved and formally named roads
- water supply pipes and sewerage/drainage systems
- crèches, leisure facilities and sports areas
- relocation for families who were currently living in high-risk areas, such as areas subject to frequent landslides
- channelled rivers to stop them changing course.

The second phase of the project aimed to bring the favela dwellers into mainstream society and keep them away from crime. This is being done by:

- generating employment, for example by creating cooperatives of dressmakers, cleaners and construction workers, and helping them to establish themselves in the labour market

- improving education and providing relevant courses such as ICT
- giving residents access to credit so that they can buy construction materials and improve their homes.

The project has been used as a model of its type. The government is also helping people to become home-owners.

The mountainous relief of Rio means there is not a great amount of building space available. Development has consequently moved out (decentralised) to create 'edge towns' such as Barra da Tijuca (see page 54). Barra is an example of decentralisation of the rich and upper classes.

Education

A number of developments have taken place to try and improve the quality of the education system. *Amigos da Escola* (school friends) encourages people from the community to volunteer their skills to improve opportunities offered by their local schools. *Bolsa Escola* (school grants) gives monthly financial incentives to low-income families to keep their children at school.

Rocinha is a central favela with a population of about 200 000 inhabitants (Figure 6). Over 90 per cent of the buildings are now constructed from brick and have electricity, running water and sewerage systems. Rocinha has its own newspapers and radio station. There are food and clothes shops, video rental shops, bars, a travel agent and there was even a McDonald's at one stage.

Figure 6 Rocinha, Rio de Janeiro

Many of these improvements and developments are the result of Rocinha's location close to wealthy areas such as São Conrado and Copacabana. Many of the residents work in these wealthy areas that surround Rocinha, and although monthly incomes are low, they are not as low as elsewhere in the city and in Brazil. These regular incomes have allowed improvements to be carried out by the residents themselves.

Transport

- An efficient **bus service**, which covers all areas of the city, has been developed and the prices are within reach of most people, being US$0.50 for each journey made. Alongside the organised bus services, vans operate along the most popular routes and charge the same price.
- A **metro system** has been built and is currently being expanded (Figure 7). A one-way ticket costs US$0.66. Currently the metro does not provide an alternative mode of transport for much of the population, as it is not extensive enough.
- The **Linha Vermelha** and the **Linha Amarela** are two major roads that have been built to try and ease traffic congestion. However, there are indications that the number of private cars on the road has increased since these roads were built, which will reduce the long-term impact they have on traffic congestion.

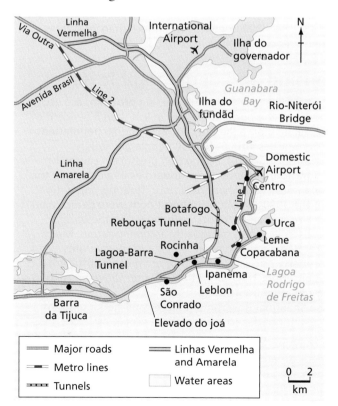

Figure 7 Transport systems in Rio de Janeiro

Water pollution

Much work has already taken place in Rio to improve the sewerage systems. Improving sanitation will also help revive the Lagoa Rodrigo de Freitas, where 4 km of new sewage pipes are currently being installed.

Case study: Urban change in Shanghai

For 700 years Shanghai has been one of Asia's major ports, and it has a varied history. It thrived until 1949, when China closed itself to trade with the west. This changed Shanghai from an international centre of production and trade to an inward-looking city. During the 1970s, China began slowly reopening its economy to the world, and Shanghai was designated one of 14 open cities. The Shanghai Economic Zone was established in 1983, and in the early 1990s an ambitious major programme of redevelopment was started, especially in the eastern hinterland around Pudong.

Since economic reforms began in China in 1978, between 150 million and 200 million Chinese have migrated from rural to urban areas. This may be the largest population movement in human history. China now has over 100 **millionaire cities**. Shanghai has a population of over 17 million and it is expected to reach 23 million by 2020.

Site

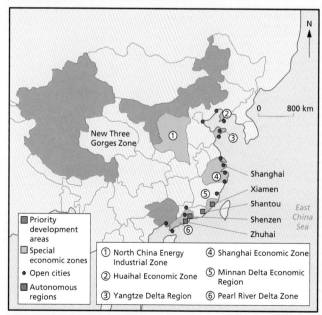

Figure 8 The situation of Shanghai in China

Shanghai developed on a flat, low-lying alluvial plain on the banks of the Yangtze river. Shanghai is also located at the confluence of the Huangpu and Suzhou rivers – so it has an excellent location for shipping and trade. From 1844 British, French, American and Japanese traders owned land in Shanghai. By 1920 it was China's largest and most important city but after the end of the Second World War and following the Communist Revolution in 1949, the foreign influence declined.

Economic change

Between 1949 and 1976 political influences, such as the Great Leap Forward (1958–60) and the Cultural Revolution (1966–76) focused attention away from rural areas, foreign influence and capitalist development. During this period one million people were returned to the countryside.

However, in 1979 the first generation of Special Economic Zones was created. Although Shanghai was not one of them, it benefited from relaxed housing restrictions such as the subdivision and subletting of housing.

In 1984, Shanghai was declared open to foreign development (Figure 8). For much of the next twenty years, Shanghai's economic growth rate was over 12 per cent per annum. In 1990, a new CBD was created in the Pudong area (Figure 9). Banks, stock exchanges and insurance companies moved in. By 2000, over 3000 skyscrapers had been built, including the Shanghai Financial Centre.

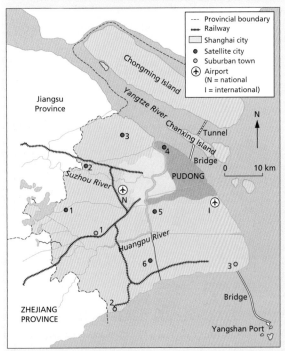

Figure 9 Land use in Shanghai

Having by 1990 established a strong industrial base, the city was well placed to take advantage of the new opportunities offered by globalisation. It became a major centre for export manufacturing based on automobiles,

biotechnology, chemicals and steel, and its service industry sector (trade, finance, real estate, tourism, e-commerce) helped to diversify its economy. Between 1990 and 2000, Shanghai began to re-emerge as a world city. Foreign investment was attracted. Over half the world's top 500 transnational corporations and 57 of the largest industrial enterprises set up in Shanghai, contributing to an annual regional growth rate of over 20 per cent, more than twice the national average. In 2009, Shanghai was ranked the seventh largest city in the world, with a population of 15 million.

Since 1990, the city's manufacturing sector has steadily contracted, shedding almost a million jobs, while the business services, finance and real estate sectors have expanded. Rising demand for highly skilled labour has led to further in-migration, resulting in an increasing disparity in wealth between rich and poor. Shanghai's experience does lend support to the general hypothesis that world city status inevitably leads to a widening gap between rich and poor.

In 2006, Yangshan deepwater port was opened in order to accommodate larger ships than could enter the Huangpu and Suzhou rivers. Yangshan was built on an island about 40 km south-east of Shanghai and connected by a 35 km bridge. It is now one of the world's largest ports.

Today, Shanghai is a city-state within China. It is part of the Yangtze River Delta, the fastest growing urban area in the world, containing 16 megacities including Shanghai. The region has 75 million people and earns 25 per cent of China's GNP – 50 per cent of its foreign direct investment. The city has been described as the largest construction site in the world: 4000 buildings with more than 24 storeys were under construction in 2010.

Housing and demographic issues

Housing shortages and overcrowding problems are acute. Almost half the population lives in less than 5 per cent of the total land area, and in central Shanghai population density reaches 40 000–160 000 people per square kilometre (Figure 10). Population pressure is caused by in-migration, overcrowding, disparities in wealth and the social insecurity of Shanghai's poor 'floating population'. From the 1990s whole neighbourhoods were demolished. Over two million residents were moved to the outer suburbs to live in better-quality accommodation. Many poorer people are unaware of their property rights. As property prices increase, they are given insufficient compensation. They cannot afford alternative housing in their old neighbourhoods.

The Shanghai government has established a series of important policies to address these problems:

- a combination of widespread family planning and medical care, which has controlled fertility levels among the young immigrant population
- compulsory work permits
- educational initiatives to improve immigrant job opportunities.

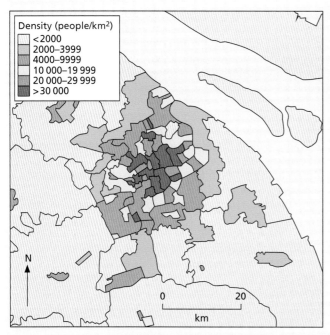

Figure 10 Population density in Shanghai

These initiatives have reduced population density in the heart of the city and increased it in the suburban satellite cities such as Songjiang. This has been a successful strategy, although for many who have been moved out under the decentralisation policy, the journey back to the centre for work is no advantage.

Economic growth has attracted an increasing number of Chinese living overseas and foreign migrants to live in Shanghai. Many of these live in luxury, gated apartments.

Water and air quality

Water quality in Shanghai is a concern: less than 60 per cent of waste water and storm water and less than 40 per cent of sewage flows are intercepted, treated and disposed of. Waste disposal is also a major problem: the Huangpu river receives 4 million cubic metres of untreated human waste every day. The construction industry generates 30 000 tonnes of building waste per day, and municipal landfill sites have almost reached capacity. Nevertheless, since the 1990s there have been marked improvements in sanitation, and almost all households have access to piped water, electricity and a means of waste disposal. Municipal organic waste is now used as fertiliser in the surrounding rural areas.

Shanghai has the highest cancer mortality rate in China, and until recently had the reputation of being the tenth most polluted city in the world. Industry generated over 72 per cent of carbon dioxide emissions, 9 per cent coming from transport systems, and the remainder from domestic use. Coal-fired power stations provide 75 per cent of China's electricity, but contributes to serious emissions of suspended particulate matter, nitrous oxides and sulfur dioxide.

Motor vehicle emissions are particularly harmful in the presence of strong sunlight, when photochemical smog is formed. A product of this is low-level ozone, a harmful irritant responsible for breathing difficulties. Efforts to reduce air pollution levels have been moderately successful: emissions of nitrous oxides have fallen from around 40 ppm to 32, and sulfur dioxide emissions from 50 ppm to around 32 over the same period. Reducing particulate matter has proved to be a much more difficult task.

The government has responded to pollution problems by upgrading the city's transport systems and attempting to limit the growth in car ownership.

Transport

The Shanghai authorities have invested heavily in transport. Eight tunnels and four bridges have been built over/under the Huangpu river. Shanghai's underground system, with a daily capacity of 1.4 million, is now linked to Pudong airport by the world's fastest commercial magnetic levitation train – MAGLEV – capable of reaching 431 km per hour. Other strategies to improve safety have been pedestrianisation and a reduction in the number of bicycles, currently estimated at 9 million and a cause of many road accidents.

Bus lanes have been introduced and over 150 km of cycle/moped lanes have been created. Shanghai developed a metro system in 1995 (Figure 11). An international airport was built in 1999 and a second terminal added in 2008. It is planned to expand fourfold by 2015.

Figure 11 The Shanghai metro system

Coastal flooding

Like many global ports, Shanghai is under threat from coastal flooding, partly due to its low elevation at only 4 metres above sea level, but also from monsoons and tropical cyclones. Future hazard events will be aggravated by climate change and the possibility of a rise in sea level. The problem is compounded by subsidence, which has been caused by over-abstraction of groundwater and the weight of high-rise buildings. Shanghai sank by 2.6 metres between 1921 and 1965, and in 2002 alone by 10.22 millimetres.

Figure 12 Shanghai landscape

Shanghai master plan

In 2000 Shanghai introduced the New Master Plan for Shanghai (2000–20). This includes the whole area and the development of three satellite cities (new towns). This is to reduce congestion and high population densities in central Shanghai.

Dongtan

Dongtan is located on the alluvial island of Chongming. It is a new city planned to produce zero waste, and using energy from clean renewable sources such as wind, solar and biofuels. Grasses will be grown on rooftops and rainfall will be harvested. The city was planned to be compact and car-free – its residents cycling or walking to school/work or to the shops and services. Links to downtown Shanghai – over 50 km away – include new bridges and tunnels.

However, these links threaten to replace the sustainable eco-city with a middle-class suburb for Shanghai workers. It could also attract holiday homes and retirement homes.

Advantages of Dongtan:

- Housing will include affordable housing as well as luxury flats.

- Waste will be treated, rather than discharged into the Yangtze. Landfills are to be allowed, and sewage will be processed for irrigation and composting.
- Renewable energy will be used – especially solar panels and wind turbines.
- Food will be locally sourced from local farmers and fishermen – about one-third of the land in Shanghai is currently used for food production and about one million people still work on the land there.
- Farming will be made more efficient by the introduction of organic fertilisers.
- Direct links between farmers and Shanghai restaurants are being developed.
- Farm tourism is being encouraged through weekend breaks.

Problems associated with Dongtan:

- There may be conflicts over water resources – planned golf courses may use vast amounts of water.
- Dongtan's first phase – by 2010 – housed mainly tourist industries: hardly a sustainable use.
- Commuting to Shanghai is inevitable, leading to a 'middle-class' ghetto.
- The extensive natural areas of freshwater marshes, saltmarshes and tidal creeks will experience major pressures for development.

The rapid development of Shanghai has presented the government and planners with some challenging problems, only some of which have been resolved. The question is whether Shanghai can maintain the principles of sustainability while growing at such a rapid pace.

Case study analysis

1 On what river is Shanghai situated?
2 What are the advantages of Shanghai's site for its economic development.
3 Describe the situation of Shanghai as shown in Figure 8.
4 Suggest reasons to help explain the rapid growth of Shanghai's population.
5 Explain why Shanghai has a problem with air quality and water quality.

● Urbanisation and the environment

Managing environmental problems

Environmental issues that most cities have to deal with include:

- water quality
- dereliction
- air quality
- noise
- environmental health of the population.

There are a range of environmental problems in urban areas (Table 7). These vary over time as economic development progresses. The greatest concentration of environmental problems occurs in cities experiencing rapid growth (Figure 13). This concentration of problems is referred to as the **Brown Agenda**. It has two main components:

- issues caused by limited availability of land, water and services
- problems such as toxic hazardous waste, pollution of water, air and soil, and industrial 'accidents' such as that at Bhopal in 1985.

Table 7 Environmental problems in urban areas

Problems (and examples)	Causes	Possible solutions
Waste products and waste disposal – 25 per cent of all urban dwellers in developing countries have no adequate sanitation and no means of sewage disposal	• Solids from paper, packaging and toxic waste increase as numbers and living standards rise • Liquid sewage and industrial waste both rise exponentially • Contamination and health hazards from poor systems of disposal, e.g. rat infestation and waterborne diseases	• Improved public awareness: recycling etc., landfill sites, incineration plants • Development of effective sanitation systems and treatment plants including recycling of brown water for industrial use • Rubbish management
Air pollution – air in Mexico City is 'acceptable' on fewer than 20 days annually!	• Traffic, factories, waste incinerators and power plants produce pollutants • Some specialist chemical pollution • Issues of acid deposition	• Closure of old factories and importation of clean technology, e.g. filters • Use of cleaner fuels • Re-siting of industrial plants, e.g. oil refineries in areas downwind of settlements
Water pollution (untreated sewage into the Ganga from cities such as Varanasi)	• Leaking sewers, landfill and industrial waste • In some developing countries, agricultural pollution from fertilisers and manure is a problem	• Control of sources of pollution at source by regulation and fining; development of mains drainage systems and sewers • Removal of contaminated land sites
Water supply (overuse of groundwater led to subsidence and flooding in Bangkok)	• Aquifer depletion, ground subsidence and low flow of rivers	• Construction of reservoirs, pipeline construction from long-distance catchment, desalination of salt water • Water conservation strategies
Transport-related issues (average speed of traffic in São Paulo is 3 km/hour)	• Rising vehicle ownership leads to congestion, noise pollution, accidents and ill-health due to release of carbon monoxide, nitrogen oxide and, indirectly, low-level ozone • Photochemical smog formation closely related to urban sprawl	• Introduction of cleaner car technology (unleaded petrol catalytic converters); monitoring and guidelines for various pollution levels; movement from private car to public transport; green transport planning • Creation of compact and more sustainable cities

Figure 13 Environmental problems: **(a)** Seoul – a city experiencing rapid economic growth

Figure 13 Environmental problems: **(b)** Castries – a city in a developing country

Attempts to turn cities green can be expensive. Increasingly local governments are monitoring the environment to check for signs of environmental stress, and then applying some form of pollution management, integrated management, or conservation order to protect the environment.

Activities

1 Describe the environmental problems shown in Figure 13.
2 a Outline the causes of environmental problems in cities.
  b Suggest why these problems might be increasing.
  c Suggest reasons why the potential solutions might not work.

● End of theme questions

Topic 1.1: Population dynamics

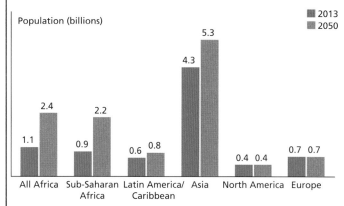

Figure 1 Population by world region 2013 and 2050

1 a Describe the distribution of the world's population by region in 2013.
  b Using the data provided calculate the total population of the world in 2013 and the world's projected population in 2050.
  c What is the total increase in population forecast to be?
2 a To what extent does the projected population growth to 2050 vary by world region?
  b Examine the reasons why population growth is forecast to be much higher in some world regions than others.
3 Suggest why these population forecasts might turn out to be inaccurate.

Topic 1.2: Migration

Destination: developing countries		
Origin	**Destination**	**2013**
Bangladesh	India	3.2
India	United Arab Emirates	2.9
Russian Federation	Kazakhstan	2.4
Afghanistan	Pakistan	2.3
Afghanistan	Iran (Islamic Republic of)	2.3
China	China, Hong Kong SAR	2.3
State of Palestine	Jordan	2.1
Myanmar	Thailand	1.9
India	Saudi Arabia	1.8
Burkina Faso	Côte d'Ivoire	1.5
Destination: developed countries		
Mexico	USA	13.0
Russian Federation	Ukraine	3.5
Ukraine	Russian Federation	2.9
Kazakhstan	Russian Federation	2.5
China	USA	2.2
India	USA	2.1
Philippines	USA	2.0
Puerto Rico	USA	1.7
Turkey	Germany	1.5
Algeria	France	1.5

Figure 2 Main international migration corridors (total migrant populations in millions)

1 a Define international migration.
  b Identify the three largest flows of migrants to destinations in developing and in developed countries.
2 a Which country is the major destination for international migrants?
  b What are the reasons for the popularity of this destination country?
  c Suggest why India appears as both a country of origin and destination.
3 a What is a refugee?
  b Suggest and justify one migration corridor that has been a route for refugees.
4 Discuss the barriers to international migration.

Topic 1.3: Population structure

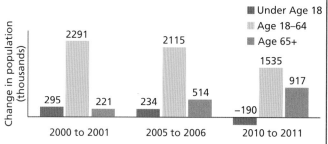

Figure 3 Changes in US population (in thousands) by age group 2000–2001, 2005–6, 2010–11

1 a By how much did the total population of the USA increase between:
    i) 2000 and 2001
    ii) 2005 and 2006
    iii) 2010 and 2011?
2 a To what extent did population change vary between the three age groups?
  b What demographic trend do the data and your analysis of it illustrate?
  c Explain the reasons for this trend.
3 a Explain a simple calculation/ratio that could be used to compare the working and non-working populations of the USA.
  b What concerns might the US government have about the changing population structure of the country?

Topic 1.4: Population density and distribution

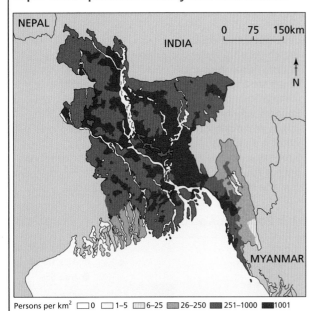

Persons per km² ☐0 ☐1–5 ☐6–25 ☐26–250 ■251–1000 ■1001

Figure 4a Population density and distribution in Bangladesh

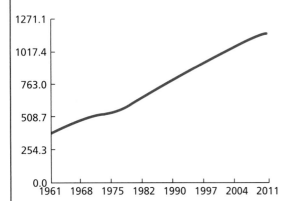

Figure 4b Increase in population density in Bangladesh

1 a Define the terms i) population density ii) population distribution.
  b Describe the population density and distribution in Bangladesh (Figure 4a).
  c Suggest likely reasons for the variations in population density.
2 a Describe the changes in the average population density in Bangladesh between 1961 and 2011.
  b Discuss the reasons for such significant change in population density.
  c Suggest where the largest increases in population density occurred.

Topic 1.5: Settlements and service provision

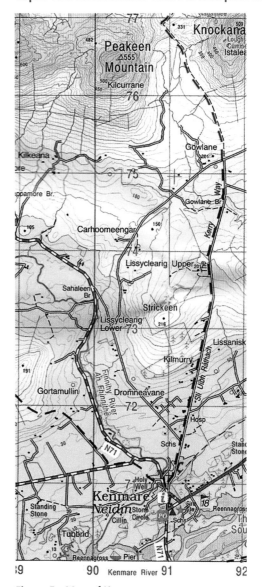

Figure 5 Map of Kenmare

1 a Suggest why a settlement developed at Kenmare (9070, 9171)
  b What type of settlement is Kenmare?
  c Describe the nature of the settlement in 9075. Suggest reasons for this.
  d Identify the services that are found in Kenmare.
  e Suggest a simple hierarchy (with three layers) of settlements in the Kenmare area.

Topic 1.6: Urban settlements

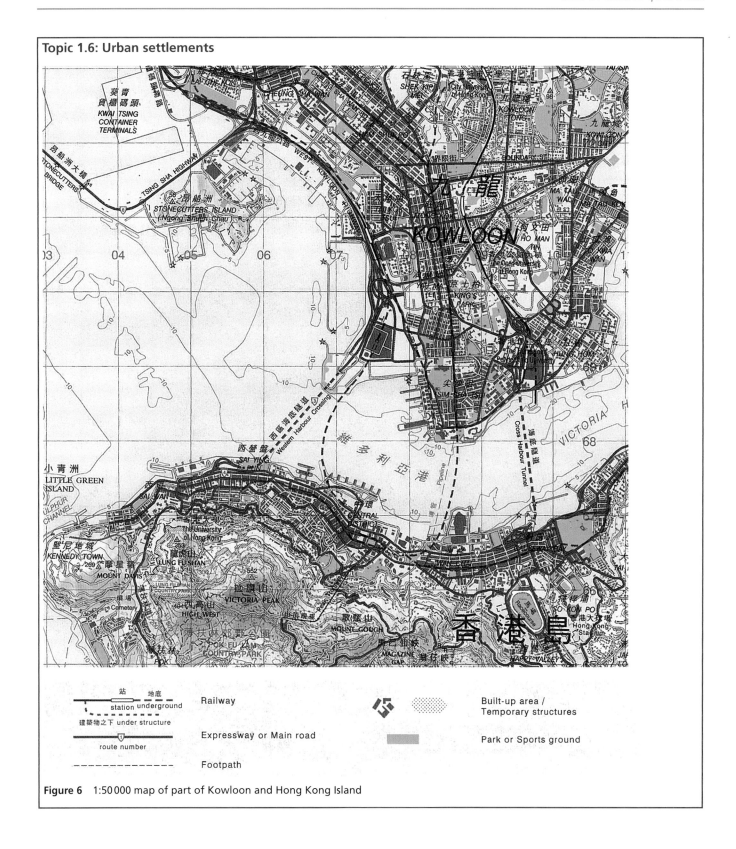

Figure 6   1:50 000 map of part of Kowloon and Hong Kong Island

Study Figure 6.

1 Give a four figure square reference for the Central Business District, Hong Kong Island.
2 Describe the location of the built environment on Hong Kong Island.
3 Describe the road pattern (network) in grid square 0869.

4 Suggest two contrasting ways in which physical geography has influenced the development of Hong Kong.
5 Identify two forms of recreational activity shown on the map.
6 a Identify an area that is likely to have industrial activity.
  b Suggest why this area has potential for economic activity.

Topic 1.7: Urbanisation

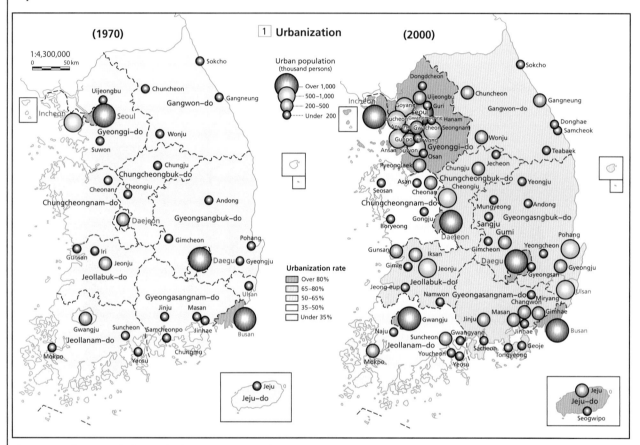

Figure 7   Urban population and urbanisation in South Korea in 1970 and 2000

1 Define the term urbanisation.
2 a Describe the changes in urbanisation in South Korea between 1970 and 2000.
  b Describe the changes in urban growth in South Korea between 1970 and 2000.
3 What factors led to an increase in urbanisation?
4 Outline the positive economic and social effects of urbanisation.
5 Comment on the negative effects of urbanisation.

Theme 2 · The Natural Environment

Topics

Impacts of the Soufrière Hills volcano, Montserrat

Key questions

- What are the main types and features of volcanoes and earthquakes?
- Where do earthquakes and volcanoes occur?
- What are the causes of earthquakes and volcanic eruptions, and what is their effect on people and the environment?
- What are the hazards and opportunities posed by volcanoes?
- What can be done to reduce the impacts of earthquakes and volcanoes?

● Types and features of volcanoes and earthquakes

Volcanoes

A volcano is an opening in the Earth's crust through which hot molten magma (lava), molten rock and **ash** are erupted onto the land. Most volcanoes are found at plate boundaries although there are some exceptions, such as the volcanoes of Hawaii, which are located at an isolated **hotspot**. Some eruptions let out so much material that the world's climate is affected for a number of years. **Magma** refers to molten materials inside the Earth's interior. When the molten material is ejected at the Earth's surface through a volcano or a crack at the surface, it is called **lava**.

Interesting note

- The greatest volcanic eruption in historic times was Tambora in Indonesia in 1815. Some 50–80 km³ of material was blasted into the atmosphere.
- In 1883 the explosion of Krakatoa was heard from as far away as 4776 km!
- The world's largest active volcano is Mauna Loa in Hawaii, 120 km long and over 100 km wide.

Types of volcano

The shape of a volcano depends on the type of lava it contains. Very hot, runny lava produces gently sloping **shield volcanoes**, while thick material produces **cone volcanoes** (Figure 1). These may be the result of many volcanic eruptions over a long period of time. Part of the volcano may be blasted away during an eruption. The shape of the volcano also depends on the amount of change there has been since the last volcanic eruption. Cone volcanoes are associated with destructive plate boundaries, whereas shield volcanoes are characteristic of constructive boundaries and hotspots (areas of weakness in the middle of a plate).

The **chamber** refers to the reservoir of magma located deep inside the volcano. A **crater** is the depression at the top of a volcano following a volcanic eruption. It may contain a lake. A **vent** is the channel which allows magma within the volcano to reach the surface in a volcanic eruption.

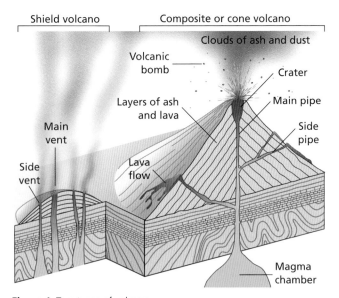

Figure 1 Two types of volcano

Active volcanoes have erupted in recent times, such as Mt Pinatubo in 1991 and Montserrat in 1997, and could erupt again. **Dormant volcanoes** are volcanoes that have not erupted for many centuries but may erupt again, such as Mt Rainier in the USA. **Extinct volcanoes** are not expected to erupt again. Kilimanjaro in Kenya is an excellent example of an extinct volcano. The Le Puys region of France is an area of extinct volcanoes which continue to influence settlements and tourism.

Earthquakes

Earthquakes involve sudden, violent shaking of the Earth's surface. They occur after a build-up of pressure causes rocks and other materials to give way. Most of this pressure occurs at plate boundaries when one plate is moving against another. Earthquakes are associated with all types of plate boundary. The **focus** refers to the place beneath the ground where the earthquake takes place. **Deep-focus earthquakes** are associated with **subduction zones**. **Shallow-focus earthquakes** are generally located along constructive boundaries and along conservative boundaries. The **epicentre** is the point on the ground surface immediately above the focus.

Some earthquakes are caused by human activity such as:

- nuclear testing
- building large dams
- drilling for oil/natural gas (fracking)
- coal mining.

Earthquake intensity: the Richter and Mercalli Scales

In 1935, Charles Richter of the California Institute of Technology developed the **Richter Scale** to measure the magnitude (strength or force) of earthquakes. These are measured on a seismometer and shown on a seismograph (Figure 2). By contrast the **Mercalli Scale** (Table 1) relates ground movement to things that you would notice happening around you. Its advantage is that it allows ordinary eyewitnesses to provide information on the strength of the earthquake. The Richter Scale is logarithmic. This means that an earthquake of 6.0 is ten times greater than one of 5.0, and one hundred times greater than one of 4.0.

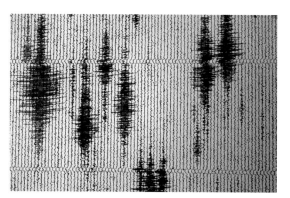

Figure 2 A seismograph reading

Table 1 The Mercalli Scale

Strength	Observations
I	Rarely felt.
II	Felt by people who are not moving, especially on upper floors of buildings. Hanging objects may swing.
III	The effects are noticeable indoors, especially upstairs. The vibration is like that experienced when a truck passes.
IV	Many people feel it indoors, a few outside. Some are awakened at night. Crockery and doors are disturbed and standing cars rock.
V	Felt by nearly everyone – most people are awakened. Some windows are broken, plaster becomes cracked and unstable objects topple. Trees may sway and pendulum clocks stop.
VI	Felt by everyone – many are frightened. Some heavy furniture moves, plaster falls. Structural damage is usually quite slight.
VII	Everyone runs outdoors. Noticed by people driving cars. Poorly designed buildings are damaged.
VIII	Damage to ordinary buildings – many collapse. Well-designed ones survive but suffer slight damage. Heavy furniture is overturned and chimneys fall.
IX	Damage occurs even to buildings that have been well designed. Many are moved from their foundations. Ground cracks and pipes break.
X	Most masonry structures are destroyed, wooden ones may survive. Railway tracks bend and water slops over banks. Landslides and sand movements occur.
XI	No masonry structure remains standing, bridges are destroyed. Large cracks occur in the ground.
XII	Total damage. Waves are seen on the surface of the ground – objects are thrown into the air.

Table 2 The world's largest earthquakes (Richter Scale)

Place	Date	Strength
Chile	1965	9.5
Alaska	1961	9.2
South-East Asia	2004	9.1
Honshu, Japan	2011	9.0
Kamchatka, Russia	1952	9.0
Chile	1960	8.9
Kansu, Japan	1920	8.6
Tokyo, Japan	1923	8.3
Mexico City	1985	8.1
Tangshen, China	1976	8.0
Erzincan, Turkey	1939	7.9
North Peru	1970	7.7
Izmit, Turkey	1999	7.2

Activities

1 Describe the main characteristics of (a) a shield volcano and (b) a cone volcano.
2 What is the difference between an active volcano and a dormant volcano?
3 What are the advantages of the Richter Scale over the Mercalli Scale? What are the advantages of the Mercalli Scale over the Richter Scale?
4 The Richter Scale is logarithmic. How much stronger is an earthquake of 7.0 compared with one of 5.0 on the Richter Scale?

● Distribution of earthquakes and volcanoes

The distribution of the world's volcanoes and earthquakes is very uneven (Figure 3). They are mostly along plate boundaries which are regions of crustal instability and tectonic activity. About 500 000 earthquakes are detected each year by sensitive instruments. Most of the world's earthquakes occur in linear chains (such as along the west coast of South America) along all types of plate boundary. Some earthquakes appear in areas away from plate boundaries such as in the mid-west of the USA. These earthquakes could still be related to plate movement as the North American plate is moving westwards. Some earthquakes are the result of human activity. The building of large dams and deep reservoirs increases pressure on the ground. Mining removes underground rocks and minerals which may cause collapse or subsidence of the overlying materials. Testing of nuclear weapons underground has been known to trigger earthquakes, too.

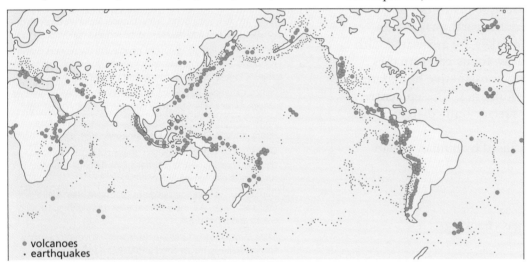

Figure 3 World distribution of volcanoes and earthquakes

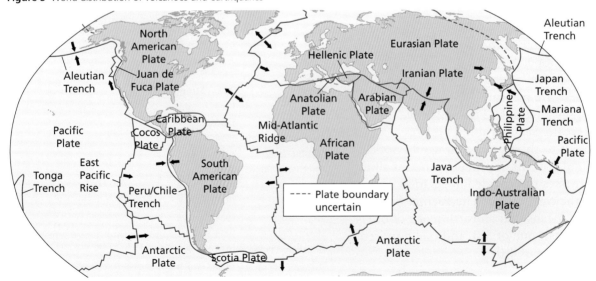

Figure 4 The world's main tectonic plates

Volcanoes

There are over 1300 active volcanoes in the world, many of them under the ocean. Three-quarters of the world's active volcanoes are located in the 'Pacific Ring of Fire', the area around the Pacific Ocean. Good examples include Mt Pinatubo (Philippines), Krakatoa (Indonesia) and Popacatapetl (Mexico). These volcanoes are related to plate boundaries, notably destructive plate boundaries (for example Mt St Helens in the USA and Soufrière in Montserrat in the Caribbean) and constructive boundaries (for example Eldfell volcano on Heimaey, Iceland). The continuing eruption of Soufrière in Montserrat occurs at the boundary of the North American and Caribbean plates. Some volcanoes, such as Mauna Loa and Kilauea in Hawaii, and Teidi on Tenerife, are located over hotspots. These are isolated plumes of rising magma that have burned through the crust to create active volcanoes.

Activities

1 Describe the global distribution of earthquakes as shown on Figure 3.
2 What is the difference between shallow-focus and deep-focus earthquakes?

● Plate tectonics

Plate tectonics is a set of ideas that describe and explain the global distribution of earthquakes, volcanoes, fold mountains and rift valleys. The cause of earth movement is huge convection currents in the Earth's interior, which rise towards the Earth's surface, drag continents apart and cause them to collide. These happen because the Earth's interior consists of semi-molten layers (magma) and the Earth's surface or crust (plates) moves around on the magma. There are seven large plates (five of which carry continents) and a number of smaller plates (Figure 4). The main plates are the Pacific, Indo-Australian, Antarctic, North American, South American, African and Eurasian plates. Smaller ones include the Caribbean, Iranian, Arabian and Juan de Fuca plates. These move relative to one another and when they collide create tectonic activity and new landforms.

The structure of the Earth

There are four main layers within the Earth (Figure 5):

● The inner core is solid. It is five times more dense than surface rocks.
● The outer core is semi-molten.
● The mantle is semi-molten and about 2900 km thick.
● The crust is a solid and is divided into two main types: oceanic crust and continental crust. The depth of the crust varies between 10 km and 70 km. Continental crust is mostly formed of granite. It is less dense than the oceanic crust. Because it is more dense the oceanic crust plunges beneath the continental one when they come together.

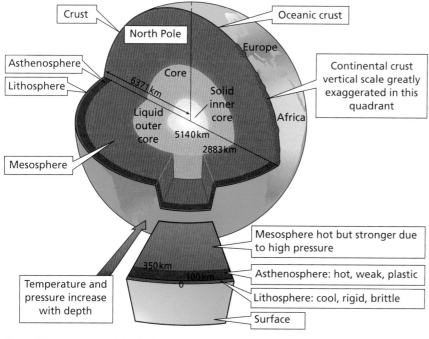

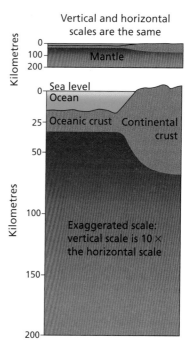

Figure 5 The structure of the Earth

Activities

1 Name the type of plate boundary located:
   a off the west coast of central America
   b in the south Atlantic Ocean
   c where the Turkish plate meets the Aegean plate.
2 Where in the world is plate movement most rapid?

constructive boundaries in which new oceanic crust is being created; **destructive boundaries** in which older crust is destroyed; **collision zones** where plates are folded and crumpled; and **conservative plates** where plates slip past each other, causing earthquakes to occur. Different plate boundaries are associated with different tectonic activities: volcanic eruptions, folding and earthquake activity (Figure 7).

Plate movement and boundaries

There are a number of different types of plate boundaries (Table 3 and Figure 6). These include:

a Constructive boundary

c Convergent boundary (collision)

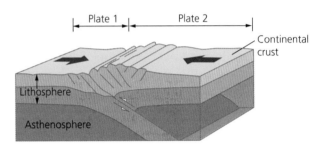

b Convergent (destructive) boundary

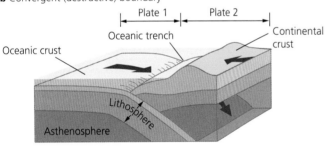

d Conservative boundary

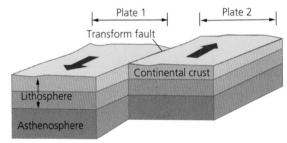

Figure 6 Types of plate boundary

Table 3 Types of plate boundary

Constructive/divergent boundary	Two plates move apart from each other causing sea-floor spreading; new oceanic crust is formed, creating mid-ocean ridges; volcanic activity is common e.g. Mid-Atlantic Ridge (Europe is moving away from North America)
Convergent (destructive) boundary	The oceanic crust moves towards the continental crust and sinks beneath it due to its greater density; deep-sea trenches and island arcs are formed; the continental crust is folded into fold mountains; volcanic activity is common e.g. Nazca plate sinks under the South American plate
Convergent (collision) boundary	Two continental crusts collide; as neither can sink they are folded up into fold mountains e.g. The Indian plate collided with the Eurasian plate to form the Himalayas
Conservative boundary	Two plates slip sideways past each other but land is neither destroyed nor created e.g. San Andreas fault in California

Figure 7 Tectonic activities
    a Folded landscape, Himalaya foothills
    b Thingvellir rift valley, Iceland
    c Volcanic eruption of Soufrière, Montserrat with the former capital city Plymouth in the foreground
    d Tourists standing by the boiling mud springs, Soufrière, St Lucia

Activities

1 Study Table 3.
    a Describe what happens at a subduction zone.
    b At what types of plate boundary are volcanoes likely to occur?
    c Which types of plate boundary produce fold mountains?

2 Study Figure 7 which shows a variety of tectonic landscapes.
    a Describe the general appearance of the land in photo (a). Suggest how it may have been formed.
    b Photo (b) shows a rift valley at Thingvellir in Iceland. At which type of plate boundary are rift valleys found? How might they be formed?

    c Photo (c) shows a volcanic eruption of Soufrière, Montserrat with the former capital city Plymouth in the foreground. Suggest the likely hazards of living close to a volcano.
    d Photo (d) shows tourists at the boiling mud springs at Soufrière in St Lucia. Suggest some of the advantages of living in a tectonically **active** region.
    e Suggest why the volcano on Montserrat and the mud springs in St Lucia have the same name: Soufrière. What does this tell us about the processes involved in these tectonic boundaries?

Causes of earthquakes and volcanoes

Earthquakes are caused by the build-up of pressure that results from plate movement. Consequently many earthquakes are found close to plate boundaries. This is illustrated clearly by tectonic activity in the Caribbean (Figure 8). The Caribbean plate is one of the smaller surface plates of the Earth. Earthquakes occur all around its periphery, and volcanoes erupt on its eastern and western sides (Figure 9). The Caribbean plate moves more slowly, at about 1–2 cm a year, while the North American plate moves westward at about 3–4 cm a year. Many earthquakes and **tsunamis** have occurred in the north-eastern Caribbean region, where the movement of plates is rapid and complicated. There are a number of hazards related to earthquakes (see Table 7 on page 88) and the impact on the death rate has been significant (Table 4).

There are 25 potentially active **volcanoes** in the Caribbean, all of them in the eastern Caribbean.

There have been 17 eruptions in recorded history: Mt Pelée (1902) accounted for most deaths and the Soufrière Hills volcano has been active since 1995. Kick 'em Jenny is an active submarine volcano, north of Grenada. All of the volcanoes are associated with subduction zones.

Figure 8 St John's Cathedral, Antigua rebuilt following the devastating earthquake of 1843

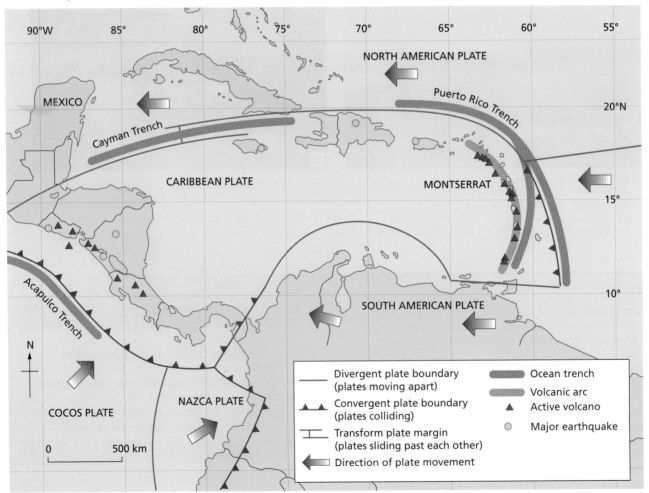

Figure 9 Distribution of plates and tectonic hazards in the Caribbean

Table 4 Major earthquakes in the Caribbean

Date	Location	Deaths	Magnitude
1902	Guatemala	2000	7.5
1907	Jamaica	1600	6.5
1918	Puerto Rico	116	7.5
1931	Nicaragua	2400	5.6
1946	Dominican Republic	100	8.0
1972	Nicaragua	5000	6.2
1976	Guatemala	23000	7.5
1985	Mexico	9500	8.0
1986	El Salvador	1000	5.5
2001	El Salvador	844	7.7
2001	El Salvador	315	6.6
2010	Haiti	300000	7.0

Volcanic eruptions eject many different types of material. **Pyroclastic flows** are superhot (700 °C) flows of ash and pumice (volcanic rock) moving at speeds of over 500 km/hr. In contrast, ash is very fine-grained but very sharp volcanic material. **Cinders** are small rocks and coarse volcanic materials. The volume of material ejected varies considerably from volcano to volcano (Table 5).

Table 5 The world's biggest volcanic eruptions

Eruption	Date	Volume of material ejected (km³)
Mt St Helens, USA	1980	1
Vesuvius, Italy	79	3
Mt Katmai, USA	1912	12
Krakatoa, Indonesia	1883	18
Tambora, Indonesia	1815	80

Volcanic strength

The strength of a volcano is measured by the Volcanic Explosive Index (VEI). This is based on the amount of material ejected in the explosion, the height of the cloud it causes, and the amount of damage caused. Any explosion above level 5 is considered to be very large and violent.

A **supervolcano** is a volcano with a VEI 8. The scale is logarithmic so a VEI 8 is 10 times more powerful than a VEI 7, 100 times more powerful than a VEI 6 and 1000 times more powerful than a VEI 5 (Mt Pinatubo, 1991). The last VEI 8 was 74 000 years ago (Mt Toba, Indonesia).

Supervolcanoes tend to be much larger than 'normal' volcanoes – the Yellowstone magma chamber is over 50 km wide. The likely impacts of a VEI 8 eruption include:

- almost complete loss of life within about 1000 km of the eruption
- destruction of all crops and livestock, leading to a global famine
- economic and social devastation.

Activities

1 Study Figure 9.
   a What are the two plates responsible for tectonic activity in Montserrat?
   b Which two plates are likely to have caused the earthquake that affected Mexico City in 1985?
   c Describe what happens when the North American plate meets the Caribbean plate.
2 a Choose a suitable diagrammatic method to show the relationship between the magnitude of an earthquake and the loss of life, as shown in Table 4.
   b Describe the relationship shown in your diagram.
   c Suggest reasons for the relationship (or lack of) in your answer to 2(a).

● Natural hazards

All natural environments provide opportunities and challenges for human activities. Some of the challenges can be described as 'natural hazards'. A **natural hazard** is a natural event that causes damage to property and/or disruption to normal life and may cause loss of life. Natural hazards involve hydrological, atmospheric and geological events. Natural hazards are caused by the impact of natural events on the social and economic environment in which people live. Some groups of people are more vulnerable to natural hazards and have greater exposure to them.

Since the 1960s more people have been affected by natural hazards. Reasons for this include:

- a rapid increase in population, especially in developing countries
- increased levels of urbanisation, including more shanty towns which are often located in hazardous environments
- changing land use in rural areas which results in flash floods, soil erosion and landslides
- increased numbers of people living in poverty who lack the resources to cope with natural hazards
- changes in the natural environment causing increased frequency and intensity of storms, floods and droughts.

A hazard refers to a potentially dangerous event or process. It becomes a disaster when it affects people and their property.

Risk suggests that there is a possibility of loss of life or damage. Risk assessment is the study of the costs and benefits of living in a particular environment.

There are two very different ways of looking at people's vulnerability:

- One view is that people choose to live in hazardous environments because they understand the environment. In this situation people choose to live in an area because they feel the benefits outweigh the risks.
- Another view is that some people live in hazardous environments because they have very little choice over where they live, as they are too poor to move.

Volcanic eruptions

People often choose to live in volcanic areas because they are useful. For example:

- Some countries, such as Iceland and the Philippines, were created by volcanic activity.
- Volcanic soils are rich, deep and fertile, and allow intensive agriculture to take place.
- Volcanic areas are important for tourism.
- Some volcanic areas are seen by people as being symbolic and are part of the national identity, such as Mt Fuji in Japan.

Figure 10 Tourists gather around the geyser at Geysir, Iceland – one of the benefits of tectonic activity

Figure 11 A building buried by a mudflow in Plymouth, Montserrat – one of the disadvantages of tectonic activity

Table 6 Hazards associated with volcanic activity

Direct hazards	Indirect hazards	Socio-economic impacts
• Pyroclastic flows • Volcanic bombs (projectiles) • Lava flows • Ash fallout • Volcanic gases • Lahars (mudflows) • Earthquakes	• Atmospheric ash fallout • Landslides • Tsunamis • Acid rainfall	• Destruction of settlements • Loss of life • Loss of farmland and forests • Destruction of infrastructure – roads, airstrips and port facilities • Disruption of communications

Earthquakes

Table 7 Earthquake hazards and impacts

Primary hazard	Secondary hazard	Impacts
• Ground shaking • Surface faulting	• Ground failure and soil liquefaction • Landslides and rockfalls • Debris flows and mudflows • Tsunamis	• Total or partial destruction of building structures • Interruption of water supplies • Breakage of sewage disposal systems • Loss of public utilities such as electricity and gas • Floods from collapsed dams • Release of hazardous material • Fires • Spread of chronic illness

The extent of earthquake damage is influenced by a number of factors:

- **Strength of earthquake** and **number of aftershocks** – the stronger the earthquake the more damage it can do. For example, an earthquake of 6.0 on the Richter Scale is 100 times more powerful than one of 4.0; the more aftershocks there are, the greater the damage that is done.
- **Population density** – an earthquake that hits an area of high population density, such as in the Tokyo region of Japan, could inflict far more damage than one that hits an area of low population and low building density.
- The **type of buildings** – developed countries generally have better-quality buildings, more emergency services and the funds to cope with disasters. People in developed countries are more likely to have insurance cover than those in developing countries.
- The **time of day** – an earthquake during a busy time, such as rush hour, may cause more deaths than one at a quiet time. Industrial and commercial areas have fewer people in them on Sundays, and homes have more people in them at night.
- The **distance from the centre (epicentre)** of the earthquake – the closer a place is to the centre of the earthquake, the greater the damage that is done.

- The **type of rocks and sediments** – loose materials may act like liquid when shaken; solid rock is much safer and buildings should be built on level areas formed of solid rock.
- **Secondary hazards** – these include mudslides and tsunamis (large sea waves), fires, contaminated water, disease, hunger and hypothermia.

The South Asian tsunami, 2004

The term *tsunami* is the Japanese for 'harbour wave'. About 90 per cent of these events occur in the Pacific Basin. Tsunamis are generally caused by earthquakes (usually in subduction zones) but can also be caused by volcanoes, for example Krakatoa (1883), and landslides, for example Alaska (1964). Tsunamis have the potential to cause widespread disaster as in the case of the South Asian tsunami 2004 (Figure 12). It became a global disaster, killing people from nearly 30 countries, many of them foreign tourists. Between 180 000 and 280 000 people were killed in the 2004 tsunami.

The cause of the tsunami was a giant earthquake and landslide caused by the sinking of the Indian plate under the Eurasian plate. Pressure had built up over many years and was released in the earthquake, which reached 9.0 on the Richter Scale.

The main impact of the 2004 tsunami was on the Indonesian island of Sumatra, the closest inhabited area to the epicentre of the earthquake. More than 70 per cent of the inhabitants of some coastal villages died. Apart from Indonesia, Sri Lanka suffered more from the tsunami than anywhere else – at least 31 000 people are known to have died there, mostly along the southern and eastern coastlines.

Figure 12 The 2004 tsunami caused widespread damage around the countries bordering the Indian Ocean

Activities

1 What is a natural hazard?
2 Suggest reasons why natural hazards appear to be increasing in frequency.
3 How may volcanic activity be a benefit to people?
4 Describe the direct and indirect hazards associated with volcanic activity in the Caribbean.
5 What are the potential impacts of volcanic activity on people's lives and livelihoods?
6 Why was the 2004 tsunami considered to be a 'global disaster'?
7 a Describe the main hazards associated with earthquakes.
  b Briefly explain any three of the impacts of earthquakes.

Managing volcanoes

There are a number of ways in which the impacts of volcanic eruptions can be reduced. These include:

- spraying lava flows with water to cool them down and cause them to solidify – this was successfully carried out in Heimaey, Iceland
- digging diversion channels to divert lava flows away from settlements – this has been successful on Mt Etna, Sicily
- adding 'cold' boulders to a lava flow in an attempt to cool the lava and stop it moving.

However, if the eruption is a pyroclastic flow, there is little that can be done to prevent the impacts apart from evacuation.

Predicting volcanoes

The main methods of predicting volcanoes include:

- seismometers to record swarms of tiny earthquakes that occur as the magma rises
- chemical sensors to measure increased sulfur levels
- lasers to detect the physical swelling of the volcano
- measurement of small-scale uplift or subsidence, changes in rock stress and changes in radon gas concentration
- ultrasound to monitor low-frequency waves in the magma, resulting from the surge of gas and molten rock, as happened at Mt Pinatubo, El Chichon and Mt St Helens.

Dealing with earthquakes

People cope with earthquakes in a number of ways. The three basic options from which they can choose are:

- do nothing and accept the hazard
- adjust to living in a hazardous environment – strengthen your home
- leave the area.

The main ways of dealing with earthquakes include:

- better forecasting and warning
- building design, building location and emergency procedures.

There are a number of ways of predicting and monitoring earthquakes. These include:

- measuring crustal movement – small-scale movement of plates
- recording changes in electrical conductivity
- noting strange and unusual animal behaviour, for example among fish (e.g. carp)
- checking historic evidence – there are possibly trends in the timing of earthquakes in some regions.

Building design

A single-storey building responds quickly to earthquake forces (Figure 13). A high-rise building responds slowly, and shock waves are increased as they move up the building. If the buildings are too close together, vibrations may be amplified between buildings and increase damage. The weakest part of a building is where different elements meet. Elevated motorways are therefore vulnerable in earthquakes because they have many connecting parts.

Certain areas are very much at risk from earthquake damage – notably areas with weak rocks, faulted (broken) rocks, and soft soils. Many oil and water pipelines in tectonically active areas are built on rollers so that they can move with an earthquake rather than fracture (Figure 14).

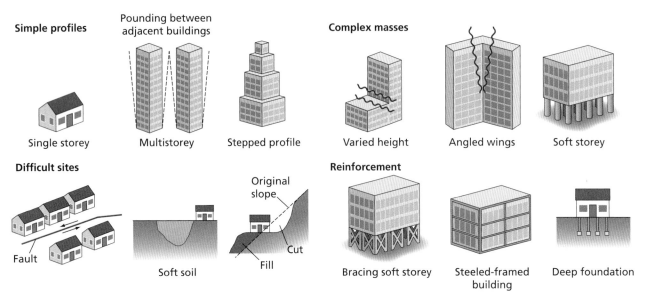

Figure 13 Buildings designed for earthquakes

Figure 14 Pipeline on rollers

Safe houses

Billions of people live in houses that cannot withstand shaking. Yet safer ones can be built cheaply, using straw, adobe or old tyres, by applying a few general principles (Figure 15).

In rich cities in fault zones, the added expense of making buildings earthquake resistant has become a fact of life. Concrete walls are reinforced with steel, for instance, and a few buildings even rest on elaborate shock absorbers. Strict building codes were credited with saving thousands of lives when a magnitude 8.8 earthquake hit Chile in February 2010. But in less developed countries, like Haiti, conventional earthquake engineering is often unaffordable – but there are some cheap solutions.

In Peru in 1970 an earthquake killed more than 70 000 people, many of whom died when their houses crumbled around them. Heavy, brittle walls of traditional adobe – cheap, sun-dried brick – cracked instantly when the ground started to move. Existing adobe walls can be reinforced with a strong plastic mesh installed under plaster. During an earthquake those walls crack but don't collapse, allowing occupants to escape. Plastic mesh could also work as a reinforcement for concrete walls in Haiti and elsewhere.

Researchers in India have successfully tested a concrete house reinforced with bamboo. A model house for Indonesia rests on ground-motion dampers – old tyres filled with bags of sand. Such a house might be only a third as strong as one built on more sophisticated shock absorbers, but it would also cost much less – and so is more likely to get built in Indonesia. In northern Pakistan, straw is available. Traditional houses are built of stone and mud, but straw is far more resilient, and warmer in winter.

	Pakistan	Haiti	Peru	Indonesia
Most destructive quake	8 October 2005	12 January 2010	31 May 1970	26 December 2004
Location	Northern Pakistan/Kashmir	Port-au-Prince area	Chimbote	Sumatra
Magnitude	7.6	7.0	7.9	9.1
Fatalities	75 000	222 500	70 000	227 900 (including the global tsunami deaths)

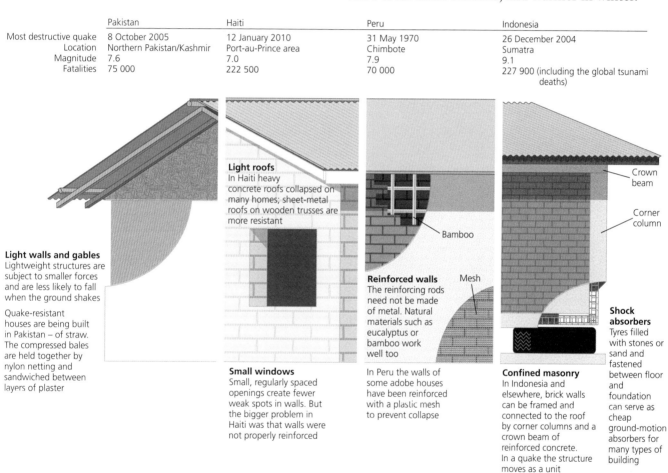

Light walls and gables
Lightweight structures are subject to smaller forces and are less likely to fall when the ground shakes

Quake-resistant houses are being built in Pakistan – of straw. The compressed bales are held together by nylon netting and sandwiched between layers of plaster

Light roofs
In Haiti heavy concrete roofs collapsed on many homes; sheet-metal roofs on wooden trusses are more resistant

Small windows
Small, regularly spaced openings create fewer weak spots in walls. But the bigger problem in Haiti was that walls were not properly reinforced

Bamboo

Mesh

Reinforced walls
The reinforcing rods need not be made of metal. Natural materials such as eucalyptus or bamboo work well too

In Peru the walls of some adobe houses have been reinforced with a plastic mesh to prevent collapse

Crown beam

Corner column

Confined masonry
In Indonesia and elsewhere, brick walls can be framed and connected to the roof by corner columns and a crown beam of reinforced concrete. In a quake the structure moves as a unit

Shock absorbers
Tyres filled with stones or sand and fastened between floor and foundation can serve as cheap ground-motion absorbers for many types of building

Figure 15 Safe houses

Activities

1 Study Figure 15. In what ways can building design reduce the impact of earthquakes?
2 What is meant by the term 'safe house'? Briefly explain how houses may be made 'safe'.
3 In what ways is it possible to predict volcanoes?
4 What is a pyroclastic flow? What are the dangers associated with pyroclastic flows?

Case study: Volcanic eruptions in Montserrat

Soufrière Hills, Montserrat

Montserrat is a small island in the Caribbean, which has been seriously affected by a volcano since 1995. The cause of the volcano is plunging of the South American and North American plates under the Caribbean plate. Rocks at the edge of the plate melt and the rising magma forms volcanic islands.

In July 1995 the Soufrière Hills volcano became active after being dormant for nearly 400 years. At first it gave off clouds of ash and steam. Then in 1996 the volcano finally erupted. It caused mudflows and finally it emitted lava flows. Part of the dome collapsed, boiling rocks and ash were thrown out and a new dome was created. Ash, steam and rocks were hurled out, forcing all the

inhabitants out of the south, the main agricultural part of the island (Figure 16). The largest settlement, Plymouth, with a population of just 4000, was covered in ash and had to be abandoned (Figure 17). This has had a severe impact on Montserrat as Plymouth was the centre for all the government offices, and most of the shops and services, such as the market, post office and cinema.

The hazard posed by the volcano was just one aspect of the risk experienced on Montserrat. For the refugees there were other hazards. For example, up to 50 people had to share a toilet. Sewage tanks in the temporary shelters were often not emptied for weeks on end. The risk of contaminated water and the spread of diseases such as cholera is greatly increased by large numbers of people living in overcrowded, unhygienic conditions.

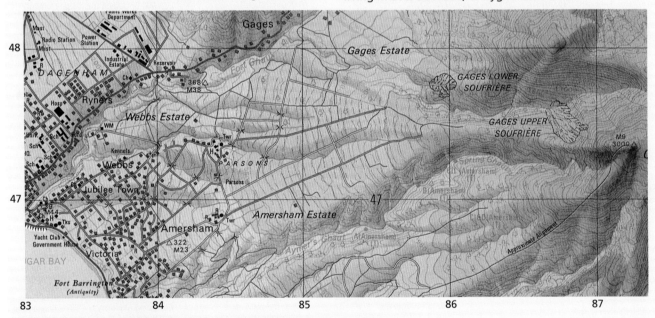

Figure 16 1:25 000 map of Soufrière and Plymouth

Figure 17 Volcanic hazards in Plymouth, Montserrat

The southern third of the island had to be evacuated (Figure 18). All public services (government, health, and education) had to be removed to the north of the island. Montserrat's population fell from 11 000 to 4500. Most fled to nearby Antigua. Some 'refugees' stayed on in Montserrat living in tents.

The northern part of the island has now been redeveloped with new homes, hospitals, crèches, schools, upgraded roads and a football pitch. The population has increased to over 9000.

The risk of eruptions continues – scientists do not know when the current activity will cease.

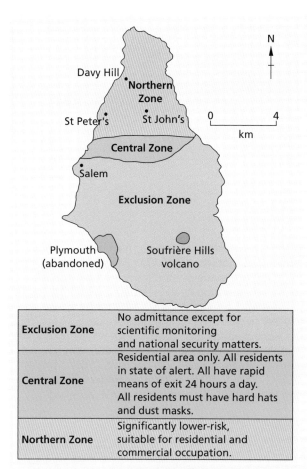

Figure 18 The exclusion zone, Montserrat

Exclusion Zone	No admittance except for scientific monitoring and national security matters.
Central Zone	Residential area only. All residents in state of alert. All have rapid means of exit 24 hours a day. All residents must have hard hats and dust masks.
Northern Zone	Significantly lower-risk, suitable for residential and commercial occupation.

By 2002 Montserrat was experiencing something of a boom. The population, which had dropped in size from over 11 000 before the eruption to less than 4000 in 1999, had risen to over 9000. The reason was very clear: there were many jobs available on the island. There were many new buildings including new government offices, a renovated theatre, new primary schools, and lots of new housing in the north of the island. There was even a new football pitch and stadium built (Figure 19). There were plans to build a new medical school and a school for hazard studies. These have not been built.

Figure 19 Montserrat's new football pitch

However, by the summer of 2009, it was very clear that conditions on Montserrat had changed. The population had fallen to a little over 5000. There are two main reasons why Montserrat's population has declined. The first is the relative lack of jobs. Although there was an economic boom in the early 2000s, once the new buildings were built, many of the jobs disappeared. There are still plans to redevelop the island – a new urban centre is being built at Little Bay but that will not be complete until 2020. The museum has been built, but not much else (Figure 20). So there are some jobs available, but not on the scale that there were previously. Second, one of the new developments on Montserrat was a new airstrip. Once this was built, the UK government and the US government stopped subsidising the ferry that operated between Antigua and Montserrat. This made it more difficult to get to Montserrat – both for visitors and for importing basic goods. Thus the number of tourists to the island fell and the price of goods on the island rose. Many Montserratians were against the airstrip and campaigned unsuccessfully for the port to be kept open. It is possible to charter a boat and sail to Montserrat but that is much more expensive than taking a ferry.

Figure 20 The new museum of Montserrat

With fewer jobs in construction, a declining tourist sector and rising prices, many Montserratians left the island for a second time. Many went to Antigua and others went to countries such as Canada, the USA and the UK. Much of the aid that was given to Montserrat following the eruptions of 1997 have dried up. The UK provided over $120 million of aid but announced in 2002 that it was phasing out any further funds. Nevertheless, in 2004 it announced a $60 million aid deal over three years.

The volcano has been relatively quiet for the last few years, although an event in May 2006 went relatively unreported. The Soufrière dome collapsed, causing a tsunami that affected some coastal areas of Guadeloupe, and English Harbour and Jolly Harbour in Antigua. The Guadeloupe tsunami was one metre high whereas the one experienced in Antigua was between 20 and 30 cm. No one was injured in the tsunami but flights were cancelled between Venezuela and Miami, and to and from Aruba, due to the large amount of ash in the atmosphere.

So volcanic activity in Montserrat may be quiet at the moment but the volcano continues to have a major impact on all those who remain on the island. The economic outlook for the island does not look good – and that is largely related to the lack of aid, the difficulty and cost of reaching Montserrat, and the small size of the island and its population.

Case study analysis

Study Figure 16.

1 What is the height of the highest point on the map? (The name of the highest point is Chance's Peak but this is not shown on the extract.)

2 How far is it from the Chance's Peak to Sugar Bay, Plymouth (the main settlement on the map)?

3 What is the average gradient of the volcano (height/distance)? Express your answer as a 1 in *x* slope.

4 Describe the shape of the volcano.

5 Suggest why a settlement was built at Sugar Bay, Plymouth.

6 Outline the hazards to Plymouth as shown in Figure 17.

Case study: The Christchurch earthquakes, 2010–12

Christchurch is New Zealand's second-largest urban area with a population of 386 000. The 2010 Christchurch earthquake (also known as the Canterbury or Darfield earthquake) was a 7.1 magnitude earthquake which struck the South Island of New Zealand at 4.35 am (local time) on 4 September 2010. Aftershocks continued into 2012. The strongest aftershock was a 6.3 magnitude earthquake, which occurred on 22 February 2011. Because this was so close to Christchurch it was much more destructive, with 185 people being killed.

2010 earthquake

The 2010 earthquake's epicentre was 40 km west of Christchurch, at a depth of 10 km. Insurance claims from the earthquake were confirmed at being between $2.3 and $2.9 billion. Private insurance and individual costs may reach as high as $3.3 billion.

Earthquakes with an intensity of VIII on the Mercalli Scale (significant property damage, loss of life possible) could occur, in the Christchurch area, on average every 55 years. Around a hundred faults have been identified in the region, some as close as 20 km to central Christchurch. However, the 2010 earthquake occurred on a previously unknown fault. The earthquake epicentre was located about 80–90 km south-east of the current surface location of the Australia–Pacific boundary.

By August 2012, over 11 000 aftershocks of magnitude 2.0 or more had been recorded, including 26 over 5.0 magnitude, and 2 over 6.0 magnitude.

Impacts

There were relatively few casualties compared with, for example, the 2010 Haiti earthquake. The Haiti earthquake also occurred in similar proximity to an urban area, also at shallow depth, and of very similar strength. The lack of casualties in New Zealand was partly due to the fact that the earthquake happened in the early hours of a Saturday morning, when most people were asleep, and many of them in timber-framed homes. Moreover, building standards in New Zealand are high. Following the 1848 Marlborough and 1855 Wairarapa earthquakes, which both seriously affected Wellington, building standards were introduced. These were further strengthened following the

1931 Hawke's Bay earthquake. In contrast, Haiti has much lower building standards, which are poorly enforced and many buildings are made of handmade non-reinforced concrete, which is extremely vulnerable to seismic damage. Ground shaking in populated areas of Canterbury was also generally less strong than for the Haiti 'quake.

Sewers were damaged, and water lines were broken. The water supply at Rolleston, south-west of Christchurch, was contaminated. Power to up to 75 per cent of the city was disrupted. Christchurch International Airport was closed following the earthquake and flights were cancelled.

Emergency response and relief efforts

Christchurch's emergency services managed the early stages of the response. Over forty search and rescue personnel and three sniffer dogs were brought from North Island to Christchurch the day of the quake.

2011 earthquake

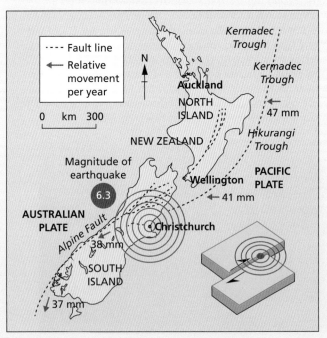

Figure 21 Plate movement and the Christchurch earthquake

The February 2011 Christchurch earthquake was a powerful natural event that severely damaged the city, killing 185 people (Figure 21). The 6.3 magnitude earthquake struck the region at 12.51pm on Tuesday 22 February local time. The earthquake was centred 2 km west of the port town of Lyttleton and 10 km south-east of the centre of Christchurch. The earthquake was probably an aftershock following the September 2010 earthquake.

This earthquake caused widespread damage across Christchurch, especially in the central city and eastern suburbs. The damage was intensified on account of the 4 September 2010 earthquake and its aftershocks. Significant liquefaction affected the eastern suburbs, producing around 400 000 tonnes of silt.

People from more than 20 countries were among the victims. Over half of the deaths occurred in the six-storey Canterbury Television (CTV) building, which collapsed and caught fire during the earthquake. Of the 185 victims, 115 people were lost in the CTV building alone, while another 18 died in the collapse of PGC House, and eight were killed when masonry fell on a bus. Between 6600 and 6800 people were treated for minor injuries.

The total cost to insurers of rebuilding was originally estimated at $12 billion. However, by April 2013, the total estimated cost had escalated to $33 billion. Some economists have estimated that it will take the New Zealand economy 50 to 100 years to completely recover.

Although smaller in magnitude than the 2010 quake, the earthquake was more damaging and more deadly for a number of reasons:

- the epicentre was closer to Christchurch
- the earthquake was shallower at 5 km underground, whereas the September earthquake was measured at 10 km deep
- the February earthquake occurred during lunchtime on a weekday when the CBD was busy
- many buildings were already weakened from previous earthquakes
- liquefaction was significantly greater than that of the 2010 earthquake, causing the upwelling of more than 200 000 tonnes of silt.

The increased liquefaction caused significant ground movement, undermining many foundations and destroying infrastructure.

Up to 80 per cent of the water and sewerage systems was severely damaged. There was damage to roads and bridges which hampered rescue efforts. Soil liquefaction and surface flooding also occurred. Around 10 000 houses would need to be demolished, and liquefaction damage meant that some parts of Christchurch could not be rebuilt on. Nevertheless, in Christchurch, New Zealand's stringent building codes limited the disaster.

Infrastructure and support

At 5pm local time on the day of the earthquake, 80 per cent of the city had no power. However, power was restored to over 80 per cent of households within five days, and to 95 per cent within two weeks. Waste water and sewerage systems had been severely damaged, so households had to establish emergency latrines. Over 2000 portaloos and 5000 chemical toilets were brought in from other parts of the country and overseas, with 20 000 more chemical toilets placed on order from the manufacturers.

Emergency management

A full emergency management programme was in place within two hours. The government response was immediate and significant, and a national emergency was declared. The New Zealand fire service coordinated the search and rescue – rescue efforts continued for over a week, then shifted into recovery mode.

Case study analysis

1 What was the strength of the earthquakes in (a) 2010 and (b) 2011?
2 What two plates are responsible for the earthquakes?
3 Comment on the relative movement of the two plates.
4 Explain why the 2011 earthquake resulted in more deaths than the 2010 earthquake.

2.2 Rivers

Waterfall, Powerscourt, Ireland

> ### Key questions
> - What are the main hydrological characteristics and processes that operate within rivers and drainage basins?
> - How do rivers erode, transport and deposit?
> - What are the main erosional and depositional features associated with rivers?
> - What are the advantages and disadvantages of rivers?
> - How can the impacts of floods be managed?

● Changing channel characteristics

Downstream changes

Rivers have three main roles: to erode the river channel, to transport materials, and to create new erosional and depositional landforms. Most rivers have three main zones – a zone of **erosion**, a zone of **transport** and a zone of **deposition**. Erosion, transport and deposition

are found in all parts of a river, although one process tends to be dominant. For example, there is more erosion in the upper part, while there is more deposition in the lower course. This is related to the changes in a river downstream. Figure 1 shows that velocity, **discharge** and load increase downstream whereas gradient and the size of load decrease downstream.

Upstream	Downstream
	Discharge
	Occupied channel width
	Water depth
	Water velocity
	Load quantity
Load particle size	
Channel bed roughness	
Slope angle (gradient)	

Figure 1 Changes in a river downstream

Energy in a river

This determines a stream's ability to erode, transport or deposit. There are two types of energy:

- potential energy, provided by the weight and elevation of the water
- kinetic energy, produced by gravity and the flow of the water.

About 95 per cent of energy is used to overcome friction with the bed and banks. The rougher the channel the more energy will be lost. In a smooth channel there is very little frictional loss and therefore there is more energy available for work.

Channel shape

The efficiency of a stream's shape is measured by its hydraulic radius – that is, the cross-sectional area divided by wetted perimeter (Figure 2). The higher the ratio the more efficient the stream and the smaller the frictional loss. The ideal form is semi-circular.

There is a close relationship between velocity, discharge and the characteristics of the channel in which the water is flowing. These include depth, width, channel roughness and hydraulic geometry. The width/depth ratio (w/d) is a good measure of comparison. The shape of the channel is also determined by the material forming the channel, and river forces. Solid rock allows only slow changes whereas alluvium allows rapid changes. Silt and clay produce steep, deep, narrow valleys (the fine material being cohesive and stable) whereas sand and gravel promote wide, shallow channels.

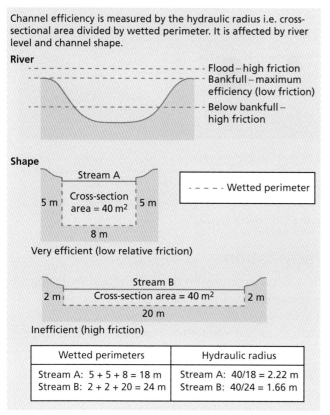

Channel efficiency is measured by the hydraulic radius i.e. cross-sectional area divided by wetted perimeter. It is affected by river level and channel shape.

River

Flood – high friction
Bankfull – maximum efficiency (low friction)
Below bankfull – high friction

Shape

Stream A
Cross-section area = 40 m²
5 m   5 m
8 m
Very efficient (low relative friction)

– – – – Wetted perimeter

Stream B
Cross-section area = 40 m²
2 m   2 m
20 m
Inefficient (high friction)

Wetted perimeters	Hydraulic radius
Stream A: 5 + 5 + 8 = 18 m	Stream A: 40/18 = 2.22 m
Stream B: 2 + 2 + 20 = 24 m	Stream B: 40/24 = 1.66 m

Figure 2 Wetted perimeter and cross-sectional area

Channel roughness

Channel roughness causes friction which slows down the velocity of the water. Friction is caused by irregularities in the river bed, boulders, trees and vegetation, and contact between the water and the bed and bank.

Discharge is the volume of water passing a given point over a set time (Figure 3). Normally it is expressed in m³/sec (cumecs). It is found by multiplying the cross-sectional area and the mean velocity. Steeper slopes should lead to higher velocities because of the influence of gravity.

Discharge (Q) normally increases downstream, as does width, depth and velocity. By contrast, channel roughness decreases. The increase downstream in channel width is normally greater than that of channel depth. Large rivers, with a higher w/d ratio are more efficient than smaller rivers with a lower w/d ratio, since less energy is spent in overcoming friction. Thus the carrying capacity increases and a lower gradient is required to transport the load. Although river gradients decrease downstream the load carried is smaller, and therefore easier to transport.

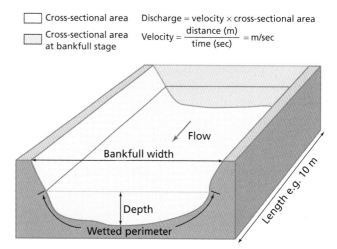

Cross-sectional area
Cross-sectional area at bankfull stage

Discharge = velocity × cross-sectional area

$$\text{Velocity} = \frac{\text{distance (m)}}{\text{time (sec)}} = \text{m/sec}$$

Flow
Bankfull width
Depth
Wetted perimeter
Length e.g. 10 m

Figure 3 Discharge

Activities

Table 1

Site	1	2	3	4	5
Gradient	1:8	1:14	1:26	1:45	1:85
Width (m)	1.3	1.6	2.4	4.1	8.3
Depth (m)	0.7	1.1	1.4	1.9	2.6
Velocity (cm/sec)	13	16	21	28	34
Discharge	...	...	...	...	...
Bedload: size (cm)	25	21	12	7	2
Shape	Angular	Angular	Sub-angular	Rounded	Rounded
Cross-sectional area	...	...	...	...	...

Study the data in Table 1.
1 a Describe how the ratio of width/depth varies with distance from the source of the river.
  b Work out (i) the cross-sectional area and (ii) the discharge of the stream for each site. How and why do these change downstream?
  c Describe the changes in bedload size and shape as you proceed downstream. What processes cause these changes to take place?
  d If the channel between sites 4 and 5 were straightened, what effect would it have on (i) the velocity of the river, (ii) the load and (iii) the work of the river?
2 Study Figure 1.
  a Describe how the amount and size of load varies downstream.
  b Suggest reasons for the changes you have identified.
  c State a reason why the channel bed roughness decreases downstream.
  d How might the nature of the load affect the type and amount of erosion carried out by the river? Give reasons for your answer.

● Drainage basins

A **drainage basin** is an area within which water supplied by **precipitation** is transferred to the ocean, a lake or larger stream. It includes all the area that is drained by a river and its **tributaries** (smaller rivers that join the larger river) and is the main unit for the study of rivers. The **confluence** is the point where a smaller river joins a larger river. Drainage basins are divided by **watersheds** – an imaginary line separating adjacent basins (Figure 4).

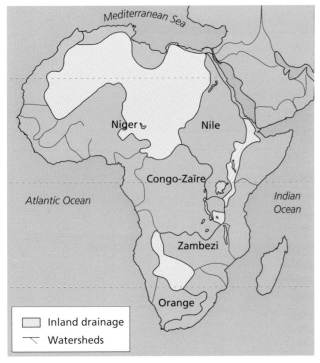

Figure 4 Drainage basins in Africa

In studying rivers, reference is made to the basin's **hydrological cycle**. In this the drainage basin is taken as the unit of study rather than the global system. The basin cycle is an open system: the main input is precipitation which is regulated by various means of storage.

The hydrological cycle refers to the movement of water between atmosphere, lithosphere and biosphere. At a global scale, it can be thought of as a closed system with no losses from the system. In contrast, at a local scale the cycle has a single input, precipitation (PPT), and two major losses (outputs), evapotranspiration (EVT) and runoff (Figure 5).

Water can be stored at a number of places within the cycle. These stores include vegetation, surface, soil moisture, groundwater and water channels. The global hydrological cycle also includes stores in the oceans and the atmosphere.

Human modifications are made at every scale. Good examples include large-scale changes of channel flow, irrigation and drainage, and abstraction of **groundwater** and surface water for domestic and industrial use.

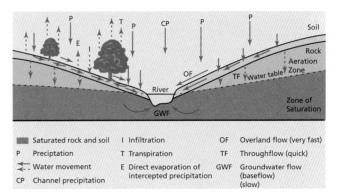

Figure 5 Basin hydrological cycle

Precipitation

The main characteristics that affect local hydrology are the amount of precipitation, the seasonality, intensity, type (snow, rain etc.), geographic distribution and variability. For rain to occur, three factors must occur:

- air is saturated – that is, it has a relative humidity of 100%
- it contains particles of soot, dust, ash, ice etc.
- its temperature is below dew point – that is, the temperature is at the level where the relative humidity is 100%, saturation is complete and clouds form.

Clouds are tiny droplets suspended in air, while rain droplets are much larger. Therefore cloud droplets must get much larger, although not necessarily by normal condensation processes. There are a number of theories to suggest how raindrops are formed.

There are three main types of rainfall:

- cyclonic – uplift of air within a low pressure area (warm air rises over cold air): it normally brings low to moderate intensity rain and may last for a few days
- orographic – a deep layer of moist air is forced to rise over a range of hills or mountains
- convectional heating which causes pockets of air to rise and cool.

Interception

Interception refers to water stored by vegetation. There are three main components:

- interception loss – water that is retained by plant surfaces and later evaporated away or absorbed by the plant

- throughfall – water that either falls through gaps in the vegetation or drops from leaves, twigs or stems
- stemflow – water that trickles along twigs and branches and finally down the trunk.

Interception loss varies with vegetation. For example, in German beech forests in summer it is up to 40 per cent of rainfall, whereas in winter only about 20 per cent. Interception is less from grasses than from deciduous woodland. Interception losses are greater from coniferous forests because:

- pine needles allow individual accumulation
- freer air circulation allows more evapotranspiration.

From agricultural crops, and from cereals in particular, interception loss increases with crop density.

Evaporation

Evaporation is the process by which a liquid or a solid is changed into a gas. Its most important source is from oceans and seas. Evaporation increases under warm, dry windy conditions.

Factors affecting evaporation include temperature, humidity, and windspeed. Of these, temperature is the most important factor. Other factors include water quality, depth of water, size of water body, vegetation cover and colour of the surface (albedo or reflectivity of the surface – see Table 2).

Table 2 Albedo values

Surface	Albedo (%)
Water (sun's angle over 40°)	2–4
Water (sun's angle less than 40°)	6–80
Fresh snow	75–90
Old snow	40–70
Dry sand	35–45
Dark, wet soil	5–15

Surface	Albedo (%)
Dry concrete	17–27
Black road surface	5–10
Grass	20–30
Deciduous forest	10–20
Coniferous forest	5–15
Crops	15–25
Tundra	15–20

Evapotranspiration

Transpiration is the process by which water vapour is transferred from vegetation to the atmosphere. The combined effects of evaporation and transpiration are normally referred to as **evapotranspiration** (EVT). EVT represents the most important aspect of water loss, accounting for the removal of nearly 100 per cent of the annual precipitation in arid areas and 75 per cent in humid areas. Only over ice and snow fields, bare rock slopes, desert areas, water surfaces and bare soil will purely evaporative losses occur.

The distinction between actual EVT and potential evapotranspiration (P.EVT) lies in the concept of moisture availability. Potential evapotranspiration is the water loss that would occur if there was an unlimited supply of water in the soil for use by the vegetation. Rates of potential and actual evapotranspiration for South Africa are shown in Figure 6.

a Potential

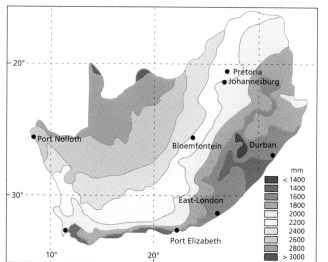

b Actual

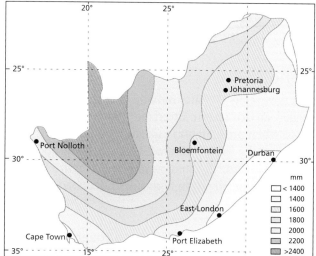

Figure 6 Rates of potential and actual evapotranspiration for South Africa

Infiltration

Infiltration is the process by which water soaks into the ground. The infiltration capacity is the maximum rate at which rain can enter the soil/ground.

Infiltration capacity decreases with time through a period of rainfall until a more or less constant value is reached (Figure 7). Infiltration rates of 0–4 mm/hour are common on clays whereas 3–12 mm/hour are common on sands. Vegetation also increases infiltration. On bare soils, infiltration rates may reach 10 mm/hour. On similar soils covered by vegetation rates of between 50 and 100 mm/hour have been recorded. Infiltrated water is chemically rich as it picks up minerals and organic acids from vegetation and soil.

Infiltration is affected by the same factors that influence overland runoff but in a different way. For example, duration (length) of rainfall decreases infiltration, as does existing moisture in the soil, raindrop size, steep slope and a lack of vegetation cover.

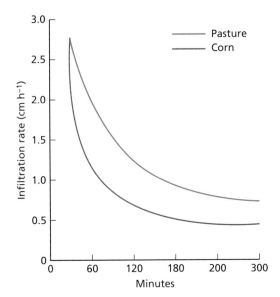

Figure 7 Infiltration and time

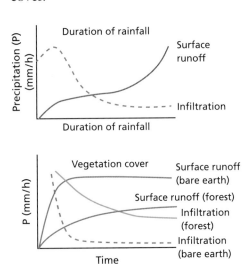
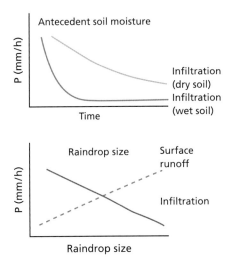
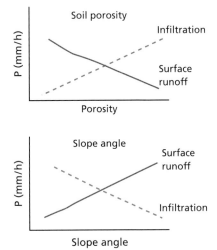

Figure 8 Factors affecting infiltration

Overland flow

Overland flow refers to water moving over the surface. It is also called surface runoff. Overland flow occurs in two main ways: when precipitation exceeds the infiltration rate, and when the soil is saturated. Overland flow and infiltration are inversely related, as shown on Figure 8. Overland flow generally has a high suspended load.

Throughflow

Throughflow refers to water flowing through the soil in natural pipes and between soil horizons.

Soil moisture

Soil moisture is the subsurface water in soil and subsurface layers above the water table. From here water may be:

- absorbed
- held
- transmitted downwards towards the water table
- transmitted upwards towards the soil surface and the atmosphere.

In coarser-textured soils much of the water is held in fairly large pores at fairly low suctions, while very little is held in small pores. In the finer-textured clay soils the range of pore sizes is much greater and, in particular, there is a higher proportion of small pores in which water is held at very high suctions.

- Field capacity refers to the amount of water held in the soil after excess water drains away – that is, saturation or near saturation.
- Wilting point refers to the range of moisture content in which permanent wilting of plants occurs.

Groundwater

Groundwater refers to subsurface water. The upper layer of the permanently saturated zone is known as the water table. The water table varies seasonally – in Britain it is higher in winter following increased levels of precipitation. Most groundwater is found within a few hundred metres of the surface but has been found at depths of up to 4 km beneath the surface (Figure 9).

a In humid regions

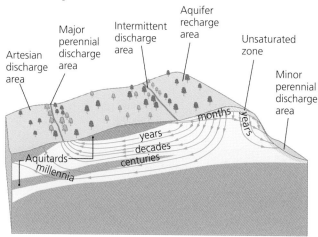

b In semi-arid regions

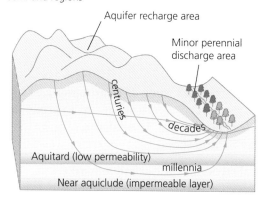

Figure 9 Groundwater

Groundwater accounts for 96.5 per cent of all freshwater on the Earth. However, while some soil water may be recycled within a matter of days or weeks, groundwater may not be recycled for as long as 20 000 years. Hence, in some places, groundwater is considered a non-renewable resource.

Aquifers (rocks that contain significant quantities of water) provide a great reservoir of water. Aquifers are permeable rocks such as sandstones and limestones. This water moves very slowly and acts as a natural regulator in the hydrological cycle by absorbing rainfall which otherwise would reach streams rapidly. In addition, aquifers maintain stream flow during long dry periods. Rocks which do not hold water are impermeable rocks which prevent large-scale storage and transmission of water, such as clay.

The groundwater balance is shown by the formula:

$$\Delta S = Qr - Qd$$

where ΔS is the change in storage (+ or −), Qr is recharge to groundwater and Qd is discharge from groundwater.

Groundwater recharge occurs as a result of:

- infiltration of part of the total precipitation at the ground surface
- seepage through the banks and bed of surface water bodies such as rivers, lakes and oceans
- groundwater leakage and inflow from adjacent aquicludes and aquifers
- artificial recharge from irrigation, reservoirs etc.

Losses of groundwater result from:

- evapotranspiration particularly in low-lying areas where the water table is close to the ground surface
- natural discharge by means of spring flow and seepage into surface water bodies
- groundwater leakage and outflow through aquicludes and into adjacent aquifers
- artificial abstraction, for example in the London basin in the UK.

● River processes

Erosion

The main types of **erosion** include:

- **abrasion** (or **corrasion**), the wearing away of the bed and bank by the load carried by a river
- **attrition**, the wearing away of the load carried by a river which creates smaller, rounder particles
- **hydraulic action**, which is the force of air and water on the sides of rivers and in cracks
- **solution** (or **corrosion**), the removal of chemical ions, especially calcium, which causes rocks to dissolve.

There are many factors affecting erosion. These include:

- **load** – the heavier and sharper the load the greater the potential for erosion
- **velocity and discharge** – the greater the velocity and discharge the greater the potential for erosion
- **gradient** – increased gradient increases the rate of erosion
- **geology** – soft, unconsolidated rocks, such as sand and gravel, are easily eroded
- **pH** – rates of solution are increased when the water is more acidic
- **human impact** – deforestation, **dams** and bridges interfere with the natural flow of a river and frequently end up increasing the rate of erosion.

Transport

The main types of transport in a river (Figure 10) include:

- **suspension** – small particles are held up by turbulent flow in the river
- **saltation** – heavier particles are bounced or bumped along the bed of the river
- **solution** – the chemical load is dissolved in the water
- **traction** – the heaviest material is dragged or rolled along the bed of the river
- **flotation** – leaves and twigs are carried on the surface of the river.

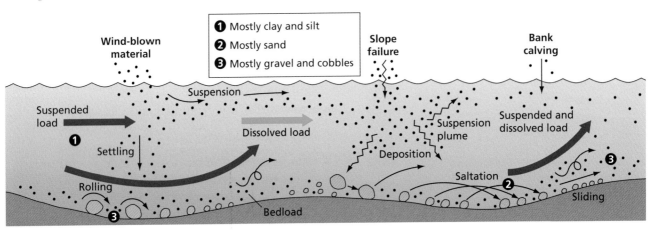

Figure 10 Types of transport in a river

Deposition

Deposition occurs as a river slows down and it loses its energy. Typically, this occurs as a river **floods** across a **floodplain** or enters the sea, or behind a dam. It is also more likely during low flow conditions (such as in a drought) than during high flow (flood) conditions – as long as the river is carrying sediment. The larger, heavier particles are deposited first, the smaller, lighter ones later. Features of deposition include deltas, **levées**, slip-off slopes (point bars), **oxbow lakes**, braided channels and floodplains.

The long profile

A number of processes, such as weathering and mass movement, interact to create variations in **cross-** and **long profiles** (Figure 11). Irregularities, or **knick-points**, may be due to:

- geological structure, for example hard rocks erode slowly, which can result in the formation of waterfalls and rapids
- variations in the load, for example when a **tributary** with a coarse load may lead to a steepening of the gradient of the main valley
- sea level changes – a relative fall in sea level will lead to renewed downcutting which enables the river to erode former floodplains and form new terraces and knick-points.

Activities

1 a Briefly describe the four main ways in which rivers erode.
  b Suggest how they will vary with (i) velocity of water (ii) rock type and (iii) pH of water.
2 a What are the main types of transport?
  b How might the type and quantity of the river's load vary between flood conditions and low flow conditions?

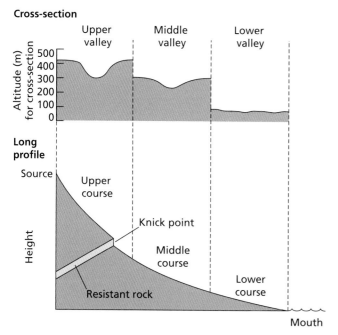

Figure 11 Long and cross-profiles

Rivers tend to achieve a condition of equilibrium, or grade, and erode the irregularities. There is a balance between erosion and deposition in which a river adjusts to its capacity and the amount of work being done. The main adjustments are in channel gradient, leading to a smooth concave profile.

Cross-profiles

The cross-profile of the upper part of a river is often described as **V-shaped** (Figure 11). Rivers in their upper course typically have a steep gradient and a narrow valley. The rivers are shallow and fast flowing. There is normally much friction with large boulders, and much energy is used to overcome friction. The processes likely to occur are vertical erosion, weathering on the slopes, mass movement and transport. Features likely to be found include **waterfalls**, rapids, **potholes**, **gorges** and interlocking spurs.

In the middle course of the river, the valley is still V-shaped but is less steep. Slopes are more gentle. A floodplain is beginning to form and **meanders** are visible. Processes in the middle course include erosion (both vertical and lateral), meandering, transport, and some deposition on the inner bends of the meanders.

In contrast, in the lower course the cross-profile is much flatter. Processes include erosion (on the outer banks), transport and deposition (especially on the inner bends and on the floodplain). Characteristic features include levées, oxbow lakes, floodplains, deltas and terraces.

Features of erosion

Localised erosion by hydraulic action and abrasion, especially by large pieces of debris, may lead to the formation of potholes (Figure 12). These are typically seen in the upper course of a river when the load is larger and more rugged. Waterfalls frequently occur on horizontally bedded rocks (Figure 13). The soft rock is undercut by hydraulic action and abrasion. The weight of the water and the lack of support cause the waterfall to collapse and retreat. Over thousands of years the waterfall may retreat enough to form a gorge of recession (Figure 14). Where there are small outcrops of hard and soft rock, rapids may develop rather than a waterfall.

Potholes formed by abrasion

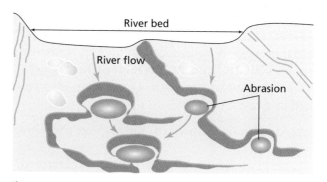

Figure 12 Formation of potholes

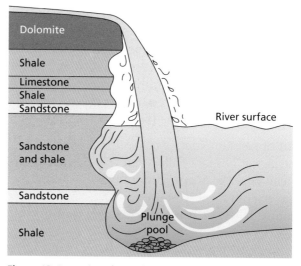

Figure 13 Formation of waterfalls

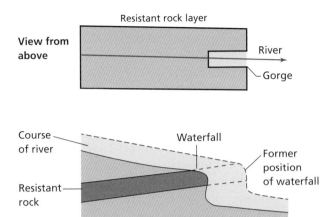

Figure 14 Formation of a gorge of recession

Gorges and waterfalls

Gorge development is common, for example where the local rocks are very resistant to weathering but susceptible to the more powerful river erosion. Similarly, in arid areas where the water necessary for weathering is scarce, gorges are formed by episodes of fluvial erosion. A rapid acceleration in downcutting is also associated when a river is rejuvenated, again creating a gorge-like landscape. Gorges may also be formed as a result of:

- antecedent drainage pattern, for example the Rhine gorge
- collapse of underground caverns in carboniferous limestone areas, for example the River Axe at Wookey Hole, UK
- retreat of waterfalls, for example Niagara Falls (Figure 15).

Plunge flow occurs where the river spills over a sudden change in gradient, undercutting rocks by hydraulic impact and abrasion, thereby creating a **waterfall**. There are many reasons for this sudden change in gradient along the river:

- a band of resistant strata, such as the resistant limestones at Niagara Falls
- a plateau edge, such as Livingstone Falls on the Congo river in D. R. Congo
- a hanging valley, such as at Glencoyne, Cumbria in the UK
- coastal cliffs, such as at Kimmeridge Bay, Dorset (UK).

The undercutting at the base of the waterfall creates a precarious overhang which will ultimately collapse. Thus a waterfall may appear to migrate upstream,

leaving a gorge of recession downstream. The Niagara Gorge is 11 km long due to the retreat of Niagara Falls (Figures 15 and 16).

Niagara Falls

Figure 15 Niagara Falls on the US/Canadian border

Most of the world's great waterfalls are the result of the undercutting of resistant cap rocks, and the retreat or recession that follows.

The Niagara river flows for about 50 km between Lake Erie and Lake Ontario. In that distance it falls just 108 metres, giving an average gradient of 1:500. However, most of the descent occurs in the 1.5 km above Niagara Falls (13 metres) and at the Falls themselves (55 metres). The Niagara river flows in a 2 km wide channel just 1 km above the Falls, and then into a narrow 400 metre wide gorge, 75 metres deep and 11 km long. Within the gorge the river falls a further 30 metres.

The course of the Niagara river was established about 12 000 years ago when water from Lake Erie began to spill northwards into Lake Ontario. In doing so, it passed over the highly resistant dolomitic (limestone) escarpment. Over the last 12 000 years the Falls have retreated 11 km, giving an average rate of retreat of about 1 metre/year. Water velocity accelerates over the Falls, and decreases at the base of the Falls. Hydraulic action and abrasion have caused the development of a large **plunge pool** at the base of the Falls, while the fine spray and eddies in the river help to remove some of the softer rock underneath the resistant dolomite. As the softer rocks are removed, the dolomite is left unsupported and the weight of the water causes the dolomite to collapse. Hence the waterfall retreats forming a gorge of recession.

Figure 16 1:25 000 map of the Niagara Falls area

In the nineteenth century rates of recession were recorded at 1.2 metres/year. However, now that the amount of water flowing over the Falls is controlled (due to the construction of hydro-electric power stations), rates of recession have been reduced. In addition, engineering works in the 1960s reinforced parts of the dolomite that were believed to be at risk of collapse. The Falls remain an important tourist attraction and local residents and business personnel did not want to lose their prized asset!

Activities

Study Figure 16.
1 Which two countries share a border at Niagara Falls?
2 In which direction is the Niagara river flowing?
3 What is the map evidence that there is a gorge below Niagara Falls?
4 Using map evidence, suggest how the Niagara Falls and river have been used for human activities.
5 Approximately how wide is Niagara Falls in squares 5870–5871, and in 5772?

Victoria Falls

Figure 17 Victoria Falls on the Zambezi river

Victoria Falls, on the border of Zambia and Zimbabwe, is one of the world's most spectacular waterfalls (Figure 17). Like Niagara it has a resistant cap rock, but unlike Niagara one of the dominant forces in the creation and development of Victoria Falls is plate tectonics. The Falls are nearly 2 km wide, up to 108 m deep, and during the rainy season over 5 million m³ per minute pass over the Falls. During the dry season the volume is much less, but the annual average is still an impressive 550 000 m³ per minute. When the Zambezi river passes over the Falls its channel narrows from over 1.7 km to just a few metres. This creates an increase in the river's velocity and causes much erosion and scouring to occur.

The evolution of the Falls is complex. About 150 million years ago molten rock formed fine-grained resistant basalt, about 300 metres thick, in the Victoria Falls area. As the lava cooled it shrank, causing cracks or fissures to appear in the rock. These fissures were later widened by weathering, and were infilled by deposits of soft clay and lime. Over time these deposits solidified to form limestone. Continued tectonic processes caused large east–west fissures to widen, allowing the limestone in the fissures to be eroded.

The river's course changed over time. Uplift caused the ancient course of the river to be blocked, diverting the river east into its present course. The river was able to erode the soft sandy deposits on the surface, but was unable to erode the harder basalt below. At the time the edge of the basalt plain was some 100 km from the present Victoria Falls. As the waterfall retreated it occupied an east–west fissure filled with relatively soft limestone. This it was able to erode easily and cut back to within 8 km of the present waterfall. Then it occupied a north–south fissure, until it cut back as far as the next east–west fissure. The present Victoria Falls are believed to be similar to the previous falls, each of which has occupied a different fissure within the resistant basalt.

Activities

1 Which two countries border Victoria Falls?
2 Identify one similarity and one difference between Victoria Falls and Niagara Falls.
3 How much water, on average, flows over Victoria Falls?
4 Why has Victoria Falls altered its course over time?

Meanders

Rivers typically **meander**. This means that the water does not follow a straight line but takes a curving route (Figure 18). As a result of this there are variations in the speed across a river. Velocity is fastest on the outside bank and slowest on the inside bank, so there is erosion on the outer bank and deposition on the inner bank. This produces a steep **river cliff** on the outer bank of a meander and a gentle slip-off slope on its inner bank.

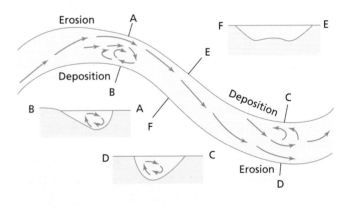

Figure 18 Cross-section through a meander

Oxbow lakes

Oxbow lakes are the result of erosion and deposition (Figure 19). Lateral erosion, caused by fast flow in the meanders, is concentrated on the outer, deeper bank. During times of flooding, erosion increases. The river breaks through and creates a new, steeper channel. In time, the old meander is closed off by deposition to form an oxbow lake.

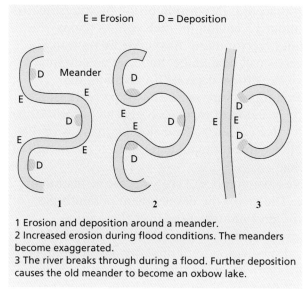

1 Erosion and deposition around a meander.
2 Increased erosion during flood conditions. The meanders become exaggerated.
3 The river breaks through during a flood. Further deposition causes the old meander to become an oxbow lake.

Figure 19 Development of an oxbow lake

Floodplains

The area covered by water when a river floods is known as its **floodplain**. When a river's discharge exceeds the capacity of the channel, water rises over the river banks and floods the surrounding low-lying area. Sometimes a floodplain will itself be eroded following a fall in sea level. When this happens, the remnants of the old floodplain are left behind as river terraces (Figure 20). These are useful for settlement as they are above the new level of the floodplain and are free from flooding.

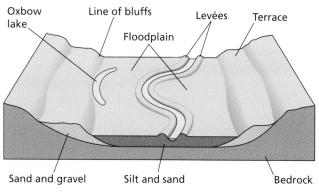

Figure 20 Formation of a floodplain and terraces

Levées

When a river floods its speed is reduced, slowed down by friction caused by contact with the floodplain. As its velocity is reduced the river has to deposit some of its load. It drops the coarser, heavier material first to form raised banks, or **levées**, at the edge of the river. This means that over centuries the levées are built up of coarse material, such as sand and gravel, while the floodplain consists of fine silt and clay (Figure 21).

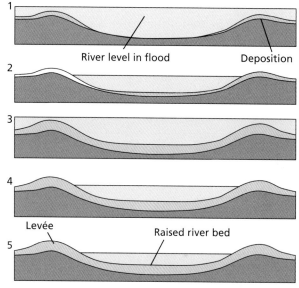

1 When the river floods, it bursts its banks. It deposits its coarsest load (gravel and sand) closest to the bank and the finer load (silt and clay) further away.
2, 3, 4 This continues over a long time, perhaps for centuries.
5 The river has built up raised banks called levées, consisting of coarse material, and a floodplain of fine material.

Figure 21 Formation of levées

Braided channels

Braiding occurs when a river transports a very heavy load in relation to its velocity. If a river's discharge falls its competence and capacity are reduced. This forces the river to deposit large amounts of its load and multi-channels, or **braided channels**, are formed. Braiding is common in rivers that experience seasonal variations in discharge. For example, in proglacial and periglacial areas such as southern Iceland, most of the discharge occurs in late spring and early summer, as snow and ice melts. This enables rivers to carry very large loads which are quickly deposited as discharge decreases.

Deltas

A **delta** is a flat, low-lying deposit of sediment that may be found at a river's **mouth** (Figure 22). For deltas to be formed a river needs to:

- carry a large volume of sediment – for example, rivers in semi-arid regions and in areas of intense human activity
- enter a still body of water which causes velocity to fall, the water loses its capacity and competence, hence deposition occurs, with the heaviest particles deposited first and the lightest last.

Deposition is increased if the water is salty, as this causes salt particles to group together, become heavier, and be deposited. Vegetation also increases the rate of deposition by slowing down the water.

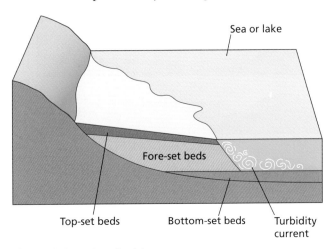

Figure 22 Formation of a delta

The Rhône delta

The Rhône river (Figure 23) divides into two main distributaries 4 km north of Arles. The east branch, the Grande Rhône, is the larger of the two, and carries 85 per cent of the Rhône's water into the Mediterranean. At Arles the river is just 2 m above sea level and takes almost 50 km to reach the sea. The delta is criss-crossed by numerous small islands, abandoned channels and active levées. Most settlements and transport routes are located close to the river, where the land is slightly higher. Further away from the river, the land is lower, swampy, and frequently covered with water. The same pattern exists along the west branch, the Petite Rhône.

Between these two limbs of the Rhône is a flat region characterised by many marshes and lakes (étangs), known as the Camargue. The largest lake is the Étang de Vaccarès, which is less than 1 m deep. The étangs receive most of their water from rainwater that becomes trapped between the slightly higher riverine locations, and the sand bars and dunes at the coast.

The Rhône delta is believed to be less than one million years old. Deposition by the river is estimated to be about 17 million m³ each year, or about 50 tonnes every minute! As the Mediterranean Sea has a very small tidal range, there are no currents to carry away these deposits. In addition, the Mediterranean is very saline. In the presence of salt water, clay and mud particles coagulate to form larger particles that cannot be held aloft by the flow of the river. Hence there is rapid deposition at the mouth of the delta.

There are a number of stages in the formation of a delta. The first is the development of sandbanks in the original mouth of the river. This causes the river to divide, and then there is a period of repeated subdivision until there are a large number of distributaries flowing towards the sea. Each of the channels develops its own set of levées, which has an impact on the human environment (settlement and transport) as well as the physical environment (affecting the development of étangs between the main branches of the river). The étangs may slowly accumulate sediment to form marshes, or they may be drained and reclaimed to form farmland or used by people in other ways.

Activities

1 What is the length of (a) the Petit Rhône and (b) the Grande Rhône between Arles and the sea?
2 How wide is the Rhône delta when it meets the sea?
3 What is the map evidence that (a) conservation and (b) tourism are important in the Rhône delta?

Interesting note

There are 76 rivers that are over 1600 km long. Four of them flow through Russia at some point and four also flow through China at some point.

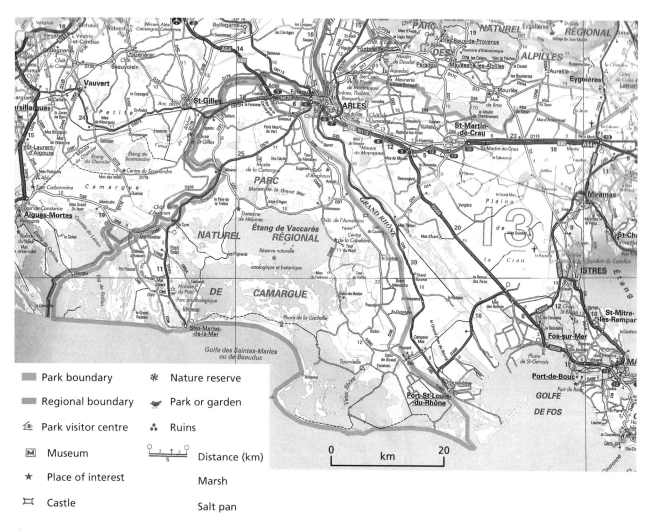

Figure 23 1:250 000 map extract of the Rhône delta

Park boundary
Regional boundary
🏛 **Park visitor centre**
Ⓜ **Museum**
★ **Place of interest**
⬚ **Castle**
✳ **Nature reserve**
🐦 **Park or garden**
∴ **Ruins**
Distance (km)
Marsh
Salt pan

River hazards and opportunities

Floods are a natural feature of all rivers. For most of the time a river is contained in its channel but at other times it may burst its bank and a flood occurs. Floods bring advantages such as water and fertile alluvium (river deposits or silt) which allow farmers to grow crops. But the problem is that they may bring too much water and too much silt. The results can be devastating, as the experience of China shows (pages 115–117). River bank erosion can cause population displacement and socio-economic impacts. For example, the River Meghna in Bangladesh caused major disruptions during the 1990s and 2000s. It eroded over 6km of land between Meghna Roads and Highway Bridge, Meghna Ghat, destroying productive land

and causing residents to lose all their possessions. Among the worst affected areas were the villages of Shikarpur, Kandargoan and Bhanipur. Many displaced farmers become day labourers or rickshaw operators. They receive little assistance from the government, although most get some assistance from friends and relatives. As over 80 per cent of people are employed in farming, the loss of land leads to widespread unemployment.

The causes of floods are natural. However, human interference intensifies many floods (Figure 24). A flood is a high flow of water which overtops the bank of a river. The primary causes of floods are mainly the result of external climatic forces. Secondary flood causes tend to be drainage basin specific. Most floods in Britain, for example, are associated with deep depressions (low pressure systems) in autumn and winter which are both long in duration and wide in areal coverage. By contrast in India, up to 70 per cent of the annual rainfall occurs in one hundred days in the summer south-west monsoon. Elsewhere, melting snow may be responsible for widespread flooding.

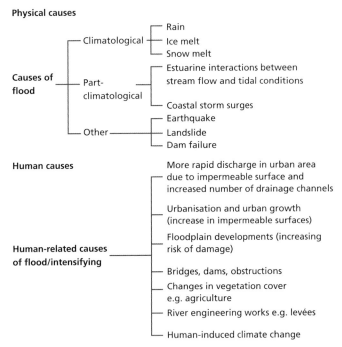

Figure 24 Natural and human causes of floods

Flood intensifying conditions cover a range of factors which alter the response of a drainage basin to a given storm. These factors include topography, vegetation, soil type, rock type, and specific characteristics of the drainage basin.

The potential for damage by flood waters increases exponentially with velocity and speeds above 3 metres per second and can undermine the foundations of buildings. The physical stresses on buildings are increased even more, probably by hundreds of times, when rough, rapidly flowing water contains debris such as rock, sediment and trees.

Other conditions that intensify floods include changes in land use. Urbanisation, for example, increases the magnitude and frequency of floods in at least four ways:

- The creation of highly impermeable surfaces, such as roads, roofs, pavements, increases runoff.
- Smooth surfaces served with a dense network of drains, gutters and underground sewers increase drainage density.
- Natural river channels are often constricted by bridge supports or riverside facilities, reducing their carrying capacity.

- Due to increased storm runoff, many sewerage systems cannot cope with the resulting peak flow.

Deforestation is also a cause of increased flood runoff and a decrease in channel capacity. This occurs due to an increase in deposition within the channel. However, there is little evidence to support any direct relationship between deforestation in the Himalayas and changes in flooding and increased deposition of silt in parts of the lower Ganges–Brahmaputra river basin. It is more likely due to the combination of heavy monsoon rains in the Himalayas, steep slopes, and the seismically unstable terrain, which ensure that runoff is rapid and sedimentation is high irrespective of the vegetation cover.

The decision to live in a floodplain, for a variety of perceived benefits, is one that is fraught with difficulties. The increase in flood damage is related to the increasing number of people living in floodplain regions.

Floods are one of the most common of all environmental hazards. This is because so many people live in fertile river valleys and in low-lying coastal areas. However, the nature and scale of flooding varies greatly. For example, 10 per cent of the US population live within a 1:100 year flood zone. In Asia floods damage about 4 million hectares of land each year and affect the lives of over 17 million people. Worst of all is China where over 5 million people have been killed in floods since 1860.

Floods account for about one-third of all natural catastrophes, cause more than half the fatalities, and are responsible for one-third of the economic losses. There are a number of reasons for the increase in the number of catastrophes and in the amount of damage they cause:

- a rising global population, including in vulnerable regions
- construction in flood-prone areas
- failure of flood protection systems
- changes in environmental conditions, for example clearance of trees and other vegetation and infilling of wetlands which reduces flood retention capacities.

Some environments are more at risk than others. The most vulnerable include the following:

- Low-lying parts of active floodplains and river estuaries, for example in Bangladesh where

110 million people live relatively unprotected on the floodplain of the Ganges, Brahmaputra and Meghna rivers. Floods caused by the monsoon regularly cover 20–30 per cent of the flat delta. In very serious floods up to half of the country may be flooded.

- Small basins subject to flash floods. These are especially common in arid and semi-arid areas. In tropical areas some 90 per cent of lives lost through drowning are the result of intense rainfall on steep slopes.
- Areas below unsafe or inadequate dams. In the USA there are about 30 000 sizable dams and 2000 communities are at risk from dams.

In most developed countries the number of deaths from floods is declining, although the number of deaths from flash floods is changing very little. By contrast, the average national flood damage has been increasing. The death rate in developing countries is much greater, partly because warning systems and evacuation plans are inadequate. It is likely that the hazard in developing countries will increase rather than decrease over time as more people migrate and settle in low-lying areas and river basins. Often newer migrants are forced into the more hazardous zones.

Managing the impacts of floods

Traditionally, floods have been managed by methods of 'hard engineering'. The main hard engineering structures include **dams** and reservoirs, levées, channel straightening and deepening (dredging), and creating flood relief channels (Figure 25). This largely means dams, levées and straight channels which are wider and deeper than the ones they replace. Although hard engineering may reduce floods in some locations, they may cause unexpected effects elsewhere in the drainage basin. Soft engineering schemes include **afforestation**, land use zoning, and river restoration.

Figure 25 Hard engineering structures
 a River defences - River Thames in London
 b Levées in Zermatt, Switzerland

Case study: Hard engineering – the Three Gorges Dam

The Three Gorges Dam is over 2 km wide and 100 metres high and the lake behind it is over 600 km long. It is an important project for China. The Yangtze basin provides 66 per cent of China's rice and contains 400 million people, while the river drains 1.8 million km² and discharges 24 000 m³/sec of water annually.

Advantages of the Three Gorges Dam

- It generates up to 18 000 megawatts, 50 per cent more than the world's largest existing HEP dam.
- It enables China to reduce its dependency on coal.

- It supplies energy to Shanghai (population over 17 million) and Chongqing (3 million), an area earmarked for economic development.
- It protects 10 million people from flooding.
- It allows shipping above the Three Gorges: the dams raised water levels by 90 m, and turned the rapids in the gorge into a lake.

Disadvantages of the Three Gorges Dam

- Over 1.25 million people were removed to make way for the dam and the lake.
- It cost $25 billion to build.
- Most floods in recent years have come from rivers that join the Yangtze below the Three Gorges Dam.
- The region is seismically active and landslides are frequent.
- The port at the head of the lake may become silted up as a result of increased deposition and the development of a delta at that point.
- Much of the land available for resettlement is over 800 m above sea level, and is cold, with thin, infertile soils on relatively steep slopes.

- Dozens of towns were evacuated, for example Wanxian (population 140 000) and Fuling (80 000), and then were drowned by the rising waters.
- Up to 530 million tonnes of silt are carried through the Gorge annually: the first dam on the river lost its capacity within seven years, and one on the Hwang He filled with silt within four years.
- To reduce the silt load, **afforestation** is needed, but resettlement of people will cause greater pressure on the slopes above the dam.
- The dam interferes with aquatic life – the Siberian crane and the white flag dolphin are threatened with extinction.
- Archaeological treasures have been drowned, including the Zhang Fei temple.

Case study analysis

1. How many people live in the Yangtze basin?
2. What is the energy potential of the Three Gorges Dam?
3. How many people could it supply with energy?
4. How much did it cost?
5. How many people had to be resettled as a result of the building of the Three Gorges Dam?

Case study: Soft engineering – the Kissimmee River Restoration Project

Between 1962 and 1971 the 165 km meandering Kissimmee river and its adjoining floodplain were channelised and thereby transformed into a 90 km, 10 m deep drainage canal (Figure 26). The river was channelised to provide an outlet canal for draining floodwaters from the developing upper Kissimmee lakes basin, and to provide flood protection for land adjacent to the river.

Impacts of channelisation

The channelisation of the Kissimee river had several unintended impacts:

- the loss of 12 000–14 000 hectares of wetlands
- a reduction in wading bird and waterfowl usage
- a continuing long-term decline in game fish populations.

Concerns about the sustainability of existing ecosystems led to a restoration study, supported by the state and national authorities. The result was a massive restoration project, on a scale unmatched elsewhere.

The project

The aim is to restore over 100 km² of river and associated floodplain wetlands. The project will benefit over 320 fish and wildlife species, including the endangered bald eagle, wood stork and snail kite. It will create over 11 000 hectares of wetlands.

Restoration of the river and its associated natural resources requires dechannelisation. This entails backfilling approximately half of the flood control channel and re-establishing the flow of water through the natural river channel. In residential areas the flood control channel will remain in place.

The costs of restoration

- It is estimated the project will cost $980 million (initial channelisation cost $20 million), the bill being shared by the state of Florida and the federal government.
- Restoration, which began in 1999, will not be completed until 2015.
- Restoration of the river's floodplain could result in higher losses of water due to evaporation during wet periods. Navigation may be impeded in some sections of the restored river in extremely dry spells. However, navigable depths should be maintained for at least 90 per cent of the time.

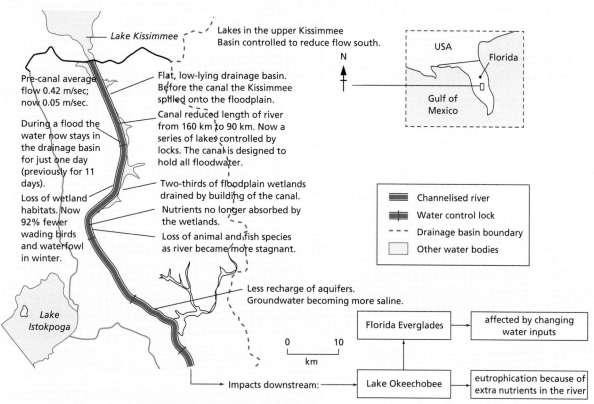

Figure 26 Kissimmee River Restoration Project – impacts of channelisation

Benefits of restoration

- Higher water levels should ultimately support a natural river ecosystem again.
- Re-establishment of floodplain wetlands and the associated nutrient filtration function is expected to result in decreased nutrient loads to Lake Okeechobee.
- It is possible that restoration of the floodplain could benefit populations of key avian species, such as wading birds and waterfowl, by providing increased feeding and breeding habitats.

Potential revenue associated with increased recreational use (such as hunting and fishing) and ecotourism on the restored river could significantly enhance local and regional economies.

Case study analysis

1 Where is the Kissimmee river?
2 Describe the changes to the Kissimmee river between 1962 and 1971.
3 Outline the benefits of river restoration.
4 How much did the Kissimmee river restoration cost?
5 When was it completed?

Event modification

Event modification includes attempts to reduce floods and to divert them away from settlements. Reducing floods involves decreasing the amount of runoff, thereby reducing the flood peak in a drainage basin. This can be achieved by weather modification and/or watershed treatment, for example to reduce flood peak over a drainage basin. There are a number of strategies including:

- reforestation
- reseeding of sparsely vegetated areas to reduce evaporative losses
- mechanical land treatment of slopes such as contour ploughing or terracing to reduce runoff
- comprehensive protection of vegetation from wildfires, overgrazing, clear-cutting of forests, land or any other practices likely to increase flood discharge and sediment load

- clearance of sediment and other debris from headwater streams
- construction of small water- and sediment-holding areas
- preservation of natural water detention zones.

Flood diversion measures, by contrast, include the construction of levées, reservoirs, and the modification of river channels (Figure 27). Levées are the most common form of river engineering. They can also be used to divert and restrict water to low value land on the floodplain. Over 4500 km of the Mississippi River has levées. Channel improvements include enlargement to increase the carrying capacity of the river. Reservoirs store excess rainwater in the upper drainage basin. Large dams are expensive and may well be causing earthquakes and siltation. It has been estimated that some 66 billion m³ of storage will be needed to make any significant impact on major floods in Bangladesh!

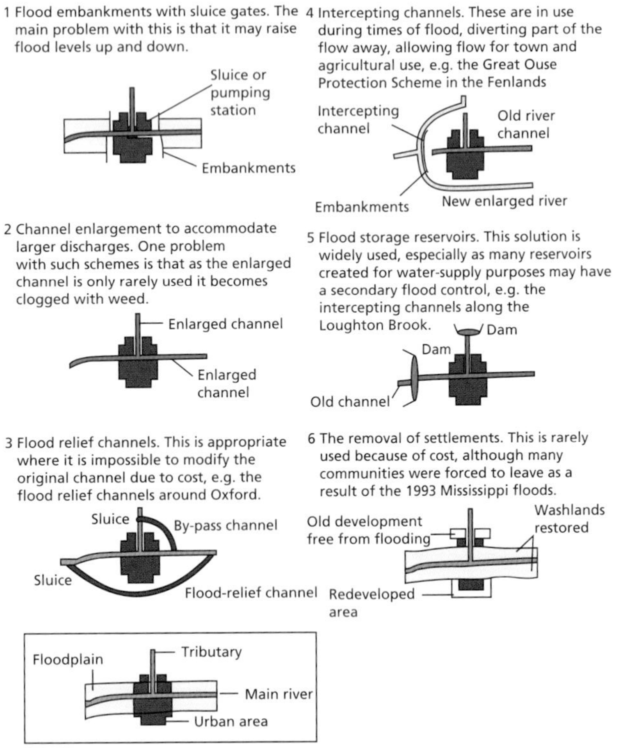

1 Flood embankments with sluice gates. The main problem with this is that it may raise flood levels up and down.

Sluice or pumping station

Embankments

2 Channel enlargement to accommodate larger discharges. One problem with such schemes is that as the enlarged channel is only rarely used it becomes clogged with weed.

Enlarged channel

Enlarged channel

3 Flood relief channels. This is appropriate where it is impossible to modify the original channel due to cost, e.g. the flood relief channels around Oxford.

Sluice   By-pass channel

Sluice

Flood-relief channel

Floodplain   Tributary

Main river

Urban area

4 Intercepting channels. These are in use during times of flood, diverting part of the flow away, allowing flow for town and agricultural use, e.g. the Great Ouse Protection Scheme in the Fenlands

Intercepting channel   Old river channel

Embankments   New enlarged river

5 Flood storage reservoirs. This solution is widely used, especially as many reservoirs created for water-supply purposes may have a secondary flood control, e.g. the intercepting channels along the Loughton Brook.

Dam   Dam

Old channel

6 The removal of settlements. This is rarely used because of cost, although many communities were forced to leave as a result of the 1993 Mississippi floods.

Old development free from flooding   Washlands restored

Redeveloped area

Figure 27 Flood relief channels

Loss sharing and insurance

Loss sharing adjustments include disaster aid and insurance. Disaster aid refers to any aid – such as money, equipment, staff and technical assistance – that is given to a community following a disaster. However, there are many taxpayers who argue that taxpayers cannot be expected to fund losses that should have been insured.

In developed countries insurance is an important loss sharing strategy. However, not all flood-prone households have insurance and many of those that are insured may be under-insured. In the floods of central England in 1998, many of the affected households had very limited flood insurance because the residents had not thought they lived in an area that was likely to flood.

Hazard resistant design

Flood proofing includes any adjustments to buildings and their contents that help reduce losses. Some are temporary, such as blocking up of entrances, use of shields to seal doors and windows, removal of damageable goods to higher levels, and the use of sandbags (Figure 28). By contrast, long-term measures include moving the living spaces above the likely level of the floodplain. This normally means building above the flood level, but could also include building homes on stilts (Figure 29).

Figure 28 Sandbags

Figure 29 Hazard-resistant design

Forecasting and warning

During the 1970s and 1980s, flood forecasting and warning became more accurate and these are now one of the most widely used measures to reduce the problems caused by flooding. In developed countries flood warnings and forecasts may reduce economic losses by as much as 40 per cent. In most developing countries there is much less effective flood forecasting. An exception is Bangladesh. Most floods in Bangladesh develop from events in the Himalayas, so authorities can give people in Bangladesh about 72 hours' warning.

Land use planning

Land use zoning involves allowing compatible land uses with land that might flood. For example, land that floods regularly (once a year) could be used for pastoral agriculture, as animals can be moved to higher ground. Alternatively it could be used for recreational purposes. However, it should not be used for industrial, commercial or residential land use, as valuable equipment and possessions would be damaged or destroyed. However, as population growth continues, people and industry may be forced to extend into floodplains.

Case study: The Hwang He river

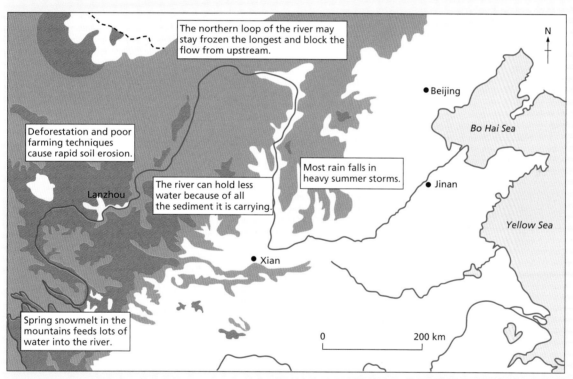

Figure 30 The Hwang He river basin

The Hwang He is the second longest river in China, and the sixth longest in the world. Originating on the Qinhai-Tibetan plateau in western China, it runs for some 5500 km across the vast North China Plain before draining into the Bo Hai Sea. Its catchment area of 795 000 km² is home to 110 million people, or about 9 per cent of China's population. (The figures increase to 189 million and 15 per cent if the floodplain surrounding the lower reach is included.) By 2000, over 25 per cent of the basin was urbanised. As the centre of China's current political, economic and social development, the river is known as 'the mother river of China'.

It is said to have killed more people than any other natural feature and for this reason it is called the 'river of sorrow'. The worst flood occurred in 1332 – over 7 million people drowned and a further 10 million died as a result of the famine that followed.

Millions of lives have been lost to floods and droughts during the long history of the Hwang He basin. From 206 BC to 1949, 1092 major floods were recorded, along with 1500 dyke failures, 26 river re-channellings and 1056 droughts. For example, in 1887 the Hwang He overtopped its bank and flooded an area of 22 000 km² to a depth of 8 m. Over 1 million people were killed by the floods and the famine that followed. It is ironic that famine should follow floods, but the silt carried by the river destroys the crops that it is dumped on. However, following the establishment of the People's Republic of China in 1949, master planning for flood control and the construction of numerous hydraulic structures significantly reduced the vulnerability of the area and losses due to floods.

During normal flow conditions the river contains a large amount of silt. In fact, 40 per cent of the river flow is sediment. An 8 m flood might be expected to dump over 3 m of material on the ground.

Now the river is 20 m higher than the floodplain and the risk of flooding is great. As China's population continues to grow, more and more people are living on the fertile floodplains. This makes the risk of a disaster even greater.

Attempts to control the river go back at least as far as 2356 BCE and there have been levées on the river for at least 2500 years! Despite this long history of engineering the Hwang He has shifted its course on at least ten occasions. When the river changes its course it can move the place where it enters the sea by as much as 1100 km.

Why does the river shift? When a river deposits its load it builds up its floodplain and also the base of the river channel. As more material is deposited in the river channel, the base of the river channel may become higher than the level of the floodplain. The rivers banks – known as levées – stop the river from flooding.

However, the river also produces some benefits:

- The sediment forms a fertile soil. As a result 13 million hectares of the drainage basin are farmed.
- Where the deposits have been dropped on the coastal plains they have formed a large alluvial fan, with a delta at the sea. This area houses over 10 million people.
- There is a huge potential for hydro-electric power, much of which is in the upper part of the river where sediment loads are low. By the mid-1980s over 150 reservoirs had been built and power stations were added to 80 of these.

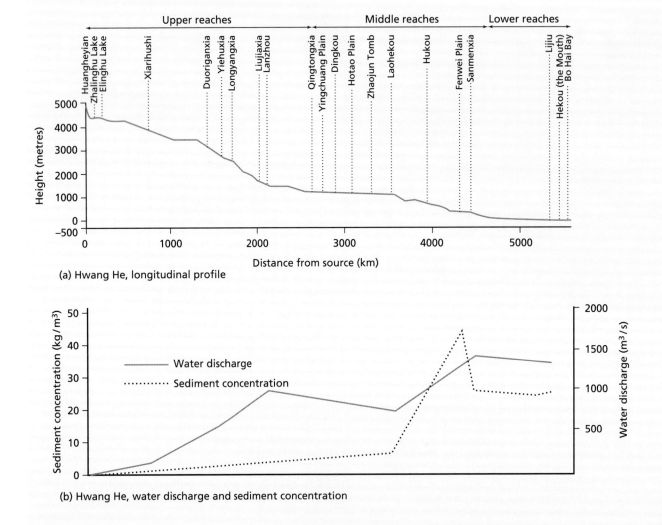

(a) Hwang He, longitudinal profile

(b) Hwang He, water discharge and sediment concentration

Figure 31 Long profile, discharge and sediment concentration on the Hwang He river

Changing demand

Water demand in the Hwang He basin increased dramatically from 10 billion m³ in 1949 to 37.5 billion m³ in 2006. **Groundwater** has been extensively exploited in the basin following the introduction of tube wells in the late 1950s. In 2000, groundwater abstraction reached 10.7 billion m³ and there were some 380 000 tube wells in the basin. Consequently, overexploitation of groundwater resources has been a serious concern, particularly in the large and mid-size cities along the Hwang He. Springs in Jinan, once known as 'the city of springs', dried up in the late 1990s. Overall, groundwater levels have dropped significantly in 65 locations due to extensive withdrawals.

One of the biggest direct impacts of rapid industrialisation and population growth has been on water quality. The amount of untreated industrial sewage being dumped into the Hwang He has doubled since the 1980s to 4.2 billion m³ per year. The river receives over 300 pollutants, and only about 60 per cent of its course is now usable for a drinking water supply.

As a result of intensive water development between 1951 and 1987, many structures were built in the basin for flood control, hydro-electric power and irrigation. In 2000, there were over 10 000 reservoirs in the basin.

The expansion of irrigation in the basin has been rapid. The irrigated area rose from 8000 km² in 1950 to 75 000 km² in 2000. Agriculture accounts for 84 per cent of total water consumption, followed by industry with 9 per cent and households with 5 per cent. When consumption exceeds water availability in the basin, the deficit is met by using groundwater resources outside the basin, and by recycling.

Case study analysis

1 Explain why the Hwang He is described as 'China's river of sorrow'.
2 Outline the benefits of the Hwang He to people.
3 Explain how human activity has affected the hydrology of the drainage basin.

Activities

1 Explain how a river can be higher than its floodplain.
2 Study Figure 31 which shows rates of erosion. How do rates of erosion vary with (a) geology and (b) the amount of water in the river?

(2.3) Coasts

Jolly Harbour, Antigua

Key questions

- How do the sea and wind erode, transport and deposit?
- What landforms do they create?
- Under what conditions do coral reefs and mangroves form?
- What are the opportunities and hazards that coasts present?
- What can be done to manage coastal erosion?

● Marine processes

The factors that affect coastal processes and coastal landforms include:

- waves and currents, including longshore drift
- local geology – that is, rock type, structure and strength
- changes in sea level
- human activity and the increased use of coastal engineering.

All of these factors interact and produce a unique set of processes that occur at the coast. These processes go on to produce different types of landform for every coastal area.

Types of wave

Wavelength is the distance between two successive crests or troughs. Wave height is the distance between the trough and the crest. Wave frequency is the number of waves per minute. Velocity is the speed of a travelling wave, and is influenced by wind, fetch and depth of water. The **fetch** is the amount of open water over which a wave has passed. **Swash** is the movement of water up the beach. **Backwash** is the movement of water down the beach. Waves are sometimes divided into **constructive** and **destructive waves** (Table 1 and Figure 1).

Table 1 Destructive and constructive waves

Destructive waves (erosional waves)	Constructive waves (depositional waves)
• Short wavelength (< 20 metres)	• Long wavelength (up to 100 metres)
• High height (> 1 metre)	• Low height (< 1 metre)
• High frequency (10–12/minute)	• Low frequency (6–8/minute)
• Low period (one every 5–6 seconds)	• High period (one every 8–10 seconds)
• Backwash > swash	• Swash > backwash
• Steep gradient	• Low gradient
• Caused by local winds and storms	• Caused by swell from distant storms
• High-energy waves	• Low-energy waves

Constructive waves

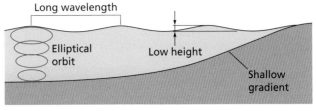

Destructive waves

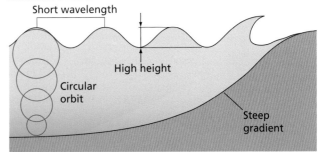

Figure 1 Destructive and constructive waves

Waves on the Palisadoes

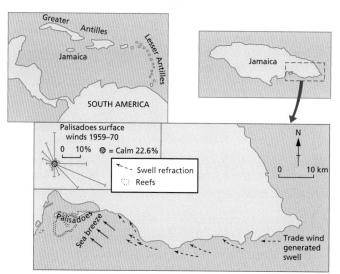

Figure 2 Constructive and destructive waves on the Palisadoes, Jamaica

Beaches are transitory features – they change shape regularly. In Jamaica, one of the most important factors for wave formation is the local sea–land breeze. The Palisadoes, on the south coast of Jamaica, has a tidal range of just 0.23 m.

Here, swell is generated by trade winds on a year-round basis, and occasionally by easterly waves and tropical cyclones (hurricanes) that originate far from the island. The sea breeze is most persistent in the summer months from May to August and strongest in June, approaching most commonly from an east-south-easterly direction (Figure 2). The breeze normally develops in the late morning, reaching velocities of 12 m/s by the early afternoon. Once the sea breeze declines, a land breeze develops from the north-west.

The sea breeze is associated with an increase in wave and breaker height. Wave heights regularly exceed 1 metre and may reach 5 m. The sea breezes provide a mechanism for shoreline erosion, caused by destructive waves during daylight hours. The change from constructive waves, where sediment is transported landwards, to destructive waves, where sediment is dragged seawards, is related to wave steepness. As steepness increases, erosion occurs.

When onshore winds occur, a return counter-current is formed, flowing seawards towards the break point. This current removes material from the front of the beach and carries it to the break point, where material is deposited to form longshore bars offshore from the Palisadoes.

With the return of the land breeze, these destructive waves decay. The trade wind generated swell becomes the dominant wave, returning sediment lost from the beach during the day.

Activities

1 What is the tidal range of the Palisadoes?
2 In which direction does the sea breeze blow?
3 From which direction does the land breeze blow?
4 What is the impact of the sea breeze on wave activity?
5 What is the impact of the land breeze on wave activity?

Wave refraction and longshore drift

Waves result from friction between wind and the sea surface. Waves in the open, deep sea are different from those breaking on shore. Sea waves are forward-moving surges of energy. Although the shape of the surface wave appears to move, the water particles follow a roughly circular path within the wave. As waves approach the shore, their speed is reduced as they touch the sea floor. **Wave refraction** causes two main changes: the speed of the wave is reduced and the shape of the wave front is altered. If refraction is completed, the wave fronts will break parallel to the shore.

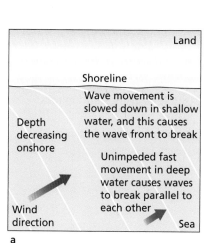

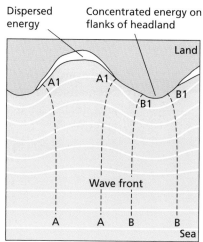

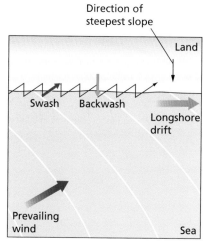

Figure 3 Wave refraction and longshore drift

Wave refraction also distributes wave energy along a stretch of coast. On a coastline with alternating **headlands** and bays, wave refraction will concentrate destructive/erosive activity on the headlands, while **deposition** will tend to occur in the bays. Irregularities in the shape of the coastline mean that refraction is not always totally achieved. This causes **longshore drift**, which is a major force in the transport of material along the coast. It occurs when waves move up to the beach (swash) in one direction, but the waves draining back down the beach (backwash) take a different route (under the effect of gravity). The net movement is *along* the shore, hence the term longshore drift. A wooden or concrete wall (groyne) may be built to prevent longshore drift from moving sand or shingle away from the beach.

Human activity and longshore drift in West Africa

The increase in coastal retreat has been blamed on the construction of the Akosombo Dam on the Volta river in Ghana. It is just 110 km from the coast and disrupts the flow of sediment from the Volta and stops it from reaching the shore. Thus there is less sand to replace that which has already been washed away by longshore drift, so the coastline retreats due to erosion by the Guinea current. Towns such as Keta, 30 km east of the Volta estuary, have been destroyed as their protective beach has been removed (Figure 4).

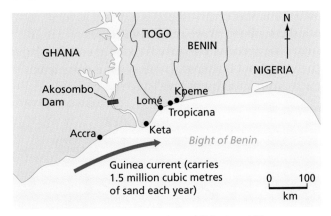

Figure 4 Human activity and longshore drift in West Africa

Activities

1 Define the following terms: swash, fetch, wave refraction, longshore drift, backwash.
2 Describe the main differences between a destructive wave and a constructive wave.
3 Describe and explain the process of longshore drift.
4 Briefly describe how human activity affected the impact of longshore drift in West Africa.

● Landscapes of erosion

There are many types of **erosion** carried out by waves:

- **Hydraulic action** occurs as waves hit or break against a cliff face. Any air trapped in cracks is put under great pressure. As the wave retreats, this build-up of pressure is released with explosive force. This is especially important in well-jointed rocks such as limestone, sandstone and granite, and in weak rocks such as clays and glacial deposits. Hydraulic action makes the most impact during storms.
- **Abrasion** is the process of a breaking wave hurling materials, such as pebbles or shingle, against a cliff face. It is similar to abrasion in a river.
- **Attrition** is the process in which eroded material, such as broken rock, is worn down to form smaller, rounder beach material.
- **Solution** occurs on limestone and chalk. Calcium carbonate, a salt found in these rocks, dissolves slowly in acidic water.

Features of erosion

On a headland, erosion will exploit any weakness, creating, at first, a **cave**. Once the cave reaches both sides of the headland, an **arch** is formed. A collapse of the top of the arch forms a **stack**, and when the stack is eroded a **stump** is created (Figure 5). Where erosion opens up a vertical crack, allowing sea water to spout up at the surface, a blowhole is formed. The sandstone of the Cape Peninsula in South Africa has been attacked by the sea, forming steep vertical **cliffs** and small-scale features such as arches and stacks (Figure 6).

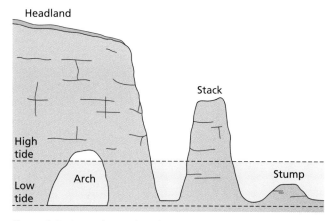

Figure 5 Features of coastal erosion

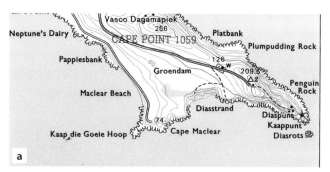

Figure 6 The Cape Peninsula, South Africa – the photo shows Cape Maclear and the Cape of Good Hope in the background

Wave action is concentrated between the high water mark (HWM) and low water mark (LWM). HWM is the level reached by the sea at high tide while LWM is the level reached by the sea at low tide. It may undercut a cliff face, creating a notch and overhang (Figure 7). As erosion continues, the notch becomes deeper and eventually the overhang collapses, causing the cliff line to retreat. The base of the cliff is left behind as an increasingly longer platform. This is sometimes called a **wave-cut**

platform, because it has been cut or eroded by wave action.

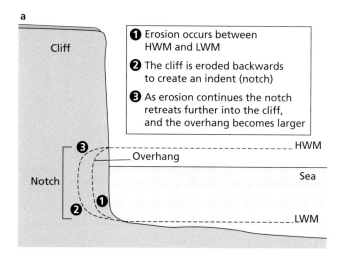

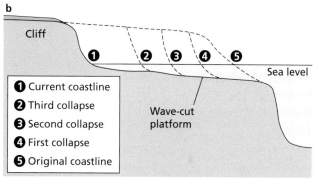

Figure 7 Formation of a wave-cut platform

Cliff profiles vary greatly, depending on the:

- rate of coastal erosion (cliff retreat)
- strength of the rock
- presence of joints and bedding planes.

In addition, cliffs change over time, from ones dominated by marine processes (like wave erosion) to ones protected from marine processes but affected by land-based processes.

On a larger scale bays may be eroded in beds of weaker rock (Figure 8). The harder rocks form headlands that protrude whereas the weaker rocks are eroded to form bays. Wave refraction in the bay spreads wave energy around the bay whereas it focuses wave energy on the flanks of the headlands. Bayhead beaches are formed when constructive waves deposit sand between two headlands, such as at Maracas Bay and Tyrico Bay in northern Trinidad.

Figure 8 Headlands and bays in Praia de Rocha, Portugal

Figure 9 Volcanic beach, Tenerife

Activities

1 What is the difference between attrition and abrasion?
2 Why does hydraulic action occur in jointed rocks?
3 What types of rock are affected by solution?
4 What types of erosion are most likely to take place:
   a during a storm
   b on beaches
   c on the face of a cliff?
5 In your own words, describe how a wave-cut platform may be formed.
6 Make a sketch of Figure 8 and label the following features: headland, bay, stack and beach.

Figure 10 An artificial beach with imported sand

● Deposition

Beaches

Excellent beach development occurs on a lowland coast (constructive waves) with a sheltered aspect/trend, composed of 'soft' rocks, which provide a good supply of material, or where longshore drift supplies abundant material.

On Tenerife, the lack of beach material other than volcanic material (Figure 9) has led to one beach, Las Terristas, being formed of sand imported from the Sahara desert (Figure 10). An artificial barrier prevents the sand from being eroded by wave action.

The term **beach** refers to the accumulation of material deposited between low spring tides and the highest point reached by storm waves at high spring tides. A typical beach will have three zones: backshore, foreshore and offshore. The backshore is marked by a line of dunes or a cliff. Above the high water mark there may be a berm or shingle ridge. This is coarse material pushed up the beach by spring tides and aided by storm waves flinging material well above the level of the waves themselves. These are often referred to as storm beaches. The seaward edge of the berm is often scalloped and irregular due to the creation of beach cusps.

The foreshore is exposed at low tide. Offshore, the first material is deposited. In this zone, the waves touch the sea bed and so the material is usually disturbed, sometimes being pushed up as offshore **bars**, when the offshore gradient is very shallow. Offshore bars are usually composed of coarse sand or shingle. Between the bar and shore, **lagoons** (often called sounds) develop (Figure 11).

If the water in the lagoon is calm and fed by rivers, marshes and mudflats can be found. Bars can be driven onshore by storm winds and **waves**. A classic area is off the coast of the Carolinas in the south-east of the USA.

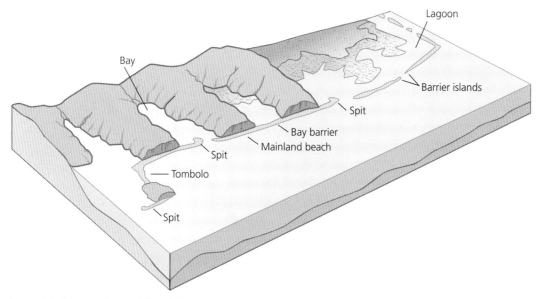

Figure 11 Features of coastal deposition

Bars and spits

These more localised features will develop where:

- abundant material is available, particularly shingle and sand
- the coastline is irregular, for example where there is a variable geology
- where there are estuaries and major rivers.

A **spit** is a beach of sand or shingle linked at one end to land. It is found where wave energy is reduced, for example along a coast where headlands and **bays** are common and near river mouths (in estuaries and **rias**).

Spits often become curved as waves undergo refraction (Figure 12). Cross-currents or occasional storm waves may assist this hooked formation. A good example is the sandspit in Walvis Bay, Namibia. The main body of the spit is curved but it has additional, smaller hooks, or recurves. Longshore drift moves sediment northwards along the coast. However, the coastline is very irregular here and there is a sudden change in the trend of the coastline. Consequently, refraction occurs, causing the waves to bend around eastwards.

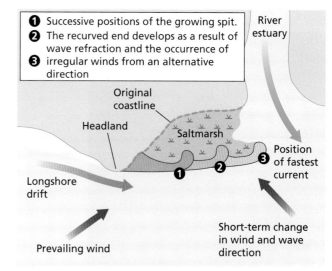

Figure 12 Development of a spit

On the seaward side, the slope to deeper water is very steep. Within the curve of the spit, the water is shallow and a considerable area of mudflat and saltmarsh is exposed at low water. These saltmarshes continue to grow as mud is trapped by the marsh vegetation.

Related features include bars. These are ridges that block off a bay or river mouth. There are many examples on the west coast of Antigua (Figure 13).

Figure 13 The west coast of Antigua

Interesting note

The longest spit in the world is the 112 km Arabat spit in the Sea of Azov, between Russia and Ukraine. There are thirteen spits in the Sea of Azov.

Tombolos are ridges that link the mainland to an island. Good examples include the Lumley area of Sierra Leone, and the Cape Verde Peninsula, Senegal. The Cape Peninsula in South Africa is a complex tombolo that has developed on a very large scale.

The Palisadoes, Jamaica: a spit or a tombolo?

The Palisadoes is one of the largest deposited coastal features in the Caribbean (Figure 14).

Figure 14 Aerial view of the Palisadoes tombolo

Located just south of Kingston in Jamaica, this 13 km long feature has been formed and re-formed many times during its history. Scientists believe that it may be 4000 years old.

Longshore drift occurs from east to west on the south coast of Jamaica. The sediment comes from rivers, cliff erosion, and offshore sediments. The Palisadoes is located at a sharp bend in the coastline. Longshore drift carries sediment westwards, and extends the length of the spit. As it grew longer, it linked up with a number of cays (small islands), turning the spit into a tombolo.

The region experiences tropical storms and **hurricanes**. These can seriously damage the coast.

For example, in 2004 Hurricane Ivan eroded up to a metre off the 2 m high **sand dunes**. Even under normal conditions, summer sea breezes cause powerful destructive waves, which are capable of eroding the seaward face of the beach, causing it to become steeper.

Activities

1 How old is the Palisadoes?
2 How long is the Palisadoes?
3 In which direction is longshore drift on the Palisadoes?
4 Where does the sediment that helps build up the Palisadoes come from?
5 What is the impact of hurricanes on the Palisadoes?

Activities

1 a Explain how a spit develops.
  b In what ways might vegetation help spits, bars and tombolos to develop?
2 a Draw a labelled sketch of photograph a in Figure 13.
  b Describe the wave conditions in the photograph.
3 Study the map in Figure 13. Name and give examples of at least two types of coastal deposit. For any one of these, describe its main characteristics and explain how it has been formed.
4 Study both the map and photograph in Figure 13.
  a What type of feature is found in Valley Church Bay and at Reeds Point?
  b What is the difference between a cove and a bay?
  c How are land-based processes affecting this area of coastline?
  d In what ways has this area of coastline influenced human activities?

Sand dunes

Sand dunes are one of the most dynamic environments in physical geography. Important changes take place in a very short space of time. Extensive sandy beaches are almost always backed by sand dunes because strong onshore winds can easily transport the sand that has dried out and is exposed at low water. The sand grains are trapped and deposited against any obstacle on land, to form dunes. Dunes can be blown inland and can therefore threaten coastal farmland and even villages. The interaction of winds and vegetation helps form sand dunes.

On the beach, conditions are very windy, dry (much water just soaks into the sand) and salty. Few plants can survive these extreme conditions but some can, including sea couch and marram grass (Figure 15). These are adapted to tolerate water with a high salt content, and high wind speeds, and they can survive burial by sand. In fact, marram grass needs to be buried by fresh sand in order to send out fresh shoots.

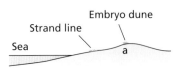

As the tide goes out, the sand dries out and is blown up the beach. A small embryo dune forms in the shelter behind the strand line.

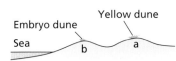

Sea couch grass colonises and helps bind the sand. Once the dune grows to over 1 m high, marram grass replaces the sea couch.

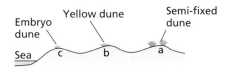

Once the yellow dune is over 10 m high, less sand builds up behind it and marram grass dies to form a thin humus layer. As the original dune a has developed, new embryo and yellow dunes have formed.

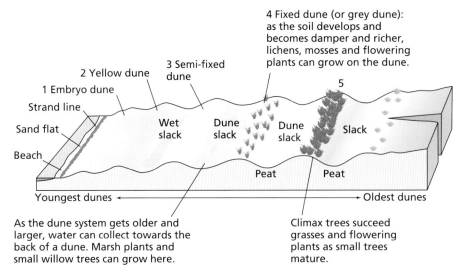

Figure 15 Formation of sand dunes

Once marram grass and sea couch are established on the beach, they reduce the wind speed and this helps trap fresh sand. As the sand builds up, these plants send out new shoots, trapping more sand and building up a dune. Increasingly, the presence of plants in the sand dune adds organic matter and moisture to the dune and allows other plants to grow, such as heather.

The growth of new plants is called succession. Plants such as heather cannot tolerate the dry, windy, salty conditions of the beach but can survive in the less windy, moister, less salty dunes. They in turn alter the environment so that other species can invade and develop (Figure 16). On a sand dune over a distance of just a few hundred metres there may be as many as four or five different types of ecosystem.

Figure 16 Sand dune vegetation

Activities

Study Figure 15 which shows the development of vegetation on a sand dune.

1 How do conditions differ between the shoreline and inland?
2 Explain why deposition occurs on the sand dunes.
3 Suggest how human activities might affect the sand dunes and/or the saltmarsh.

Coral reefs

Coral reefs are calcium carbonate structures, made up of reef-building stony corals. **Coral** is limited to the depth that light can reach, so reefs develop in shallow water, ranging to depths of 60 m. This dependence on light also means that reefs are only found where the surrounding waters contain relatively small amounts of suspended material. Reef-building corals live only in tropical seas, where temperature, salinity and clear water allow them to develop.

There are many types of coral reef (Figure 17).

- **Fringing reefs** are those that fringe the coast of a landmass (Figures 18 and 19). Many fringing reefs grow along shores that are protected by barrier reefs and are thus characterised by organisms that are best adapted to low wave-energy conditions.
- **Barrier reefs** occur at a greater distance from the shore than fringing reefs and are commonly separated from it by a wide, deep lagoon. Barrier reefs tend to be broader, older, and more continuous than fringing reefs. For example, the Beqa barrier reef off Fiji stretches unbroken for more than 37 km, and that off Mayotte in the Indian Ocean for around 18 km. The largest barrier reef system in the world is the Great Barrier Reef, which extends 1600 km along the east Australian coast, usually tens of kilometres offshore. Another long barrier reef is located in the Caribbean off the coast of Belize between Mexico and Guatemala.
- **Atoll reefs** rise from submerged volcanic foundations. Atoll reefs are essentially indistinguishable in form and species composition from barrier reefs except that they are confined to the flanks of submerged oceanic islands, whereas barrier reefs may also flank continents. Over 300 atolls are present in the Indo-Pacific but only 10 are found in the western Atlantic.

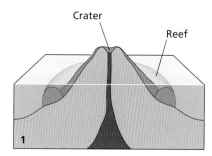

Rocky volcanic islet encircled by fringing coral reef

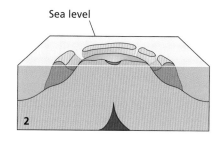

Reef enlarges as land sinks (or sea rises)

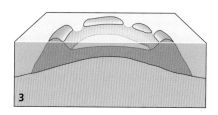

Circular coral reef or atoll (with further change in level)

Figure 17 Formation of coral reefs

about 25 per cent of the world's sea fish breed, grow, and evade predators in coral reefs. Some of the world's best coral reefs include Australia's Great Barrier Reef, many of the reefs around the Philippines and Indonesia, Tanzania and the Comoros, and the Lesser Antilles in the Caribbean (Figure 19).

Coral reefs face many pressures. The fishing industry now uses dynamite to flush out fish and cyanide solution to catch live fish. Destruction takes many forms – collection of specimens, trampling, berthing of boats, oil spills, mining, and the cement industry. Indirect pressures include sedimentation from rivers, and waste disposal from urban areas. Coastal development, especially for tourism, is taking its toll too. Dust storms from the Sahara have introduced bacteria into Caribbean coral, while global warming may cause coral bleaching. Bleaching occurs when high temperatures kill the algae in coral, removing their colour so the coral appears bleached. Many areas of coral in the Indian Ocean were destroyed by the 2004 tsunami.

Coral reefs are of major biological and economic importance. Countries such as Barbados, the Seychelles and the Maldives rely on tourism based on their reefs. Florida's reefs attract tourism worth $1.6 billion annually. The global value of coral reefs in terms of fisheries, tourism and coastal protection is estimated to be $375 billion.

Figure 18 A fringing reef on the south coast of Antigua

Figure 19 A fringing reef on the west coast of Antigua

Coral reefs are often described as the 'rainforests of the sea' on account of their rich biodiversity. Some coral is believed to be 2 million years old, although most is less than 10 000 years old. Coral reefs contain nearly a million species of plants and animals, and

Activities

1 Under what conditions does coral grow?
2 What is the difference between a fringing reef and a barrier reef?
3 How are atolls formed?
4 Why are coral reefs so valuable?
5 What are the main threats to coral reefs?

Mangroves

Mangroves are salt-tolerant forests of trees and shrubs that grow in the tidal estuaries and coastal zones of tropical areas (Figure 20). The muddy waters, rich in nutrients from decaying leaves and wood, are home to a great variety of sponges, worms, crustaceans, molluscs and algae. Mangroves cover about 25 per cent of the world's tropical **coastline**, the largest being the 570 000 ha Sundarbans in Bangladesh.

Figure 20 Mangrove swamps

The value of mangroves

Mangroves have many uses, such as providing large quantities of food and fuel, building materials and medicine. One hectare of mangrove in the Philippines can yield 400 kg of fish and 75 kg of shrimp. Mangroves also protect coastlines by absorbing the force of hurricanes and storms. They also act as natural filters, absorbing nutrients from farming and sewage disposal.

Pressures on mangroves

Despite their value, many mangrove areas have been lost to rice paddies and shrimp farms. As population growth in coastal areas is set to increase, the fate of mangroves looks bleak. Already most Caribbean and South Pacific mangroves have disappeared, and India, West Africa and South-East Asia have lost half of theirs (Table 2).

Table 2 Mangrove losses

Thailand	185 000 ha (1960–91) to shrimp ponds
Malaysia	235 000 ha (1980 and 1990) to shrimp ponds and farming
Indonesia	269 000 ha (1960–90) to shrimp ponds
Vietnam	104 000 ha (1960–74) due to action by the US army
Philippines	170 000 ha (1967–76) mostly to shrimp ponds
Bangladesh	74 000 ha (since 1975) largely to shrimp ponds
Guatemala	9500 ha (1965–84) to shrimp ponds and salt farming

Mangroves and coral reefs

Mangroves and coral reefs are fundamentally connected ecosystems. Mangroves protect coral reefs from sedimentation from land-based sources, as well as helping to keep the water clear of particles and nutrients. Both of these functions are necessary to maintain reef health. Mangroves also provide spawning and nursery areas for many animal species that spend their adult lives on the reefs. In return, the coral reefs provide shelter for the mangroves and their inhabitants, while the calcium carbonate eroded from the reef provides sediment in which the mangroves grow.

● Coastal hazards and opportunities

Coastal areas offer many opportunities to people. However, actions in coastal areas may cause new problems. These are summarised in Table 3.

Table 3 Relationships between human activities and coastal zone problems

Human activity	Agents/consequences	Coastal zone problems
Urbanisation and transport	Land use changes, e.g. for ports, airports; road, rail, and air congestion; dredging and disposal of harbour sediments; water abstraction; waste water and waste disposal	Loss of habitats and species diversity; visual intrusion; lowering of groundwater table; salt water intrusion; water pollution; human health risks; eutrophication; introduction of alien species
Agriculture	Land reclamation; fertiliser and pesticide use; livestock densities; water abstraction	Loss of habitats and species; diversity; water pollution; eutrophication; river channelisation
Tourism, recreation and hunting	Development and land use changes, e.g. golf courses; road, rail and air congestion; ports and marinas; water abstraction; waste water and waste disposal	Loss of habitats and species diversity; disturbance; visual intrusion; lowering of water table; salt water intrusion in aquifers; water pollution; eutrophication; human health risks
Fisheries and aquaculture	Port construction; fish processing facilities; fishing gear; fish farm effluents	Overfishing; impacts on non-target species; litter and oil on beaches; water pollution; eutrophication; introduction of alien species; habitat damage and change in marine communities
Industry (including energy production)	Land use changes; power stations; extraction of natural resources; process effluents; cooling water; windmills; river impoundment; tidal barrages	Loss of habitats and species diversity; water pollution; eutrophication; thermal pollution; visual intrusion; decreased input of fresh water and sediment to coastal zones; coastal erosion

Case study: Erosion of the USA's eastern seaboard

Many beaches along the east coast of America have disappeared since 1900, such as that at Marshfield, Massachusetts. As the sea level rises, the beaches and barrier islands (barrier beaches) that line the coasts of the Atlantic Ocean and the Gulf of Mexico, from New York to the Mexican border, are in retreat.

The problem is that much of the shore cannot retreat naturally because industries and properties worth billions of dollars have been built here. Many important cities and tourist centres, such as Miami, Atlantic City and Galveston (Texas), are sited on barrier islands. Consequently, many shoreline communities have built sea walls and other protective structures to protect them from the power of destructive waves.

- **Relief** – the flat topography of the coastal plains from New Jersey southward means that a small rise in sea level can allow the ocean to advance a long way inland.
- **Changing sea levels** – much of the North American coast is sinking relative to the ocean, so local sea levels are rising faster than global averages. The level of tides along the coasts shows that subsidence varies between 0.5 and 19.5 mm a year. By contrast, the west coast, in particular Alaska, is rising.
- **Coastal development** – extensive coastal development has accelerated erosion. While sea level rises, apartment blocks, resorts and second homes have developed rapidly along the shoreline. By 1990, 75 per cent of Americans lived within 100 km of a coast (including the Great Lakes).
- **Erosion and tourism-related developments** – erosion is evident at many places along the coasts of the Atlantic and the Gulf of Mexico. Major resorts such as Miami Beach and Atlantic City have pumped in dredged sand to replenish eroded beaches. Erosion threatens islands to the north and south of Cape Canaveral, although the Cape itself appears safe. Resorts built on barrier beaches in Virginia, Maryland and New Jersey have also suffered major erosion.
- **Rates of erosion** – overall losses are not well known. Massachusetts loses about 26 hectares a year to rising seas. Nearly 10 per cent of that loss is from the island of Nantucket, south of Cape Cod. However, these losses are minimal compared with Louisiana, which is losing 40 hectares of wetlands a day – about 15 500 hectares a year.

Case study analysis

1 Why are beaches on the eastern seaboard of the USA retreating?
2 What has been done to reduce erosion of barrier beaches?
3 What proportion of Americans live within 100 km of the coast?
4 By how much is the east coast subsiding each year?
5 How much land is being lost to rising sea levels each year?

Tropical storms

A hurricane is one of the most dangerous natural hazards to people and the environment. Damage is caused by high winds, floods and storm surges. In Asia, hurricanes are also known as tropical cyclones.

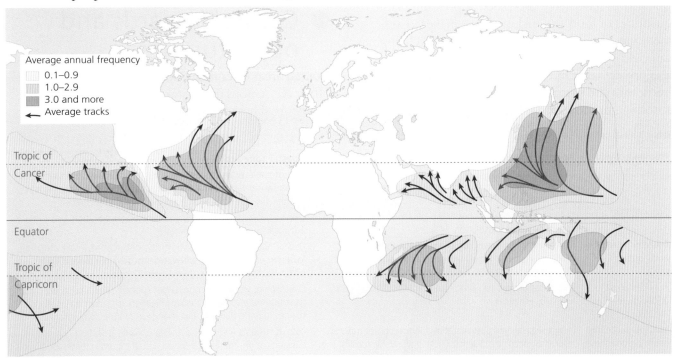

Figure 21 The distribution of hurricanes

Hurricanes are intense hazards that bring heavy rainfall, strong winds and high waves, and they cause other hazards such as flooding and mudslides. Hurricanes are also characterised by enormous quantities of water. This is due to their origin over moist tropical seas. High-intensity rainfall with totals of up to 500 mm in 24 hours invariably cause flooding. The path of a hurricane is erratic, so it is not always possible to give more than 12 hours' notice. This is insufficient for proper evacuation measures.

Hurricanes develop as intense low-pressure systems over tropical oceans. Winds spiral rapidly around a calm central area known as the eye. The diameter of the whole hurricane may be as much as 800 km, although the very strong winds that cause most of the damage are found in a narrower belt up to 300 km wide. In a mature hurricane, pressure may fall to as low as 880 millibars. This very low pressure, and the strong contrast in pressure between the eye and outer part of the hurricane, lead to strong gale-force winds.

Hurricanes move excess heat from low latitudes to higher latitudes. They normally develop in the westward-flowing air just north of the equator (known as an easterly wave). They begin life as small-scale tropical depressions, localised areas of low pressure that cause warm air to rise. These trigger thunderstorms that persist for at least 24 hours and may develop into tropical storms, which have greater wind speeds of up to 117 km/hr (73 mph). However, only about 10 per cent of tropical disturbances ever become hurricanes – storms with wind speeds above 118 km/hr (above 74 mph).

For hurricanes to form, a number of conditions are needed:

- Sea temperatures must be over 27 °C to a depth of 60 m (warm water gives off large quantities of heat when it is condensed; this is the heat that drives the hurricane).
- The low pressure area has to be far enough away from the equator so that the Coriolis force (the force caused by the rotation of the Earth) creates rotation in the rising air mass. If it is too close to the equator there is insufficient rotation and a hurricane will not develop.
- Conditions must be unstable: some tropical low-pressure systems develop into hurricanes, but not all of them, and scientists are unsure why some do but others do not.

Impacts of hurricanes

The Saffir-Simpson Scale, developed by the National Oceanic and Atmospheric Administration, assigns hurricanes to one of five categories of potential disaster (Table 4). The categories are based on wind intensity: in order to be classified as a hurricane a tropical cyclone must have maximum sustained winds of at least 118 km/hr (74 mph). The classification is used for hurricanes forming in the Atlantic and northern Pacific – other areas use different scales.

Table 4 The Saffir-Simpson Scale

Type	Hurricane category	Damage	Pressure (mb)	Wind-speed (km/hr)	Storm surge (metres above normal)
Depression	–	–	–	< 56	–
Tropical storm	–	–	–	57–118	–
Hurricane	1	Minimal	> 980	119–53	1.2–1.5
Hurricane	2	Moderate	965–79	154–77	1.6–2.5
Hurricane	3	Extensive	945–64	178–209	2.6–3.6
Hurricane	4	Extreme	920–44	210–49	3.7–5.5
Hurricane	5	Catastrophic	< 920	> 250	> 5.5

- **The unpredictability of hurricane paths makes the effective management of hurricanes difficult**. It was fortunate for Jamaica that Hurricane Ivan (2004) suddenly changed course away from the most densely populated parts of the island, where it had been expected to hit. In contrast, it was unfortunate for Florida's Punta Gorda when Hurricane Charley (2004) moved away from its predicted path.
- **The strongest storms do not always cause the greatest damage**. Only six lives were lost to Hurricane Frances in 2004, but 2000 were taken by Jeanne when it was still categorised as just a 'tropical storm' and had not yet reached full hurricane strength.
- **The distribution of the population throughout the Caribbean islands increases the risk associated with hurricanes**. Much of the population lives in coastal settlements and is exposed to higher sea levels and the risk of flooding.

- **Hazard mitigation depends on the effectiveness of the human response to natural events**. This includes urban planning laws, emergency planning, evacuation measures and relief operations, such as rehousing schemes and the distribution of food aid and clean water (Figure 22).
- **Developing countries continue to lose more lives to natural hazards as a result of inadequate planning and preparation**. By contrast, insurance costs continue to be greatest in American states such as Florida, where multi-million-dollar waterfront homes proliferate.

Figure 22 Hurricane management

Case study: Cyclone Nargis

In May 2008 Cyclone Nargis struck Burma. Winds exceeding 190 km/hr (118 mph) and torrential rain devastated the area, killing some 134 000 people. As many as 95 per cent of all buildings in the affected area were demolished by the cyclone and the resulting floods (Figure 23). The Burmese government identified 15 townships in the Irrawaddy delta that had suffered the worst. Seven of them had lost 90–95 per cent of their homes, with 70 per cent of their population dead or missing. International frustration mounted as disaster management experts failed to get the necessary visas to enter the country.

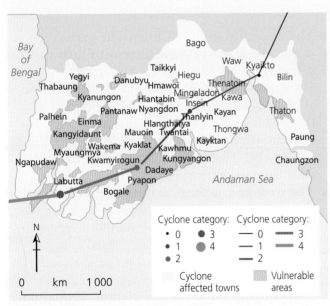

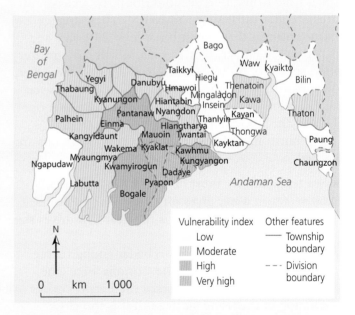

Figure 23 Cyclone Nargis

The low-lying land in the Irrawaddy delta is home to an estimated 7 million of Burma's 53 million people. Nearly 2 million of the densely packed area's inhabitants live on land that is less than 5 m above sea level, leaving them extremely vulnerable to flooding. As well as the cost in terms of lives and homes, valuable agricultural land was lost in the fertile delta.

Case study analysis

1   What were the maximum wind speeds reached by Cyclone Nargis?
2   How many people were killed by Cyclone Nargis?
3   How many people live in the Irrawaddy delta?
4   What is the height above sea level of the Irrawaddy delta?

Case study: Typhoon Haiyan

The term 'hurricane' is used in the Atlantic and north-east Pacific, 'typhoon' in the north-west Pacific, and 'cyclone' in the South Pacific and Indian Ocean.

At least 6000 people were killed in the central Philippine province of Leyte when Typhoon Haiyan, one of the strongest storms ever to make landfall, struck the Philippines in November 2013. The super-typhoon brought winds of up to 315 km/hr (195 mph) and tore roofs off buildings, turned roads into rivers full of debris, and knocked out electricity pylons.

About 70–80 per cent of the buildings in the area in the path of Haiyan in Leyte province were destroyed. Tacloban, the provincial capital of Leyte, had a population of over 200 000. The storm surge caused sea waters to rise by over 6 m when the typhoon hit. Power was knocked out and there was no mobile phone signal, making communication possible only by radio.

With many provinces left without power or telecommunications, and airports in the hardest-hit areas such as Tacloban closed, it was impossible to know the full extent of the storm's damage – or to provide badly needed aid. Government figures showed that more than 4 million people had been directly affected. The World Food Programme mobilised some $2 million in aid and aimed to deliver 40 tonnes of fortified biscuits to victims within days. Estimates of the economic cost are about $15 billion. Many countries pledged aid to the Philippines including the UK (US$131m), Japan (US$52m), Canada (US$40m) and USA (US$37m).

Satellite images showed normally green patches of vegetation ripped up into brown squares of debris in Tacloban, where a local TV station broadcast images of huge storm surges, flattened buildings and families wading through flooded streets with their possessions held high above the water. Those living in the hardest-hit areas, such as the eastern Visayas, are among the poorest in the Philippines. Many have little or no savings, so the typhoon put an already vulnerable population at even greater risk of future food and job insecurity. On Bohol Island, where a 7.3 magnitude earthquake had killed some 200 people in October 2013, residents were successfully evacuated ahead of the storm. However, because the island's main power supply comes from neighbouring Leyte, residents were left without electricity or water. In Tacloban, the sheer force of the storm was just too much for some evacuation centres, which collapsed.

The Philippines experiences about 20 typhoons every year. In 2012 Typhoon Bopha killed more than 1100 people and caused over $1 billion in damage.

How does Typhoon Haiyan compare with other tropical cyclones?

Typhoon Haiyan, described as the strongest tropical cyclone to make landfall in recorded history, hit the Philippines with winds of 314 km/hr mph and gusts of up to 378 km/hr – the fourth strongest typhoon ever recorded (Table 5) but the strongest to reach landfall.

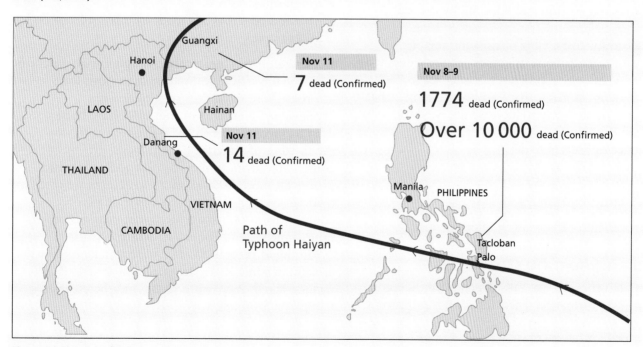

Figure 24 The path of Typhoon Haiyan

Table 5 The strongest tropical cyclones in world history

Super-typhoon	Year	Windspeed (km/hr)	Pressure (mb)	Landfall
Nancy	1961	346	882	Made landfall in Japan as a category 2 storm, killing 191 people
Violet	1961	330	886	Made landfall in Japan as a tropical storm, killing 2 people
Ida	1958	322	877	Made landfall in Japan as a category 1 storm, killing 1269 people
Haiyan	2013	314	895	Made landfall in the Philippines at peak strength
Kit	1966	314	880	Did not make landfall
Sally	1961	314	895	Made landfall in the Philippines as a category 4 storm

Note: Some of the data from the 1940s and the 1960s may have over-estimated wind speeds.

Activities

1 Describe the distribution of hurricanes shown in Figure 21.
2 Outline the main changes that occur as hurricane intensity increases.
3 Suggest reasons why the impacts of hurricanes vary from place to place.

● Coastal management

Human pressures on coastal environments create the need for a variety of **coastal management strategies** (Table 6). Coastal defence protects against coastal erosion and flooding by the sea. Coastal management strategies may be long-term or short-term, sustainable or non-sustainable. Successful management strategies require a detailed knowledge of coastal processes. Rising sea levels, more frequent storm activity, and continuing coastal development are likely to increase the need for coastal management.

Defence options include:

- do nothing
- maintain existing levels of coastal defence
- improve the coastal defence
- allow retreat of the coast in selected areas.

Hard engineering structures

The effectiveness of sea walls depends on their cost and their performance. Their function is to prevent erosion and flooding but much depends on whether they are:

- sloping or vertical
- permeable or impermeable
- rough or smooth
- what material they are made of (clay, steel or rock, for example).

In general, flatter, permeable, rougher walls perform better than vertical, impermeable smooth walls.

Cross-shore structures such as groynes, breakwaters, piers and strongpoints have been used for decades. Their main function is to stop the drifting of material. Traditionally, groynes were constructed from timber, brushwood and wattle. However, modern cross-shore structures are often made from rock. They may be part of a more complex form of management that includes beach nourishment and offshore structures.

Managed retreat allows nature to take its course – erosion in some areas, deposition in others. Benefits include less money being spent, and the creation of natural environments.

Table 6 Different forms of coastal management

Type of management	Aims/methods	Strengths	Weaknesses
Hard engineering	To control natural processes		
Cliff base management	To stop cliff or beach erosion		
Sea walls	Large-scale concrete curved walls designed to reflect wave energy	Easily made; good in areas of high density	Expensive; life span about 30–40 years; foundations may be undermined
Revetments	Porous design to absorb wave energy	Easily made; cheaper than sea walls	Life span limited
Gabions	Rocks held in wire cages absorbs wave energy	Cheaper than sea walls and revetments	Small scale
Groynes	To prevent longshore drift	Relatively low cost; easily repaired	Cause erosion on downdrift side; interrupt sediment flow
Rock armour	Large rocks at base of cliff to absorb wave energy	Cheap	Unattractive; small-scale; may be removed in heavy storms
Offshore breakwaters	Reduce wave power offshore	Cheap to build	Disrupt local ecology
Rock strongpoints	To reduce longshore drift	Relatively low cost; easily repaired	Disrupt longshore drift; erosion downdrift
Cliff face strategies	To reduce the impacts of sub-aerial processes		
Cliff drainage	Removal of water from rocks in the cliff	Cost-effective	Drains may become new lines of weakness; dry cliffs may produce rockfalls
Vegetation	To increase interception and reduce overland runoff	Relatively cheap	May increase moisture content of soil and lead to landslides
Cliff regrading	Lowering of slope angle to make cliff safer	Useful on clay (most other measures are not)	Uses large amounts of land – impractical in heavily populated areas
Soft engineering	Working with nature		
Offshore reefs	Waste materials, e.g. old tyres weighted down, to reduce speed of incoming wave	Low technology and relatively cost-effective	Long-term impacts unknown
Beach nourishment	Sand pumped from sea bed to replace eroded sand	Looks natural	Expensive; short-term solution
Managed retreat	Coastline allowed to retreat in certain places	Cost-effective; maintains a natural coastline	Unpopular; political implications
'Do nothing'	Accept that nature will win	Cost-effective!	Unpopular; political implications
Red-lining	Planning permission withdrawn; new line of defences set back from existing coastline	Cost-effective	Unpopular; political implications

Case study: Miami Beach

Miami Beach is a barrier island with a long history of human intervention. Human interference in the Miami area resulted in the almost complete removal of its beach. Channels through the beach, groynes, dredging and sea walls all affected the beach so that, by the 1950s, very little of the beach remained. Miami is a very popular place with tourists and the elderly. Given the importance of tourism to the area, it was crucial that the beach was replenished and protected. During the late 1970s and 1980 the US Army Corps of Engineers built an 18 km long, 200 m wide beach. Essentially it resembled a natural beach with a shallow, shelving seaward edge and a rampart at the landward side (Figure 26). Over 18 million m³ of sand were needed to make the beach, and up to 750 000 m³ of sand have to be replenished each year.

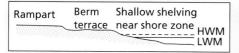

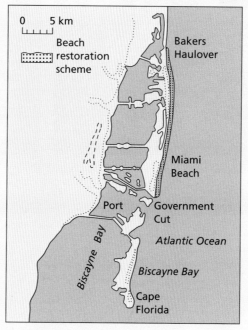

Figure 25 Beach nourishment, Miami

Figure 26 Coastal defence schemes, Pantai Berakas, Brunei

Case study: **The Palisadoes Peninsula Shoreline Development Plan**

Over the years, consistent storm surges which occur as a result of natural disasters such as flood rains, including those accompanying tropical storms and hurricanes, have led to massive erosion of the Palisadoes peninsula's natural dunes.

The Jamaican government partnered with the China Harbour Engineering Company (CHEC) to repair and protect the extensively degraded shoreline of the Palisadoes peninsula. At a cost of over $65 million, CHEC constructed rock revetment walls on the seaward side of the peninsula, and at the harbour.

The road was raised from its previous levels of 0.6–1.0 m, to 2.4 –3.2 m above **sea level**. Additional drainage facilities were placed along the roadway – this was needed to remove excess water from rainfall and from over-topping by waves. A 10 metre wide boardwalk was constructed on the harbour side of the peninsula. These works along the peninsula have been designed for a 100-year return period – that is, the shoreline is expected to withstand storm surges that are only anticipated to re-occur, on average, once in every 100 years.

In October 2012 Kingston was affected by Hurricane Sandy. This caused massive storm surges. Following 'Superstorm Sandy' as it came to be known, experts claimed that businesses along the Kingston Harbour could have suffered significant damage during the passage of the hurricane, but that the protection scheme had proved very effective.

Case study analysis

1   Which natural hazards does the Palisadoes experience?
2   a   Which country provided financial aid to Jamaica to build the defences?
    b   Describe the coastal protection measures that were put in place.
3   What level of protection is provided by the new defences?

Activities

1 Identify the coastal defence strategies illustrated in Figure 25.
2 Define the term 'coastal management'.
3 Distinguish between hard engineering and soft engineering.
4 a What is a groyne?
   b Using a sketch diagram, suggest the likely distribution of sediment around groynes 50 years after the groynes' construction. Suggest reasons to support your answer.
5 a What are the benefits of sea walls?
   b Outline some of the disadvantages of using sea walls as a form of coastal management.
6 For a coastal area you have studied, describe how the coastline is being protected, and comment on the effectiveness of the measures used.

● Opportunities and impacts – tourism

Case study: **St Lucia**

Tourism is extremely important to the economic growth of many countries. St Lucia and Antigua in the Caribbean are two such countries. The tourist industry generates a large number of jobs, for example in hotels and restaurants, as tour guides, as well as in supporting services: farmers, retailers, taxi operators and so on. It is therefore an important money earner for national governments, businesses and for individual workers and their dependents.

There are many conflicts between those who wish to develop tourism and, for example, environmentalists, fishermen and local people who risk losing their access to water supplies. The construction of buildings has the greatest impact on the environment (Figure 27). Most of the hotels built in St Lucia are on the beach front, and the clearing of land has led to slope instability, erosion and sedimentation of the shallow offshore environment. These developments have had a negative impact on the nesting grounds of endangered turtles, for example. Over-exploitation of sand has led to significant increases in beach erosion and environmental degradation. Although many hotels are artificially replenishing the beaches, the introduced sand is rapidly eroded and causes problems for offshore coral reefs which require clear water.

The building of the Pointe Seraphine cruise facility altered wave and swell patterns significantly in the harbour. At Gros Inlet, wetlands were destroyed to make way for buildings and an artificial lagoon to expand the Rodney Bay resort (Figure 28). The results of the reclamation were unforeseen: the ebb and flood tide patterns were modified, creating stronger currents which increased erosion on nearby beaches. In addition, local fisheries declined as the offshore waters became murkier – and the problem of sand flies remained.

Figure 27 The effects of tourism in St Lucia

Figure 28 Development for tourism in St Lucia

Solid and liquid waste disposal are now amongst the greatest environmental challenges facing St Lucia. This reduces the attractiveness of the tourist experience and raises issues about standards of health, the freshwater and marine environment, and the aesthetics of the island.

The discharge of poorly treated wastewater into coastal waters poses environmental and health risks. Nutrient loading in the sea has led to the decline and loss of corals. The discharge of sewage by yachts also contributes to inshore marine pollution.

Scuba diving is increasing and efforts have been made to minimise the impact of divers on coral reefs. The development of marine leisure craft facilities has led to the loss of mangrove swamps. Boating activities also damage the marine environment. The lack of holding tanks causes water pollution through sewage disposal from craft. Dropping anchors on coral is also a problem.

Case study analysis

1   Outline the advantages of tourism to St Lucia.
2   Explain why fisheries have declined as a result of tourist developments.
3   Outline the impacts of tourist developments on natural vegetation.
4   Explain briefly how tourist development may affect offshore developments.

● Coastal development

Case study: Dubai

Coastal reclamation in the United Arab Emirates has been developing on a large scale since 2001. Two palm-shaped artificial islands, Palm Jumeirah (Figure 29) and Palm Jebel Ali were completed in 2007, and in 2003 plans were unveiled for a third palm-shaped island, Palm Deira, and 'The World', a collection of over 300 islands, each one in the shape of a country.

Figure 29 Coastal development in Dubai

Palm Jumeirah not only created a new shoreline, it became the centre for world-class hotels, over 200 shopping outlets, and a range of luxury housing and leisure and entertainment developments. An Environmental Impact Assessment (EIA) was carried out to investigate likely environmental impacts. Water circulation and quality studies were investigated, to ensure that the project did not lead to a deterioration in environmental quality.

According to a report in the journal *Nature*, uncontrolled development, weak regulatory oversight and a lack of scientific monitoring are seriously threatening ecosystems along this coast. Sea-front projects ranging from desalination plants to artificial islands in the gulf between the Arabian peninsula and Iran have transformed the entire coastline in the past few decades. More than 40 per cent of the shores of some countries in the region are now developed. The change is happening more quickly, and with greater environmental impact, than in any other coastal region.

To create the islands for Palm Jumeirah, some 94 million m³ of sediment were dredged from the sea. Such large-scale projects are changing the ecology in ways that will become clear in the coming decades.

One of the problems is water circulation. Water around some parts of the islands can remain almost stationary for several weeks. This increases the risk of algal blooms. In addition, the fish that have colonised the new environment are invasive species (species from outside the area).

The Gulf region has already lost 70 per cent of its coral reefs since 2001, and most of the remaining reefs are threatened or degraded. Construction of Dubai's Palm Jebel Ali, an even larger artificial archipelago, has already destroyed 8 km² of natural reef.

Case study analysis

1 When were Palm Jumeirah and Palm Jebel Ali completed?
2 Describe the developments on Palm Jumeirah.
3 Outline the environmental impacts of Palm Jumeirah.

Activities

1 Outline the range of opportunities in a named coastal area.
2 Describe how coastal activities can have unwanted impacts on the coastal environment.
3 Using Table 6 on page 137, suggest ways in which unwanted impacts can be managed.

2.4 Weather

Clouds formed by convectional uplift

Key questions
- How are weather data collected?
- How can the data be used to describe the weather?
- How do graphs and other diagrams show weather data?

● Measuring the weather

The weather station

A weather station is a place where the elements of **weather** such as temperature, **rainfall**, **humidity**, **air pressure**, wind direction and velocity, sunshine and cloud cover are measured and recorded as accurately as possible. The weather station is placed on an open piece of land and it contains the following instruments: thermometers (ideally kept in a Stevenson screen – see Figure 1), a **rain gauge**, barometer, wind vane, anemometer and sunshine recorder.

Whatever instruments are used, ideally they should have good exposure (that is, they should be sited away from buildings, fences, trees and other obstacles). The flat top of a science block is a favoured location in many schools.

- Thermometers should be placed in the shade. Ideally, they should be in a **Stevenson screen** or slatted box. If this is not available they could be hung on a shaded wall or fence.
- Rain gauges should be away from walls, fences and bushes as they affect the amount of rain caught in the rain gauge.
- Wind instruments should be well clear of walls, fences and houses as these cause eddies that spoil the reading and make the direction difficult to assess.

It is important that readings are taken at the same time each day.

The Stevenson screen is a wooden box standing on four legs at a height of about 120 cm. The screen is built so that the shade temperature of the air can be measured. The sides of the box are slatted to allow free entry of air, and the roof is made of double boarding to prevent the sun's heat from reaching the inside of the screen. Insulation is further improved by painting the outside of the screen white so as to reflect much of the sun's energy. The screen is usually placed on a grass-covered surface, thereby reducing the radiation of heat from the ground.

Figure 1 A Stevenson screen

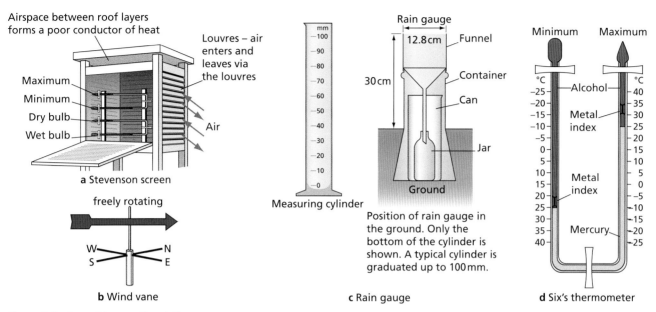

Figure 2 Equipment in a weather station

Instruments kept inside the Stevenson screen include a **maximum-minimum thermometer** and a wet- and dry-bulb thermometer (also called a hygrometer) – see Figure 2d and Figure 4 on page 142.

Instruments kept outside the Stevenson screen include a rain gauge, a wind vane to determine wind direction, and an anemometer to assess wind speed.

Measuring temperature

Variations in temperature represent responses to differences in insolation, or the amount of energy received from the sun at different times.

Meteorologists measure shade temperature. This is less variable than air temperature which is affected by cloud cover and direct insolation. Temperature is measured using a thermometer. A continuous temperature reading is given by a thermograph.

- **Maximum thermometer** – When the temperature rises, the mercury in the thermometer expands and pushes the index along the tube. When the temperature falls, the mercury contracts but the index stays where it was pushed to by the mercury. The maximum temperature is obtained by reading the scale at the point where the index is. The index is then drawn back to the mercury by a magnet for measuring the next reading.
- **Minimum thermometer** – When the temperature falls, the alcohol contracts and its meniscus pulls the index along the tube. When the temperature rises, the alcohol expands. It is read in the same way as the maximum thermometer.

A Six's thermometer (see Figure 2d) can be used to measure maximum and minimum temperatures at the same time.

The daily readings of the maximum and minimum thermometers are used to work out the average or mean temperature for one day (this is called the mean daily temperature) and the temperature range for one day (the daily or diurnal temperature range).

To find the mean daily temperature, the maximum and minimum temperatures for one day are added together and then halved. For example: maximum temperature (35 °C + minimum temperature 25 °C) ÷ 2 = mean daily temperature 30 °C. The sum of the daily mean temperatures for one month divided by the number of days for that month gives the mean monthly temperature. The sum of the mean monthly temperatures divided by 12 gives the mean annual temperature.

The daily or diurnal temperature range is found by subtracting the minimum temperature from the maximum temperature for any one day. For example: maximum temperature 35 °C – minimum temperature 25 °C = daily or diurnal temperature range 10 °C.

The highest mean monthly temperature minus the lowest mean monthly temperature gives the mean annual temperature range. For example, Lagos has a mean maximum temperature of 27.5 °C (March), and a mean minimum temperature of 24.5 °C (August). Its mean annual temperature range is therefore 3 °C.

Interesting note

The highest temperature recorded was at Furnace Creek, California, USA in 1913 when it reached 56.7°C. In contrast, the lowest temperature recorded was –89.2°C in Antarctica in 1983.

Measuring rainfall

A rain gauge is used to measure rainfall. It consists of a cylinder in which there is a collecting can containing a glass or plastic jar, and a funnel that fits in the top of the container. The gauge is placed in an open space so that only raindrops enter the funnel of the gauge, and no runoff from trees, buildings or other objects can get into the funnel. The gauge is sunk into the ground so that the top of the funnel is about 30 cm above ground level (Figures 2c and 3). This is to prevent the sun's heat from evaporating any water collected and to ensure no rain splashes up from the ground into the funnel.

Figure 3 Rain gauge

Rain falling over the funnel collects in the jar. This is emptied, usually every 24 hours, and measured in a tapered glass measure, graduated in millimetres. The tapered end of the jar enables very small amounts of rain to be measured accurately.

The rainfall recorded for a place, either for a day or for a week, month or year, can be shown on a map. This is done by using lines called isohyets. An isohyet is a line on a map that joins areas of equal rainfall.

It is important to check the rain gauge every day, preferably at the same time, even if there has not been any rainfall. This is because small amounts of dew may accumulate in the gauge, leading to false readings when it does rain.

Interesting note

The highest rainfall over a 24-hour period was in Foc-Foc, Réunion, when 1.825 m of rain fell. The largest one-minute burst of rainfall was 31.2 mm in Unionville, Maryland, USA in 1956.

Measuring relative humidity

Wet- and dry-bulb thermometers are used to measure relative humidity. The dry-bulb is a glass thermometer which records the actual air temperature. The wet-bulb is a similar thermometer, but with the bulb enclosed in a muslin bag which is dipped into a bottle of water (Figure 4). This thermometer measures the wet-bulb temperature which, unless the relative humidity is close to 100%, is generally lower than the dry-bulb temperature.

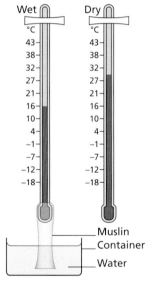

Figure 4 Wet- and dry-bulb thermometer

Measuring air pressure, wind speed and direction

Because air has weight it exerts a pressure on the Earth's surface. At sea level the pressure is about 1.03 kg/cm². Pressure varies with temperature and altitude, and the instrument that measures pressure is called a barometer (Figure 5). Air pressure is usually measured in millibars.

Figure 5 A simple barometer

A mercury barometer is a hollow tube from which the air is extracted before the open end is placed in a bath of mercury. Mercury is forced up the tube by the pressure of the atmosphere on the mercury in the bath. When the pressure of the mercury in the tube balances the pressure of the air on the exposed mercury, the mercury in the tube stops rising. The height of the column of mercury changes as air pressure changes: it rises when air pressure increases and falls when air pressure decreases.

An aneroid barometer is a vacuum chamber in the form of a small metal cylinder. Inside, a strong metal spring prevents the chamber from collapsing. The spring contracts and expands with changes in atmospheric pressure. These changes are magnified by a series of levers and the movements are conveyed to a pointer which moves across a calibrated scale.

A barograph is a tracing from an aneroid barometer which records continuously for one week. Changes in pressure are recorded by a flexible arm which traces an ink line on a rotating paper-covered drum. The paper is divided by vertical lines at two-hour intervals.

The atmospheric pressure is recorded at numerous weather stations for a region and these are plotted on a map of the region. First, though, the pressures are 'reduced' to sea level – that is, they are adjusted to what they would be if the stations were at sea level. The pressures are plotted on a map. Lines are then drawn through points where pressure is the same. These lines are called isobars.

The wind vane is used to indicate wind direction. It consists of a horizontal rotating arm pivoted on a vertical shaft. The rotating arm has a tail at one end and a pointer at the other. When the wind blows, the arm swings until the pointer faces the wind. The directions north, east, south and west are marked on the arms which are rigidly fixed to the shaft.

The speed of the wind is measured by an anemometer (Figure 6), which consists of three or four metal cups fixed to metal arms that rotate freely on a vertical shaft. When there is a wind, the cups rotate. The stronger the wind, the faster the rotation. The number of rotations are recorded on a meter to give the speed of the wind in km/hr.

The wind vane and anemometer are placed well away from any buildings or trees that may interfere with the free movement of air. Buildings may channel air through narrow passages between two buildings, or decrease the flow of air by blocking its path. Trees have a similar effect.

Figure 6 An anemometer

Winds are shown by arrows on a weather map. The shaft of the arrow shows wind direction and the feathers on the shaft indicate wind velocity. The tip of the arrow, at the opposite end from the feathers, points to the direction in which the wind is blowing.

Wind direction for a specific place can be shown on a wind rose (Figure 7). It is made up of a circle from which rectangles radiate. The directions of the rectangles represent the points of the compass.

The lengths of the rectangles are determined by the number of days/times the wind blows from that direction. The number of days/times (hours) when there is no wind is recorded in the centre of the rose.

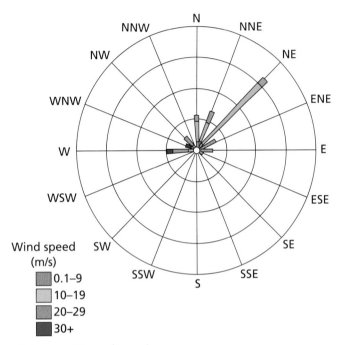

Wind speed (m/s)
- 0.1–9
- 10–19
- 20–29
- 30+

Calm conditions shown in centre of wind rose

Figure 7 A wind rose

Measuring sunshine hours

The number of hours and minutes of sunshine received at a place can be measured and recorded by a sunshine recorder. This is a glass sphere partly surrounded by a metal frame (Figure 8). A strip of special card, divided up into hours and minutes, is placed below the sphere. When the sun shines, the sphere focuses the sun's rays on the card. As the sun moves, the rays burn a trace on the card. At the end of the day, the card is removed and replaced. The length of the trace represents the amount of sunshine that the location received.

Activities

1 Describe and explain the main characteristics of a Stevenson screen.
2 What information does a Six's thermometer show?
3 Why are weather readings taken at the same time each day?
4 Where is the best place to locate a rain gauge? Briefly explain why.
5 How are wind speed and wind direction measured?

Figure 8 Campbell-Stokes sunshine recorder

Recording the weather

Clouds

The ten main types of **cloud** can be separated into three broad categories according to the height of their base above the ground: high clouds, medium clouds and low clouds (Figure 9).

High clouds are usually composed solely of ice crystals and have a base between 5500 and 14000 m. These are described as:

- cirrus – white filaments
- cirrocumulus – small rippled elements
- cirrostratus – a transparent sheet, often with a halo.

Medium clouds are usually composed of water droplets or a mixture of water droplets and ice crystals, and have a base between 2000 and 7000 m:

- altocumulus – layered, rippled elements, generally white with some shading
- altostratus – a thin layer, grey, allows sun to appear as if through ground glass.

Low clouds are usually composed of water droplets, although cumulonimbus clouds include ice crystals, and have a base below 2000 m:

- stratocumulus – layered, a series of rounded rolls, generally white with some shading
- stratus – layered, uniform base, grey
- nimbostratus – a thick layer with a low base, dark, and rain or snow may fall from it
- cumulus – individual cells, vertical rolls or towers with a flat base
- cumulonimbus – large cauliflower-shaped towers, often with 'anvil tops', and sometimes giving thunderstorms or showers of rain or snow.

Cloud cover is measured in oktas (eights). This is made by a visual assessment of how much of the sky is covered by cloud. For example, in Figure 9b approximately 5/8 of the sky is covered by cloud, where as in Figure 9c, the whole sky is covered in cloud, so 8/8 cloud cover.

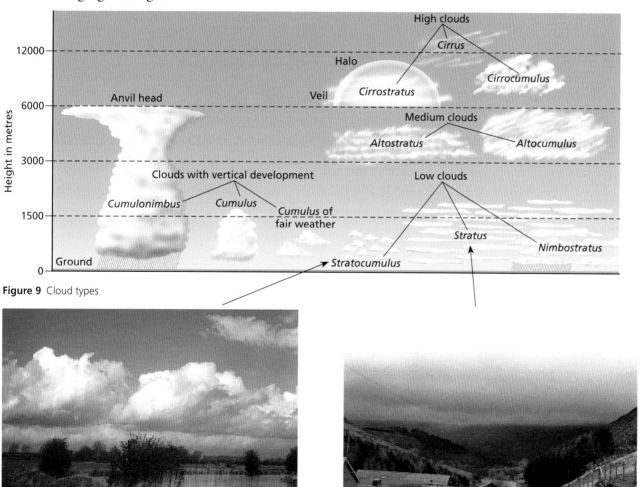

Figure 9 Cloud types

Table 1 Daily weather observations at Frankston, Victoria (Australia), 1–7 August 2007

Date	Day	Temperature		Rainfall (mm)	Wind direction	Wind speed (km/hr)	Air pressure (mb)
		Max. (°C)	Min. (°C)				
1 August	W	14.2	9.7	4.0	N	22	1006
2	Th	13.4	11.5	0	N	37	1004
3	F	9.9	8.1	0	WNW	33	1011
4	S	11.5	7.2	0	WNW	31	1016
5	S	11.6	8.2	0	W	28	1019
6	M	12.7	9.5	20.2	W	20	1023
7	T	14.5	9.2	0	N	30	1019

Table 2 Daily weather observations at Frankston, Victoria (Australia), 1–7 February 2008

Date	Day	Temperature		Rainfall (mm)	Wind direction	Wind speed (km/hr)	Air pressure (mb)
		Max. (°C)	Min. (°C)				
1 February	F	25.6	11.7	6.8	SSE	15	1020
2	S	25.7	16.9	0	NNW	9	1016
3	S	27.6	17.9	0	SE	9	1016
4	M	29.1	19.9	0	ENE	11	1013
5	T	23.2	19.7	0	SW	13	1012
6	W	23.1	19.2	0	SW	19	1004
7	Th	17.9	15.7	8.4	SW	19	1005

Daily weather

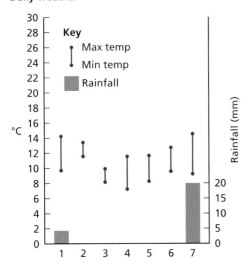

Wind direction and frequency

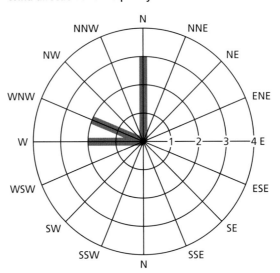

Figure 10 Daily weather, wind direction and frequency at Frankston, August 2007

Activities

The results recorded by a school in Victoria are shown in Tables 1 and 2. The data for the first week (August) are plotted in Figure 10.

1 Plot the data for February using the same methods as in Figure 10.
2 State the maximum and minimum temperature of the seven-day period in February.
3 Work out the mean minimum temperature and the mean maximum temperature for the seven days.
4 How much rain fell during the seven days?
5 Compare the weather in February with that in August.

Tropical rainforest, Sarawak

Key questions

- What are the main characteristics of equatorial and hot desert climates?
- What are the factors that influence equatorial and desert climates?
- What are the characteristics of tropical rainforest and hot desert ecosystems?
- What are the causes and impacts of the deforestation of tropical rainforest?

● Equatorial and hot desert climates

The main characteristics of an equatorial climate include:

- hot conditions – generally above 26 °C – throughout the year
- high levels of rainfall, often over 2000 mm
- a lack of seasons – the temperatures are high throughout the year
- the difference between daytime and night-time temperatures (known as the diurnal range) is, in fact, higher than the seasonal differences in temperature
- rainfall is mainly **convectional** and may fall on as many as 250 days each year

- cloud cover varies – in the morning it may be limited but by the afternoon, towering cumulonimbus clouds mark the start of the convectional rains
- the presence of clouds tends to reduce the amount of heat that is lost at night – hence the diurnal range is less than in hot desert areas
- the humidity (moisture in the atmosphere) is high, and relative humidities of 100% are often reached in the late afternoon
- wind speeds within the rainforest are reduced by the large numbers of trees present.

The data for Manaus in Brazil (Table 1) show that the warmest months are September and October with a mean monthly temperature of 34 °C. In contrast, all of the months from December to September share the mean minimum monthly temperature of 24 °C. Thus the mean annual temperature range is 10 °C.

Rainfall in Manaus is high, nearly 2100 mm. There is a definite wet season between November and May whereas the months of June to October are relatively dry.

In contrast, the main characteristics of hot desert climates include:

- very hot days and cold nights, caused by the lack of cloud cover
- low and irregular amounts of rainfall, which lack any seasonal pattern
- low levels of humidity for much of the year
- warm dry winds, sometimes causing sandstorms.

The data for Cairo (Table 2) show that the highest mean monthly temperatures are between June and August when the temperature reaches 35 °C. In contrast, the lowest mean monthly temperature is in January, reaching just 9 °C. Thus the temperature range is 26 °C. There is a seasonal pattern to temperature, with the highest values in the summer and lowest readings in the winter. Rainfall figures are very low, just 27 mm. Sunshine levels are lower

Table 1 Climate data for Manaus

	J	F	M	A	M	J	J	A	S	O	N	D	Av/Total
Temperature													
Daily max (°C)	31	31	31	31	31	31	32	33	34	34	33	32	32
Daily min (°C)	24	24	24	24	24	24	24	24	24	25	25	24	24
Average monthly (°C)	28	28	28	27	28	28	28	29	29	29	29	28	28
Rainfall													
Monthly total (mm)	278	278	300	287	193	99	61	41	62	112	165	220	2096
Sunshine													
Sunshine (hours)	3.9	4	3.6	3.9	5.4	6.9	7.9	8.2	7.5	6.6	5.9	4.9	5.7

during the months when there is more rain (winter months between November and March). In general, sunshine levels are much higher in Cairo – 9.5 hours per day compared with 5.7 hours per day in Manaus.

Table 2 Climate date for Cairo

	J	F	M	A	M	J	J	A	S	O	N	D	Av/Total
Temperature													
Daily max (°C)	19	21	24	28	32	35	35	35	33	30	26	21	28
Daily min (°C)	9	9	12	14	18	20	22	22	20	18	14	10	16
Average monthly (°C)	14	15	18	21	25	28	29	28	26	24	20	16	22
Rainfall													
Monthly total (mm)	4	4	3	1	2	1	0	0	1	1	3	7	27
Sunshine													
Sunshine (hours)	6.9	8.4	8.7	9.7	10.5	11.9	11.7	11.3	10.4	9.4	8.3	6.4	9.5

Activities

1 In which months is the average temperature in Cairo higher than in Manaus?
2 How much rain falls in Manaus in April?
3 In which months is the minimum temperature in Cairo higher than that in Manaus? How do you explain this?
4 Describe the variations in monthly sunshine levels in Manaus.
5 Suggest why there is a link between sunshine levels and rainfall.
6 What is the mean monthly temperature range in Manaus and Cairo in (a) July and (b) December?

● Factors affecting climate

Many factors affect the temperature of a place. These include latitude, distance from the sea, the nature of nearby ocean currents, altitude, dominant winds, cloud cover, and aspect. Differences in pressure systems also affect whether it rains or whether it is dry.

Latitude

On a global scale latitude is the most important factor determining temperature (Figure 1). Two factors affect the temperature: the angle of the overhead sun, and the thickness of the atmosphere. Firstly, at the equator the overhead sun is high in the sky, hence high-intensity insolation is received. By contrast, at the poles the overhead sun is low in the sky, hence the quantity of energy received is low. Secondly, the thickness of the atmosphere affects temperature. Radiation has more atmosphere to pass through near the poles, due to its low angle of approach. Hence more energy is lost, scattered or reflected here than over equatorial areas, making temperatures lower over the poles.

Equatorial climates have high temperatures throughout the year on account of their location. They also receive high levels of rainfall due to the daily convection. Hot deserts are hot due to their tropical location, but receive low rainfall for a variety of reasons, including the presence of the subtropical high pressure belt.

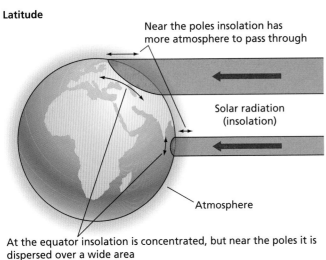

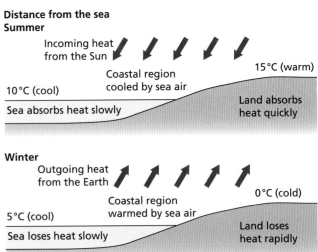

Figure 1 Factors that affect climate

Proximity to the sea

The specific heat capacity is the amount of heat needed to raise the temperature of a body by 1 °C. Land heats and cools more quickly than water. It takes five times as much heat to raise the temperature of water by 1 °C as it does to raise land temperatures.

Water also heats more slowly because:

- it is clear, so the sun's rays penetrate to great depth, distributing energy over a wider area
- tides and currents cause the heat to be distributed further.

Therefore a greater volume of water is heated for every unit of energy than land, so water takes longer to heat up.

Distance from the sea therefore has an important influence on temperature. Water takes up heat and emits it much more slowly than the land. In winter, in mid latitudes air over sea is much warmer than over land, so onshore winds bring heat to the coastal lands. By contrast, during the summer coastal areas remain much cooler than inland sites. Areas with a coastal influence are termed maritime or oceanic whereas inland areas are called continental. Areas that are very far from the sea may be extremely arid, such as parts of central North Africa.

Ocean currents

The effect of ocean currents on temperatures depends on whether the current is cold or warm. Warm currents from equatorial regions raise the temperature of polar areas (with the aid of prevailing westerly winds). However, the effect is only noticeable in winter. Areas that lie close to cold upwelling ocean currents, such as Namibia in Africa, may contains hot deserts, such as the Namib desert. This is because the cold current cools the air above it, reducing the amount of evaporation from the ocean, and producing dry conditions.

Altitude

In general, air temperature decreases with increasing **altitude**. This is because air under the greater pressure of lower altitudes is denser and therefore warmer. As altitude increases so the pressure on the air is reduced and the air becomes cooler. The normal decrease of temperature with height is on average 10 °C/km.

Winds

The effects of wind on temperature depend on the initial characteristics of the wind. In temperate latitudes **prevailing** (dominant) **winds** from the land lower the winter temperatures, but raise them in summer. This is because continental areas are very hot in summer but very cold in winter. Prevailing winds from the sea do the opposite – they lower the summer temperatures and raise them in winter.

Cloud cover

Cloud cover decreases the amount of insolation reaching the surface by reflecting some of it. Clouds also reduce the amount of insolation leaving the surface by absorbing the radiation. If there is limited cloud then incoming shortwave radiation and outgoing longwave radiation are at a maximum. This is the norm in many hot deserts.

Pressure

In low pressure systems air is rising. Low pressure produces rain as the air may rise high enough, cool, condense and form clouds and rain. This can happen in very warm areas, such as in equatorial areas, at mountain barriers and at weather fronts, when warm air is forced over cold air. In contrast, where there is high pressure air is sinking, and rain formation is prevented. The world's great hot deserts are located where there is high pressure caused by sinking air.

Activities

1 How does latitude affect the amount of heat a place receives?
2 Why are equatorial areas not getting any hotter nor polar areas any colder?
3 What is meant by the term 'specific heat capacity'?
4 Explain why temperature decreases with height.
5 Why is there a large temperature difference between day and night in hot deserts, but not in equatorial areas?

● Tropical rainforests

Tropical evergreen rainforests are located in equatorial areas, largely between 10 °N and 10 °S (Figure 2). There are, nevertheless, some areas of rainforest that are found outside these areas but these tend to be more seasonal in nature. The main areas of rainforest include the Amazon rainforest in Brazil, the Congo rainforest in central Africa, and the Indonesian-Malaysian rainforests of South-East Asia. There are many small fragments of rainforest, such as those on the island of Madagascar and in the Caribbean. Everywhere tropical rainforests are under increasing threat from human activities such as farming and logging. The result is that rainforests are disappearing and those that remain are not only smaller, but broken up into fragments.

Interesting note

Tropical rainforests cover 6 per cent of the world's land surface but hold 50 per cent of the world's species. The Amazon rainforest alone is home to 10 per cent of the world's known species.

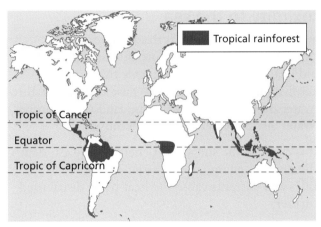

Figure 2 World distribution of tropical rainforests

Vegetation

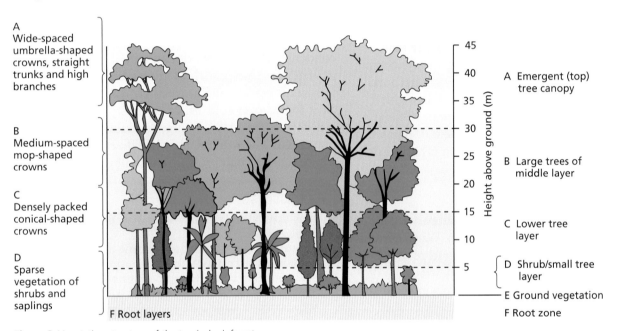

Figure 3 Vegetation structure of the tropical rainforest

The vegetation is evergreen, enabling photosynthesis to take place all year round. This is possible due to the high temperatures all year, and the presence of water throughout the year. The vegetation is layered, and the shape of the crowns varies at each layer. Species at the top of the canopy receive most of the sunlight whereas species that are located near the forest floor are adapted to darker conditions, and generally have a darker pigment so as to photosynthesise at low light levels. There is a great variety in the number of species in a rainforest – this is known as **biodiversity**. A rainforest may contain as many as 300 different species in a single hectare. Typical rainforest species include figs, teak, mahogany and yellow woods.

Tropical vegetation has many adaptations. Some trees have leaves with drip-tips (Figure 4a), which are designed to get rid of excess moisture. In contrast, other plants have saucer shaped leaves in order to collect water. Pitcher plants have developed an unusual means of getting their nutrients. Rather than taking nutrients from the **soil**, they have become carnivorous and get their nutrients from insects and small frogs that are trapped inside the pitcher (Figure 4b). This is one way of coping with the very infertile soils of the rainforest. Other plants are very tall. To prevent being blown over by the wind, very large trees have developed buttress roots that project out from the main trunk above the ground, which gives the plant extra leverage in the wind.

Figure 4 Adaptations of rainforest plants
a Drip-tip **b** Pitcher plant

Rainforests are the most productive land-based **ecosystems**. Ironically, the soils of tropical rainforests are quite infertile. This is because most of the nutrients in the rainforest are contained in the biomass (living matter). Rainforest soils are typically deep due to the large amount of weathering that has taken place, and they are often red in colour, due to the large amounts of iron present in the soil. Nevertheless, there are some areas in which tropical soils may be more fertile: in floodplains and in volcanic areas the soils may be enriched by flooding or the weathering of fertile lava flows.

The nutrient cycle is easily disrupted (Figure 5). Tropical rainforests have been described as 'deserts covered by trees'. Once the vegetation is removed, nutrients are quickly removed from the system, creating infertile conditions and even deserts.

Rainforest are found only in areas with over 1700 mm of rain and temperatures of over 25 °C

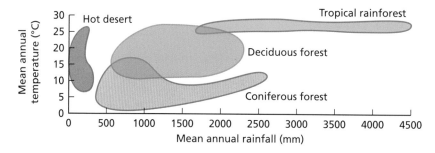

The links between climate, soils and vegetation are very strong

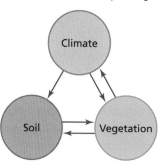

Figure 5 Conditions required for the growth of rainforest, deciduous forest, coniferous forest and hot desert

Activities

Study Figure 5.

1. What is the minimum temperature required for the growth of tropical rainforests?
2. What is the minimum amount of rainfall needed for a tropical rainforest?
3. Suggest how a rainforest with a mean annual temperature of 30 °C and an annual rainfall of 3500 mm might differ from one with a mean annual rainfall of 1700 mm and a mean annual temperature of 25 °C.
4. Suggest how the vegetation in Figure 4 is adapted to conditions in the rainforest.
5. What is biodiversity? Suggest reasons why it may be important to protect biodiversity.
6. Why are rainforests described as 'deserts covered with trees'?

● Hot deserts

The world's hot **deserts** are largely found in subtropical areas between 20° and 30° north and south of the equator (Figure 6). The largest area of hot desert is the Sahara but there are other important deserts such as the Great Victoria Desert and Great Sandy Desert in Australia, the Kalahari and Namib deserts in southern Africa, the Atacama desert in South America, and the Arabian desert. The Gobi desert in Mongolia and China lies outside the tropics and therefore is not a hot desert.

The main factors influencing the vegetation are that it is hot throughout the year and there is low and unreliable rainfall (≤ 250 mm per year).

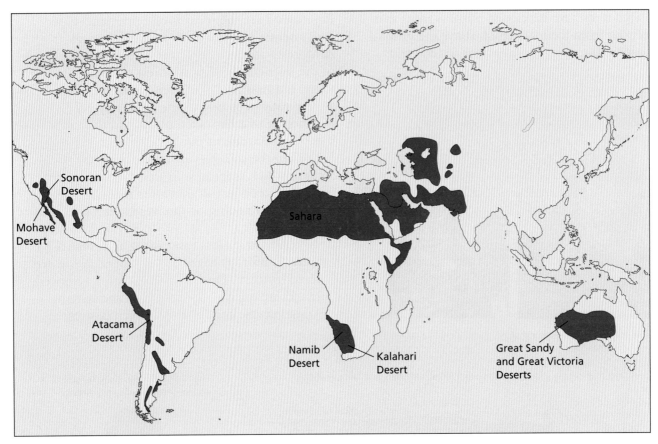

Figure 6 World distribution of hot deserts

Vegetation

The vegetation responds to this in a number of ways. There are two main types of desert plant. Perennials (plants that grow over a number of years) may be succulent (they store lots of water), they are often small (to reduce water loss by evaporation and transpiration) and they may be woody. Annuals or ephemerals are plants that live for a short time but these may form a dense covering of vegetation immediately after rain.

Ephemerals evade **drought**. During the infrequent wet periods they develop rapidly, producing a large number of flowers and fruits. These help produce seeds which remain dormant in the ground until the next rains.

Many plants are adapted to drought (Figure 7) – these are called **xerophytes**. Water loss is minimised in a number of ways:

- Leaf hairs reduce windspeed and therefore reduce transpiration.
- Thick waxy cuticles and the rolling-up or shedding of leaves at the start of the dry season reduce water loss.
- Some plants have the bulk of their biomass (living material) below the ground surface.
- Others have very deep roots to reach the water table.

Figure 7 Plant adaptations to hot desert environments

- In woody species the wood prevents the collapse of the plant even when the plant is wilting.

Vegetation from desert margins is often referred to as scrub. Tropical scrub on the margins of hot deserts includes acacias, cacti, succulents, tuberous-rooted plants and herbaceous plants that only grow with rain. Special types are mulga in Australia (dense acacia thickets), spinifax in Australia ('porcupine grass'), and chanaral in Chile (spiny shrubs).

Soils in desert areas are very infertile. As a result of the low rainfall there is little organic or moisture content in the soil. Due to the lack of chemical weathering (largely due to the lack of moisture), soils contain few nutrients.

> **Interesting note**
> Although it is a hot desert, the Sahara contains some 300 plant species and around 70 animal species.

Animals

Animals are adapted to living in the desert in a number of ways (Figure 8). Different animals may:

- be nocturnal (active only at night) in order to avoid the heat of the day
- use panting and/or have large ears help to reduce body heat
- remain in underground burrows during the day
- may secrete highly concentrated uric acid in order to reduce water loss
- migrate during the hottest season to escape the heat
- adopt a strategy of long-term aestivation (dormancy, or sleep), which ends only when moisture and temperature conditions become more favourable.

Kangaroo rat
Lives in south-west USA and Mexico; weighs 35–180 g and measures 100–200 mm

Fennec fox
Lives in Sahara desert: weighs 1.5 kg and is 200 mm tall

Highly efficient kidneys which concentrate urine and produce dry droppings

Reduces water lost by respiration by cooling exhaled air in nasal passage

Light-coloured coat reflects heat

Lives in burrows during the day to avoid heat

Does not perspire

Licks fur to keep cool

Does not need to drink – gets water from its food

Large ears with blood vessels close to the surface to lose body heat

Gets most of its moisture from its prey

Excretes highly concentrated urine

Lives in burrows during the day to avoid heat

Soles of feet protected by thick fur to run across hot sand

Figure 8 Animal adaptations to hot desert environments

> ## Activities
>
> 1 Why is it difficult to live in a hot desert?
> 2 Study Figure 5 on page 151 which shows the conditions required for the growth of hot deserts.
>   a What is the maximum rainfall in a hot desert, as suggested by Figure 5?
>   b What is the range of mean annual temperatures in hot deserts?
> 3 Suggest how a hot desert with a mean annual temperature of 30 °C and an annual rainfall of 250 mm might differ from one in which the mean annual rainfall is 250 mm and the mean annual temperature is 20 °C.
> 4 How have plants adapted to survive in the desert?
> 5 How do animals survive in the desert?
> 6 Visit www.bbc.co.uk/nature/life/Camel to find out how camels are adapted to living in hot deserts.

⬤ Impacts of deforestation of the tropical rainforest

About 200 million people live in areas that are or were covered by tropical rainforests. These areas offer many advantages for human activities such as farming, hydro-electric power, tourism, fishing and food supply, mineral development, and forestry (Figure 9). Rainforests also play a vital role in regulating the world's climate, and they account for 50 per cent of the world's plants and animals. They are vital too for the protection of soil and water resources.

The year-round growing season is very attractive for farmers, although the poor quality of the soil results in the land being farmed for only a few years before the land is abandoned (Figure 10). Nevertheless, large-scale **plantations** occur in areas of tropical rainforest, producing crops such as palm oil which is increasingly being used for the **biofuels** industry. High rainfall totals, especially in hilly areas, favour the development of HEP, such as at Batang Ai in Sarawak, Malaysia. Areas of rainforest have a long history of commercial farming. Tropical hardwoods, such as teak and mahogony, are prized by furniture manufacturers. Mineral developments, such as iron ore at Carajas in Brazil and ilminite on the south-east coast of Madagascar, are also developed in some rainforest areas.

Figure 9 Tropical rainforest along with **shifting cultivation** – rice growing in Sarawak

Figure 10 Rainforest at Batang Ai affected by flooding, shifting cultivation and soil erosion

There are a large number of effects of deforestation including:

- disruption to the circulation and storage of nutrients
- surface erosion and compaction of soils
- sandification

- increased flood levels and sediment content of rivers
- climatic change
- loss of biodiversity.

Table 3 The value of tropical rainforests

Industrial uses	Ecological uses	Subsistence uses
• Charcoal • Saw logs • Gums, resins and oils • Pulpwood • Plywood and veneer • Industrial chemicals • Medicines • Genes for crops • Tourism	• Watershed protection • Flood and landslide protection • Soil erosion control • Climate regulation e.g. balancing levels of carbon dioxide and oxygen • Special woods and ashes • Fruit and nuts	• Fuelwood and charcoal • Fodder for agriculture • Building poles • Pit sawing and saw milling • Weaving materials and dyes • Rearing silkworms and beekeeping

Deforestation disrupts the closed system of **nutrient cycling** within tropical rainforests. Inorganic elements are released through burning and are quickly flushed out of the system by the high intensity rains.

Soil erosion is also associated with deforestation. As a result of soil compaction, there is a decrease in infiltration, an increase in overland runoff and surface erosion.

Sandification is a process of selective erosion. Raindrop impact washes away the finer particles of clay and humus, leaving behind the coarser and heavier sand. Evidence of sandification dates back to the 1890s in Santarem, Rondonia.

As a result of the intense surface runoff and soil erosion, rivers have a higher **flood peak** and a shorter time lag. However, in the dry season river levels are lower, the rivers have greater turbidity (murkiness due to more sediment), an increased bed load, and carry more silt and clay in suspension.

Other changes relate to **climate**. As deforestation progresses, there is a reduction of water that is re-evaporated from the vegetation hence the recycling of water must diminish. Evaporation rates from savanna grasslands are estimated to be only about one-third of that of the tropical rainforest. Thus, mean annual rainfall is reduced, and the seasonality of rainfall increases.

Causes of deforestation in Brazil

There are five main causes of deforestation in Brazil:

- agricultural colonisation by landless migrants and speculative developers along highways and agricultural growth areas
- conversion of the forest to cattle pastures especially in eastern and south eastern Para and northern Mato Grosso

- mining, for example the Greater Carajas Project in south eastern Amazonia, which includes a 900 km railway and extensive deforestation to provide charcoal to smelt the iron ore. Another threat from mining are the small-scale informal gold mines, *garimpeiros*, causing localized deforestation and contaminated water supplies
- large-scale hydroelectric power schemes such as the Tucurui Dam on the Tocantins River
- forestry taking place in Para, Amazonas and northern Mato Grosso.

Deforestation in Brazil shows five main trends:

- it is a recent phenomenon
- it has partly been promoted by government policies
- there are a wide range of causes of deforestation

- deforestation includes new areas of deforestation as well as the extension of previously deforested areas
- land speculation and the granting of land titles to those who 'occupy' parts of the rainforest is a major cause of deforestation.

Activities

1 Comment on the value of tropical rainforests to human population.
2 Outline the main impacts of deforestation on the natural environment.
3 Explain the main causes of deforestation in Brazil.
4 Comment on the trends of deforestation in Brazil.

Case study: Danum Valley Conservation Area, Malaysian Borneo

Figure 12 Danum Valley Conservation Area

The Danum Valley Conservation Area (DVCA) contains more than 120 mammal species including 10 species of primate. The DVCA and surrounding forest is an important reservation for orang-utans. These forests are particularly rich in other large mammals including the Asian elephant, Malayan sun bear, clouded leopard, bearded pig and several species of deer. The area also provides one of the last refuges in Sabah for the critically endangered Sumatran rhino. Over 340 species of bird have been recorded at Danum, including the argus pheasant, Bulwer's pheasant, and seven species of pitta bird.

The DVCA covers 43 800 hectares, comprising almost entirely lowland dipterocarp forest (dipterocarps are valuable hardwood trees). It is the largest expanse of pristine forest of this type remaining in Sabah, north-east Borneo (Figure 12).

Until the late 1980s, the area was under threat from commercial logging. The establishment of a long-term research programme between Yayasan Sabah and the Royal Society in the UK created local awareness of the conservation value of the area and provided important scientific information about the forest and what happens to it when it is disturbed through logging. Danum Valley is controlled by a management committee containing all the relevant local institutions – wildlife, forestry and commercial sectors are all represented. To the east of the DVCA is the 30 000 hectare Innoprise-FACE Foundation Rainforest Rehabilitation Project (INFAPRO), one of the largest forest rehabilitation projects in South-East Asia, which is replanting areas of heavily disturbed logged forest.

Because all areas of conservation and replantating are embedded within the larger commercial forest, the value of the whole area is greatly enhanced. Movement of animals between forest areas is enabled and allows the continued survival of some important and endangered Borneo animals such as the Sumatran rhino, the orang-utan and the Borneo elephant. In the late 1990s, a hotel was established on the north-eastern edge of the DVCA. It has established flourishing ecotourism in the area and exposed this unique forest to a wider range of visitors than was previously possible. As well as raising revenue for the local area, it has raised the international profile of the area as an important centre for conservation and research.

Case study analysis

1 What was the main threat to the Danum Valley before the late 1980s?
2 Why is the DVCA important for the conservation of species?
3 What are the main interest groups in the forest?

Case study: The Sonoran desert

The Sonoran Desert is located in southern USA (southern California and southern Arizona) and northern Mexico. Its vegetation includes the saguaro cactus, which can grow to a height of 15 m and may live for up to 175 years. Its ribbed stem expands as it fills with water during the winter wet season. Its stem also reduces wind speed and water loss from the plant, while sunken stomata reduce water loss. It has shallow roots to catch water from storms before it evaporates. Other species with similar adaptations include the prickly pear (Figure 15) and barrel and hedgehog cacti.

Figure 15 Prickly pear: a plant well adapted to desert conditions

The Palo Verde is a small, drought-tolerant tree. It loses its leaves in the dry season, but its green bark allows it to photosynthesise without leaves. Creosote bushes have small, dark leaves to reduce transpiration. Plant density depends upon water availability.

Soils in the Sonoran desert are typically thin, relatively infertile and alkaline. Seasonal rains carry soluble salts down through the soil. However, during the dry season they are drawn up to surface by evaporation. Concentrations may be toxic for some plants. In addition, flash flooding can compact the soil, leaving the surface impermeable.

There has been considerable human impact in the area. Some cities, notably Phoenix in Arizona, have expanded rapidly at the expense of the desert. The increased demand for and abstraction of water has lowered water tables. Mesquite bushes and cottonwood trees which were growing along water courses have died back. Road construction and pipelines have affected the movement of mammals, and fenced highways have prevented pronghorn antelopes, for example, from reaching water supplies. Off-road vehicles have compacted soils and made them less able to hold water. Overgrazing by cattle has removed more palatable species. Domesticated animals have escaped into the wild, and reduced grazing availability for wild mammals. The introduction of exotic plant species, such as tamarisk, has displaced native species such as cottonwoods and desert willows. Removal of native species has speeded up the spread of exotic species.

Case study analysis

1 Where is the Sonoran desert?
2 How is the saguaro cactus adapted to life in the desert?
3 Describe the main features of the prickly pear as shown in Figure 15.

● Vegetation distribution in Death Valley

The distribution of vegetation in Death Valley (average rainfall less than 50 mm per year), depends very much on ground and soil conditions. These determine the quantity and quality of water supply. Three main zones can be recognized: the central salt pan, the lower sandy slopes and the upper gravel slopes.

The central salt pans occupy depressions, into which rainfall runs off from the upper slopes and collects. This rain contains dissolved chemicals. As it seeps into the ground it raises the water table, in some places quite close to the surface. Due to the high temperatures of the valley, groundwater containing chemicals in solution is drawn up by evaporation and capillary action. As the water evaporates, salts are deposited on the surface forming thick crusts. These crusts are high in salt, and so no flowering plants can grow here.

Around the edge of the salt pans, however, the ground is not so salty. The groundwater is still quite high and the ground sandy. In this zone, phraetophytes (plants with very deep tap roots) grow. Their roots penetrate as far down as the water table. The various plants in this zone are found in a regular order depending on their tolerance of salinity. The most salt-tolerant is pickleweed, which can tolerate levels as high as 6 per cent salt (twice as salty as seawater). This plant is found closest to the salt pan. Next is arrow-weed, which can tolerate 3 per cent salt, and finally honey mesquite, which can only tolerate a maximum of 0.5 per cent salt content.

Between the sandy zone and the valley sides are the gravel deposits. This is where the xerophytes (a species of plants that have adapted to survive in an environment with little water) are located. They are too far above the water table to reach it and survive through being drought-resistant. Xerophytes, such as the creosote bush, may receive some water from dew, infrequent rainfall, and some occasional overland flow. Once again, a sequence can be observed. On the slightly wetter, upper slopes, burrow-weed is found. With increasing water shortage, creosote bushes are found, and finally, in the lowest, driest gravel, desert holly bush is found.

Activities

1 Suggest reasons why hot deserts offer limited opportunities for human activities.
2 Explain how plants are adapted to desert environments.
3 Explain how human activities have impacted upon desert ecosystems.

● End of theme questions

Topic 2.1: Earthquakes and volcanoes

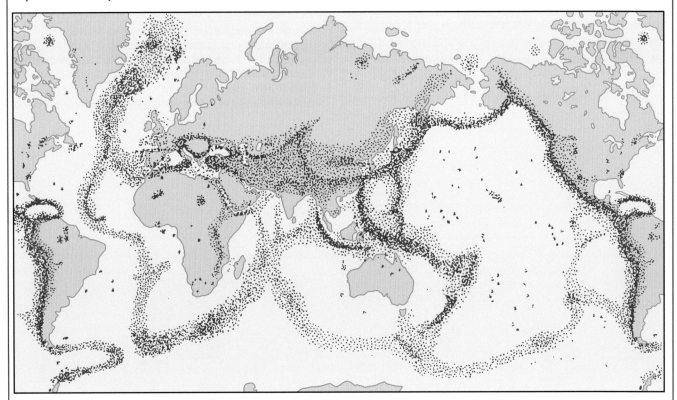

Figure 1   Global distribution of earthquakes

1   Study Figure 1 which shows the global distribution of
    earthquakes
    a   Describe the global distribution of earthquakes.
    b   Suggest reasons for the distribution of earthquakes.
    c   Dedine the terms 'focus' and 'epicentre'.
    d   Using examples, explain the main factors that
        increase the impacts of earthquakes.

Topic 2.2: Rivers

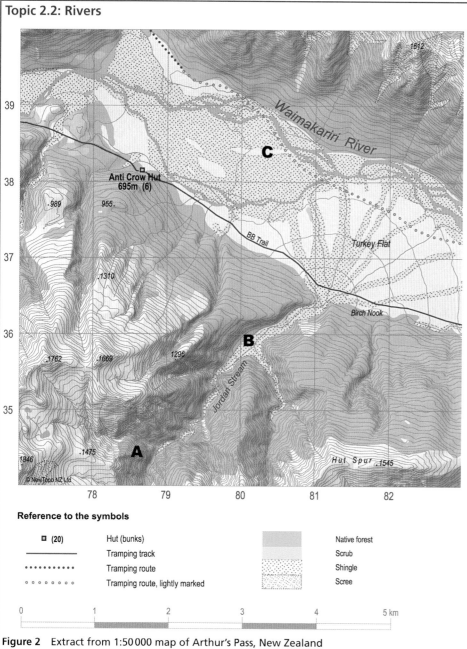

Reference to the symbols

☐ (20)	Hut (bunks)
——	Tramping track
••••••••••••	Tramping route
○ ○ ○ ○ ○ ○ ○ ○	Tramping route, lightly marked

	Native forest
	Scrub
	Shingle
	Scree

0 1 2 3 4 5 km

Figure 2 Extract from 1:50 000 map of Arthur's Pass, New Zealand

1 Study Figure 2.
  a Approximately how high is the source of the Jordan Stream?
  b How much does it fall in order to reach the Waimakarari river?
  c How far is the Waimakarari river from the source of the Jordan Stream?
  d Describe the changes in the river valley between point A and point B.
  e Describe the valley of the Jordan Stream and its valley at point C.
  f Outline the likely human activities that might occur in this area. Give reasons for your choices.

Topic 2.3: Coasts

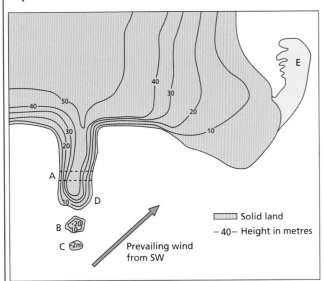

Figure 3 Sketch map of coastal features

1 Study Figure 3.
  a Identify the features A, B, C and D.
  b Explain how landform C may be formed.
  c What feature is located at E?
  d Explain the meaning of the term 'prevailing wind'.
  e Suggest how landform E may be formed.

Topic 2.4: Weather

Table 1 Daily weather observations Frankston, Victoria (Australia), 16–22 September 2013

Date	Day	Maximum Temperature (°C)	Minimum Temperature (°C)	Rainfall (mm)	Wind direction	Wind speed (km/hour)	Air pressure (mb)
16 September	Monday	16.5	12.7	0.4	E	24	1008
17 September	Tuesday	16.7	12.4	26.0	ESE	28	1003
18 September	Wednesday	16.4	12.7	14.6	WNW	57	999
19 September	Thursday	14.4	9.9	19.8	WNW	61	1005
20 September	Friday	14.6	10.7	1.0	SW	48	1007
21 September	Saturday	17.2	8.1	0.0	NNE	24	1016
22 September	Sunday	18.1	9.9	0.0	N	30	1012

1 Plot the data for 16–22 September 2013 using the same methods as in Figure 10 (for August 2007, see page 146).
2 State the maximum and minimum temperature of the 7-day period.
3 Work out the mean minimum temperature and the mean maximum temperature for the seven days.
4 How much rain fell during the seven days?
5 Compare the weather of August 2007 and September 2013.

Topic 2.5 Climate and natural vegetation

1 Study Figure 6 on page 152.
  a Comment on the distribution of the world's hot deserts.
  b Why is it difficult to live in a desert?
  c How have plants in the desert adapted in order to survive?
  d How do animals survive in the desert?

Development

An open pit toilet: this is as far as sanitation goes in many parts of the developing world.

Key questions

- How can the level of economic development of a country be measured?
- What are the reasons for inequalities between and within countries?
- How can economic production be classified into different sectors?
- How do the proportions employed in each sector of an economy vary according to the level of development?
- What is globalisation?
- How important are technology, transnational corporations and other economic factors in the process of globalisation?
- What are the impacts of globalisation at local, national and global scales?

● Indicators of development

Development, or improvement in the quality of life, is a wide-ranging concept. It includes wealth, but it also includes other important aspects of our lives (Figure 1). For example, many people would consider good health to be more important than wealth. Development occurs when there are improvements to individual factors making up the quality of life. For example, development occurs in a low-income country when:

- local food supply improves due to investment in farm machinery and fertilisers
- the electricity grid extends outwards from the main urban areas to rural areas
- levels of literacy improve throughout the country.

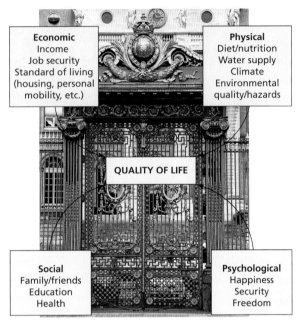

Economic
Income
Job security
Standard of living
(housing, personal
mobility, etc.)

Physical
Diet/nutrition
Water supply
Climate
Environmental
quality/hazards

QUALITY OF LIFE

Social
Family/friends
Education
Health

Psychological
Happiness
Security
Freedom

Figure 1 Factors comprising the quality of life

Gross National Product

One of the traditional indicators of a country's wealth is the **Gross National Product** (GNP). You will also be able to find data for Gross Domestic Product (GDP) and Gross National Income (GNI). It is reasonable at this level of study to regard these as broadly similar measures. The Gross National Product is:

- the total value of goods and services produced by a country in a year
- plus income earned by the country's residents from foreign investments
- minus income earned within the domestic economy by overseas residents.

To take account of the different populations of countries the **Gross National Product per capita** is often used. Here, the total GNP of a country is divided by the total population. Per capita figures allow for better comparisons between countries when their total populations are very different. For example, the total GNP of China is greater than that of the UK, but GNP per capita is much higher in the UK.

However, such data do not take into account the way in which the cost of living can vary between countries. For example, a dollar buys much more in China than it does in the USA! To account for this the GNP per capita at **purchasing power parity** (PPP) can be calculated. Figure 2 shows GNP per capita for 2013. It is clear to see where regions of high and low GNP per capita are located.

Figure 2 World map showing GNP per capita in 2013

Table 1 shows the top and bottom 12 countries in GNP per capita for 2013. The **development gap** between the world's wealthiest and poorest countries is huge. However, a major limitation of GNP and other national data is that these are 'average' figures for a country which tell us nothing about:

- the way in which wealth is distributed within a country – in some countries the gap between rich and poor is much greater than in others
- how government invests the money at its disposal; for example, Cuba has a low GNP per capita but high standards of health and education because these have been government priorities for a long time.

Development not only varies between countries, it can also vary significantly within countries. Most of the measures that can be used to examine the contrasts between countries can also be used to look at regional variations within countries.

Table 1 Top 12 and bottom 12 countries in GNP per capita ($), 2013

Top		Bottom	
Qatar	102 800	Togo	1100
Liechtenstein	89 400	Afghanistan	1000
Bermuda	86 000	Madagascar	1000
Luxembourg	80 700	Niger	900
Monaco	70 700	Malawi	900
Singapore	60 900	Eritrea	800
Norway	55 300	Central African Rep.	800
Switzerland	54 600	Liberia	700
Hong Kong	50 700	Somalia	600
Brunei	50 500	Burundi	600
USA	49 800	Zimbabwe	500
United Arab Emirates	49 000	Congo Dem. Rep.	400

Literacy

Education is undoubtedly the key to socio-economic development. It can be defined as the process of acquiring knowledge, understanding and skills. Education has always been regarded as a very important individual indicator of development and it has figured prominently in aggregate measures. Adult literacy is one of the main ways in which differences in educational standards between countries can be shown. In 2010, the global adult literacy rate was 84 per cent, but many African countries have rates below 50 per cent. About 775 million adults worldwide are illiterate! A low adult literacy rate is a great obstacle to development.

The World Bank has concluded that improving female literacy is one of the most fundamental achievements for a developing nation to attain, because so many aspects of development depend on it. For example, there is a very strong relationship between the extent of female literacy and infant and child mortality rates. People who are literate are able to access medical and other information that will help them to a higher quality of life compared with those who are illiterate.

Life expectancy

Life expectancy is viewed as a very important measure of development as it is to a large extent the end result of all the factors contributing to the quality of life in a country. It is important for international and national government agencies to know about variations in life expectancy as this is a key measure of inequality. It helps development programmes to target those in most need. The main influences on life expectancy are:

- the incidence of disease (for example, malaria)
- physical environmental conditions (for example, very low rainfall)
- human environmental conditions (for example, pollution)
- personal lifestyle (for example, smoking).

Rates of life expectancy have converged significantly between rich and poor countries during the last 50 years, in spite of a widening wealth gap. Figures for life expectancy by world region are given in Theme 1.

Infant mortality

The infant mortality rate is regarded as one of the most sensitive indicators of socio-economic progress (Table 2). It is an important measure of health inequality both between and within countries.

There are huge differences in the infant mortality rate around the world, despite the wide availability of public health knowledge. Fortunately infant mortality rates have fallen sharply in many developing countries over the last 20 years. However, there is still a considerable gap between the richer and poorer world regions. The infant mortality rate in Africa is more than thirteen times that of Europe. Infant mortality generally compares well to other indicators of development, which is a good indication of its value as a measure of development.

Table 2 Infant mortality rate by world region, 2012

Region	Infant mortality rate 2012
World	41
More developed world	5
Less developed world	45
Africa	67
Asia	37
Latin America/Caribbean	20
North America	6
Oceania	21
Europe	5

Other individual indicators of development

Other measures of development include:

- school enrolment
- doctors per 100 000 people
- food intake (calories per capita per day)
- energy consumption per capita
- percentage of the population with access to an improved water supply
- percentage of the population living in urban areas
- number of motor vehicles per 1000 population
- internet penetration rate.

Interesting note
In 2012, seven countries had infant mortality rates above 100/1000. All were in Africa, apart from Afghanistan which had the world's highest rate at 129/1000.

The Human Development Index: a broader measure of development

In 1990 the **Human Development Index (HDI)** was devised by the United Nations as a measure of the disparities between countries. The HDI is a composite index which has changed slightly in composition in recent years. The current index contains four indicators of development (Figure 3):

- life expectancy at birth
- mean years of schooling for adults aged 25 years
- expected years of schooling for children of school entering age
- GNI per capita (PPP$).

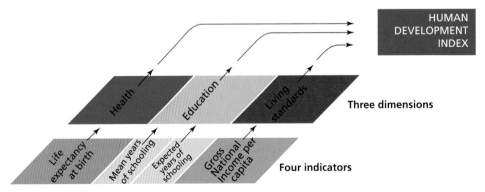

Figure 3 The components of the Human Development Index

The actual figures for each of these measures are converted into an index which has a maximum value of 1.0 in each case. The index values are then combined and averaged to give an overall HDI value. This also has a maximum value of 1.0. Every year the United Nations publishes the *Human Development Report* which uses the HDI to rank all the countries of the world in their level of development. The countries of the world are divided into four groups:

- Very high human development
- High human development
- Medium human development
- Low human development.

Figure 4 shows the global distribution of these four groups in 2011, while Table 3 lists the 'Very high human development' countries in rank order.

Every measure of development has merits and limitations. No single measure can provide a complete picture of the differences in development between countries. This is why the United Nations combines four measures of different aspects of the quality of life to arrive at a figure of human development for each country. Although the development gap can be measured in a variety of ways it is generally taken to be increasing. Many people are concerned about this situation, either because they see it as very unfair, or because it can create political instability.

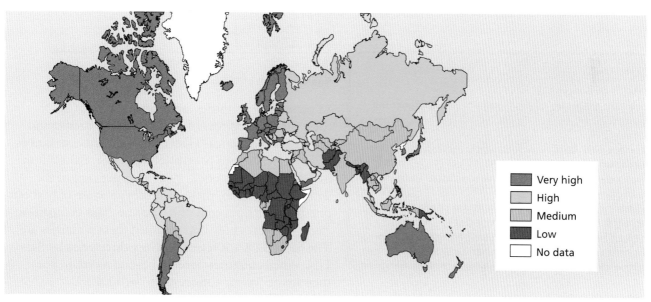

Figure 4 Map of the Human Development Index, 2011

Table 3 The 47 'Very high human development' countries: 2012 rankings

1	Norway	17	Belgium	33	Andorra
2	Australia	18	Austria	33	Estonia
3	USA	18	Singapore	35	Slovakia
4	Netherlands	20	France	36	Qatar
5	Germany	21	Finland	37	Hungary
6	New Zealand	21	Slovenia	38	Barbados
7	Ireland	23	Spain	39	Poland
8	Sweden	24	Liechtenstein	40	Chile
9	Switzerland	25	Italy	41	Lithuania
10	Japan	26	Luxembourg	41	United Arab Emirates
11	Canada	26	United Kingdom	43	Portugal
12	South Korea	28	Czech Republic	44	Latvia
13	Hong Kong	29	Greece	45	Argentina
13	Iceland	30	Brunei Darussalam	46	Seychelles
15	Denmark	31	Cyprus	47	Croatia
16	Israel	32	Malta		

Figure 5 The Waterfront, Vancouver, Canada has a very high level of human development

Activities

1 What is development?
2 Give two examples of development in a low-income country.
3 Define GNP.
4 Why are GNP data now frequently published at purchasing power parity (PPP)?
5 Briefly discuss the merits of one individual indicator of development.
6 Which indicators of development are combined to form the Human Development Index?
7 Look at Figure 4 and briefly describe the distribution of countries according to the Human Development Index.

● Explaining inequalities between countries

Stages of development

Over the years there have been a number of descriptions and explanations of levels of development between countries and how countries have moved from one level of development to another. A reasonable division of the world in terms of stages of economic development is shown in Figure 6. You will be familiar with the concept of developed and developing countries, but you may not be so sure about least developed countries and newly industrialised countries (NICs).

The concept of **least developed countries** (LDCs) was first identified in 1968 by the United Nations Conference on Trade and Development (UNCTAD). These are the poorest of the developing countries. They have major economic, institutional and human

resource problems. These are often made worse by geographical handicaps such as very low rainfall and natural and man-made disasters.

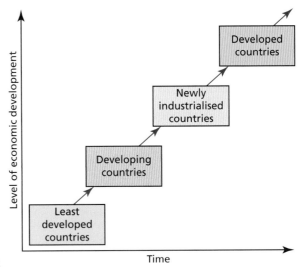

Figure 6 Stages of development

At present 49 countries are identified as LDCs. Of these 34 are in Africa, 14 in Asia and the Pacific, and one in Latin America (Haiti). You can find case studies of LDCs relating to population (Gambia, Bangladesh) in Theme 1 of this book. With 10.5 per cent of the world's population, these countries generate only about one-tenth of 1 per cent of its income. This comparison clearly indicates how impoverished these countries are! The list of LDCs is reviewed every three years by the UN. When countries develop beyond a certain point they are no longer considered to be LDCs.

Newly industrialised countries (NICs) are nations that have undergone rapid and successful industrialisation since the 1960s. They have moved up the development ladder, having previously been considered developing countries. The first countries to become newly industrialised countries (in the 1960s) were South Korea, Singapore, Taiwan and Hong Kong. The media referred to them as the 'four Asian tigers'. A 'tiger economy' is one that grows very rapidly. The reasons for the success of these countries were:

- a good initial level of infrastructure
- a skilled but relatively low-cost workforce
- cultural traditions that revere education and achievement
- governments that welcomed foreign direct investment (FDI) from transnational corporations
- all four countries had distinct advantages in terms of geographical location
- the ready availability of bank loans, often extended at government behest and at attractive interest rates.

The success of these four countries provided a model for others to follow, such as Malaysia, Brazil, China and India. In the last 20 years the growth of China has been particularly impressive. South Korea and Singapore have developed so much that many people now consider them to be developed countries.

Explaining the development gap

There has been much debate about the causes of development. Detailed studies have shown that variations between countries are due to a variety of factors.

Physical geography

- Landlocked countries have generally developed more slowly than coastal ones.
- Small island countries face considerable disadvantages in development.
- Tropical countries have grown more slowly than those in temperate latitudes, reflecting the cost of poor health and unproductive farming. However, richer non-agricultural tropical countries such as Singapore do not suffer a geographical deficit of this kind.
- A generous allocation of natural resources has spurred economic growth in a number of countries.

Economic policies

- Open economies that welcomed and encouraged foreign investment have developed faster than closed economies.
- Fast-growing countries tend to have high rates of saving and low spending relative to GDP.
- Institutional quality in terms of good government, law and order and lack of corruption generally result in a high rate of growth.

Demography

- Progress through demographic transition is a significant factor, with the highest rates of growth experienced by those nations where the birth rate had fallen the most.

Figure 7 combines a range of factors to consider some of the most basic points in explaining differences in development. For example, in Figure 7a Brazil would satisfy all three criteria. It is by far the largest country in South America. It has abundant natural resources and it is clearly a newly industrialised country. In contrast, countries such as Haiti and Niger would be affected by all three negative factors in Figure 7b.

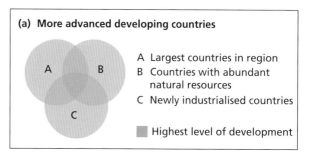

(a) **More advanced developing countries**

A Largest countries in region
B Countries with abundant natural resources
C Newly industrialised countries

■ Highest level of development

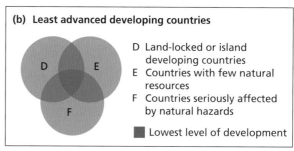

(b) **Least advanced developing countries**

D Land-locked or island developing countries
E Countries with few natural resources
F Countries seriously affected by natural hazards

■ Lowest level of development

Figure 7 Fast and slow development in developing countries

Figure 8 Low-income housing on the banks of the river Nile, Egypt

Consequences of the development gap

The development gap has significant consequences for people in the most disadvantaged countries. The consequences of poverty can be economic, social, environmental and political (Table 4). Development may not bring improvements in all four areas at first, but over time all four categories should witness advances.

Table 4 The consequences of poverty

Economic	Global integration is spatially selective: some countries benefit, others it seems do not. One in five of the world's population live on less than a dollar a day, almost half on less than two dollars a day. Poor countries frequently lack the ability to pay for (1) food (2) agricultural innovation and (3) investment in rural development.
Social	More than 775 million people in poor countries cannot read or write. Nearly a billion people do not have access to clean water and 2.4 billion to basic sanitation. Over 3.5 million children under the age of 5 die each year from conditions that could be prevented/treated with access to simple, affordable interventions. The impact of HIV/AIDS in poor countries can be devastating.
Environmental	Poor countries have increased vulnerability to natural disasters. They lack the capacity to adapt to droughts induced by climate change. Poor farming practices lead to environmental degradation. Often, raw materials are exploited with very limited economic benefit to poor countries and little concern for the environment. Landscapes can be devastated by mining, vast areas of rainforest destroyed by logging and clearance for agriculture, and rivers and land polluted by oil exploitation.
Political	Poor countries that are low on the development scale often have non-democratic governments or they are democracies that function poorly. There is usually a reasonably strong link between development and improvement in the quality of government. In general, the poorer the country the worse the plight of minority groups.

Activities

1 Look at Figure 4 and name two countries in each of the four levels of development.
2 a Define a 'least developed country'.
 b What are the general problems facing least developed countries?
3 a Define a 'newly industrialised country'.
 b Briefly discuss the factors responsible for the development of the first newly industrialised countries.
4 Consider the physical, economic and demographic factors that help explain the inequalities between countries.
5 Explain the two scenarios shown in Figure 7.

● Explaining inequalities within countries

The scale of disparities within countries is often as much an issue as the considerable variations between countries.

The **Gini coefficient** is a technique frequently used to show the extent of income inequality. It allows:

● analysis of changes in income inequality over time in individual countries and
● comparison between countries.

Figure 9 shows global variations in the Gini coefficient for 2009. It is defined as a ratio, with values between 0 and 1.0. A low value indicates a more equal income distribution while a high value shows more unequal income distribution. A Gini coefficient of zero would mean that everyone in a country had exactly the same income (perfect equality). At the other extreme a Gini coefficient of 1 would mean that one person had all the income in a country (perfect inequality). In general, more affluent countries have a lower income gap than lower income countries. Southern Africa and South America show up clearly as regions of very high income inequality. Europe is the world region with the lowest income inequality.

Legend:
<0.25
0.25–0.29
0.30–0.34
0.35–0.39
0.40–0.44
0.45–0.49
0.50–0.54
0.55–0.59
>0.60
No data

Figure 9 World map showing variations in the Gini coefficient, 2009

In China, the income gap between urban residents and the huge farm population reached its widest level ever in recent years as rural unemployment in particular rose steeply. The ratio between more affluent urban dwellers and their rural counterparts reached 3.36 to 1. This substantial income gap is a very sensitive issue in China as more and more rural people feel they have been left behind in China's economic boom.

A theory of regional disparities

The model of **cumulative causation** helps to explain regional disparities. Figure 10 is a simplified version of the model. The overall scenario is that there are three stages of regional disparity:

- the pre-industrial stage when regional differences are minimal
- a period of rapid economic growth characterised by increasing regional economic divergence
- a stage of regional economic convergence when the significant wealth generated in the most affluent region(s) spreads to other parts of the country.

Figure 11 shows how the regional economic divergence of the earlier stages of economic development can eventually change to regional economic convergence as regional differences narrow.

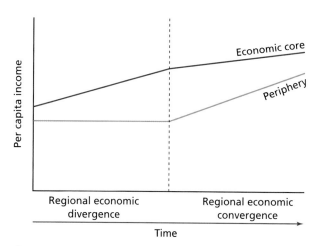

Figure 11 Regional economic divergence and convergence

In the model (Figure 10), economic growth begins with the location of new manufacturing industry in a region with a combination of advantages greater than elsewhere in the country. Once growth has been initiated in this 'core' region, flows of labour, capital and raw materials develop to support it and the growth region undergoes further expansion by the cumulative causation process. A detrimental negative effect (the backwash effect) is transmitted to the less developed regions (the periphery) as skilled labour and locally generated capital is attracted away. Manufactured goods and services produced and operating under the economies of scale of the core region flood the market of the relatively underdeveloped periphery, undercutting smaller-scale enterprises in such areas. In Figure 11, the wealth gap between the core and the periphery widens and regional inequality increases (regional economic divergence).

However, increasing demand for raw materials from resource-rich parts of the periphery may stimulate growth in other sectors of the economies of such regions. If the impact is strong enough to overcome local negative effects, a process of cumulative causation may begin. This may lead to the development of new centres of self-sustained economic growth (spread effects). If the process is strong enough and significant economic growth occurs in the periphery, the inequality between core and periphery may begin to narrow. This is the second stage on Figure 11 – regional economic convergence.

Many developing countries are in the first stage of Figure 11, where the wealth gap between core and periphery is still widening. Thus, they have a high

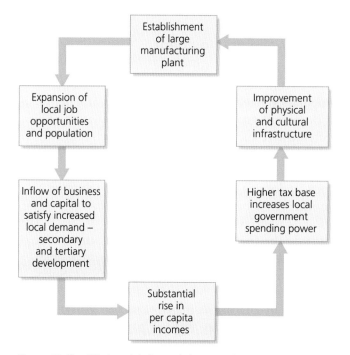

Figure 10 Simplified model of cumulative causation

Gini coefficient. Movement from stage 1 to stage 2 is usually attained by a combination of:

- market forces – the cost of doing business in the core region of a country may become so high that investment in the periphery becomes increasingly popular
- government regional development policies – government investment to improve conditions in peripheral regions, such as improvements in infrastructure, can help attract business investment.

- Urban/rural disparities with urban areas generally attracting much greater investment. This results in higher per capita incomes in urban areas.
- Intra-urban contrasts – low, middle and high income areas often exist close together in the same city. Large numbers of people live in slum conditions in cities in developing countries and find it extremely difficult to break out of this situation.

The UN has recognised that the focus of global poverty is moving from rural to urban areas, a process known as the **urbanisation of poverty**.

Figure 12 A village in eastern Siberia – the standard of living in most parts of Asiatic Russia (the periphery) is lower than in European Russia

Figure 13 Manholes in Ulaanbaatar, Mongolia – people in poverty sometimes live down these manholes because of access to the underground hot water pipes that can provide warmth in the harsh winters

Activities

1 a What is the Gini coefficient?
  b Briefly describe the global differences shown in Figure 9.
2 Define the terms (a) economic core region (b) periphery.
3 Explain the process shown in Figure 10 in your own words.
4 Describe and explain the trends shown in Figure 11.
5 What is the evidence in Figure 12 that this region is part of the economic periphery of Russia?

Factors affecting inequalities within countries

Residence

Where people are born and where they live can have a very significant impact on their quality of life. This includes:

- Regional differences within countries – the wealth gap between core and periphery, which is explained above.

Ethnicity and employment

The development gap often has an ethnic and/or religious dimension whereby some ethnic groups in a population have income levels significantly below the dominant group(s) in the same population. This is often the result of discrimination which limits the economic, social and political opportunities available to the disadvantaged groups. Examples include South Africa, Indonesia and Bolivia. Ethnic or religious minorities may be heavily concentrated in a particular region or regions of a country.

Jobs in the **formal sector** of the economy will be known to the government department responsible for taxation and to other government offices. Such jobs generally provide better pay and much greater security than jobs in the informal sector. Fringe benefits such as holiday and sick pay may also be available. Formal sector employment includes health and education service workers, government workers, and people working in established manufacturing and retail companies.

In contrast, the **informal sector** is that part of the economy operating outside official recognition. Employment is generally low-paid and often temporary and/or part-time in nature. While such employment is outside the tax system, job security will be poor with an absence of fringe benefits. About three-quarters of those working in the informal sector are employed in services. Typical jobs are shoe-shiners, street food stalls, messengers, repair shops and market traders. Informal manufacturing tends to include both the workshop sector, making for example cheap furniture, and the traditional craft sector. Many of these goods are sold in bazaars and street markets.

Figure 14 The informal sector – shoe-shiners in Cairo, Egypt

Education

Education is a key factor in explaining disparities within countries. Those with higher levels of education invariably gain better-paid employment. In developing countries there is a clear link between education levels and family size, with those with the least education having the largest families. Maintaining a large family usually means that saving is impossible and varying levels of debt likely. In contrast, people with better educational opportunities have smaller families and are thus able to save and invest more for the future. Such differences serve to widen rather than narrow disparities. Standards of education can vary significantly by region and as a result of other factors in a country.

Land ownership

The distribution of land ownership (tenure) has had a major impact on disparities in many countries. The greatest disparities tend to occur alongside the largest inequities in land ownership. The ownership of even a very small plot of land provides a certain level of security that those in the countryside without land cannot possibly aspire to. Households headed by women are often the most disadvantaged in terms of land tenure.

Regional contrasts in Brazil

South-east Brazil (Figure 15) is the economic core region of Brazil. Over time the south-east has benefited from flows of raw materials, capital and labour (Figure 16a). The last two inputs have come from abroad as well as from internal sources. The region grew rapidly through the process of cumulative causation. This process not only resulted in significant economic growth in the core, but also had a considerable negative impact on the periphery. The overall result was widening regional disparity. However, more recently some parts of the periphery, with a combination of advantages above the level of the periphery as a whole, have benefited from spread effects from the core (Figure 16b). The south has been the most important recipient of spread effects from the south-east, but the other regions have also benefited to an extent. This process has caused the regional gap to narrow at times, but often not for very long. In Brazil income inequality still remains very wide.

Figure 15 South-east Brazil

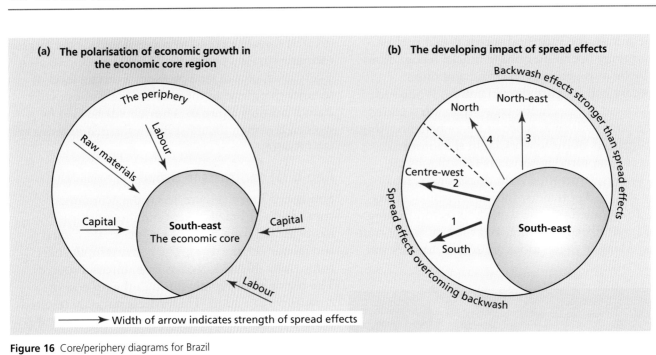

Figure 16 Core/periphery diagrams for Brazil

The south-east has benefited from a number of advantages:

- The natural environment of the south-east provided the region with several advantages for the development of its economy.
- The south-east is the centre of both foreign and domestic investment.
- The region is the focus of the country's road and rail networks. It contains the main airports and seaports.
- More transnational companies are located in the south-east than in the rest of Brazil. With the highest population density in Brazil, the labour supply here is plentiful.
- The region also has the highest educational and skill levels in the country.

The success of the first large wave of investment by foreign companies in the south-east encouraged others to follow. For the last 50 years the area has experienced an upward cycle of growth (cumulative causation).

Activities
1 Briefly explain two factors that can affect income inequality in a country.
2 a Where is the economic core region of Brazil?
  b Why is this region the most highly developed in the country?
3 a What are the two diagrams in Figure 16 attempting to show?
  b How do these diagrams relate to Figure 11?

● Classifying production into different economic sectors

In all modern economies of a significant size, people do hundreds – and in some cases thousands – of different jobs, all of which can be placed into four broad economic sectors:

- The **primary sector** exploits raw materials from land, water and air. Farming, fishing, forestry, mining and quarrying make up most of the jobs in this sector. Some primary products are sold directly to the consumer but most go to secondary industries for processing.
- The **secondary sector** manufactures primary materials into finished products. Activities in this sector include the production of processed food, furniture and motor vehicles. Secondary products are classed either as consumer goods (produced for sale to the public) or capital goods (produced for sale to other industries).
- The **tertiary sector** provides services to businesses and to people. Retail employees, drivers, architects, teachers and nurses are examples of occupations in this sector.
- The **quaternary sector** uses high technology to provide information and expertise. Research and development is an important part of this sector. Jobs in this sector include aerospace engineers, research scientists, computer scientists and biotechnology workers.

The **product chain** can be used to illustrate the four sectors of employment. The product chain is the full sequence of activities needed to turn raw materials into a finished product. The food industry provides a good example (Figure 17). Some companies are involved in all four stages of the food product chain.

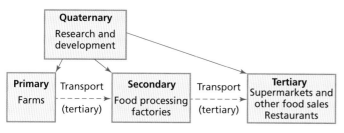

Figure 17 The food industry's product chain

Figure 18 The primary sector: an oil well in Dorset, UK

Figure 19 The secondary sector: grain processing factories, Chicago, USA

Figure 20 The tertiary sector: a street market in Nabul, Tunisia

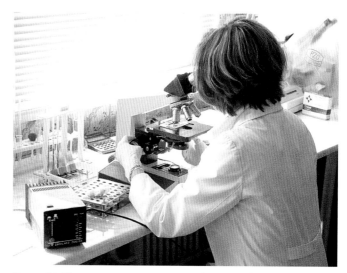

Figure 21 The quaternary sector: a research scientist

Activities

1 Define the terms (a) primary sector (b) secondary sector (c) tertiary sector (d) quaternary sector.
2 Give three examples of jobs in each of the four sectors of an economy.
3 Describe the food industry product chain shown in Figure 17.
4 What job do you want to do when you complete your education? In which sector of employment is this job?

How employment structure varies

As an economy develops, the proportion of people employed in each sector changes (Figure 22). Countries such as the USA, Japan, Germany and the UK are 'post-industrial societies' where the majority of people are employed in the tertiary sector. Yet in 1900, 40 per cent of employment in the USA was in the primary sector. However, the mechanisation of farming, mining, forestry and fishing drastically reduced the demand for labour in these industries. As these jobs disappeared, people moved to urban areas where secondary and tertiary employment was expanding. Less than 4 per cent of employment in the USA is now in the primary sector.

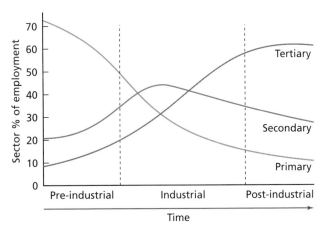

Figure 22 The sector model

Human labour has been steadily replaced in manufacturing too. In more and more factories, robots and other advanced machinery handle assembly-line jobs which once employed large numbers of people. The most advanced forms of manufacturing are in the developed world. In 1950 the same number of Americans were employed in manufacturing as in services. By 1980 two-thirds

were working in services. Today 79 per cent of Americans work in the tertiary sector.

The tertiary sector is also changing. In banking, insurance and many other types of business, computer networks have reduced the number of people required. But elsewhere service employment is often rising, such as in health, education and tourism. In developed countries employment in the quaternary sector has become more and more important. Employment in the quaternary sector is a significant measure of how advanced an economy is.

People in the poorest countries of the world are heavily dependent on the primary sector for employment. Such countries are often **primary product dependent** because they rely on one or a small number of primary products for all their export earnings. In newly industrialised countries employment in manufacturing has increased rapidly in recent decades. Table 5 compares the employment structure of a developed country, a newly industrialised country (NIC) and a developing country. The contrasts are very considerable indeed and a good fit with the sector model presented above.

Table 5 Employment structure of a developed country, an NIC and a developing country

Country	% primary	% secondary	% tertiary
Australia (developed)	4	21	75
Malaysia (NIC)	11	36	53
Bangladesh (developing)	45	30	25

There is a very clear link between employment structure and indicators of development. Compare the data in Table 5 with those in Table 6, which shows development indicators for the same three countries. Such a comparison could be conducted with a much larger number of countries and the results of the comparison would be very similar.

Table 6 Development indicators for Australia, Malaysia and Bangladesh

Country	GNI per capita, 2010 $PPP	% of population urban 2012	Infant mortality rate 2012
Australia	36910	82	3.9
Malaysia	14220	63	7
Bangladesh	1810	25	43

A graphical method often used to compare the employment structure of a large number of countries is the triangular graph (Figure 23). One side (axis) of the triangle is used to show the data for each of the primary, secondary and tertiary sectors. Each axis is scaled from 0 to 100 per cent. The indicators on the graph show how the data for the UK can be read. Figure 23 shows data for two developing countries, two NICs and two developed countries.

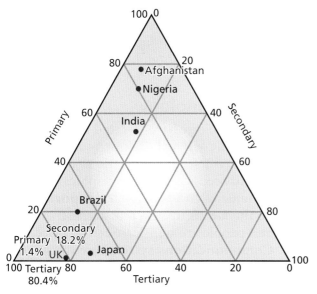

Figure 23 Triangular graph

Activities

1 Look at Figure 22. Describe and explain how the employment structure of developed countries has changed over time.
2 Describe and explain the different employment structures of the three countries in Table 5.
3 Comment on the relationship between the data presented in Tables 5 and 6.
4 a How would you classify each of the six countries shown in Figure 23?
  b On a copy of Figure 23, insert the employment structure positions of Australia, Malaysia and Bangladesh (Table 5).

● The process of globalisation

Figure 24 City of London emblem marking the boundary of the City of London, one of the world's great financial centres

Globalisation is the increasing interconnectedness and interdependence of the world, economically, culturally and politically. Most political borders are not the obstacles they once were and as a result goods, capital, labour and ideas flow more freely across them than ever before.

Transnational corporations

A **transnational corporation (TNC)** is a firm that owns or controls productive operations in more than one country through foreign direct investment (FDI). TNCs can exploit raw materials, produce goods such as cars and oil, and provide services such as banking. Table 7 shows the 10 largest TNCs in the world according to the business journal *Fortune*. Every year *Fortune* publishes a list of the 'Global 500' – the 500 largest TNCs in the world by revenue (the value of what they sell). The 2013 list shows that the five largest TNCs had revenues over $400 billion. The 100 largest TNCs represent a significant proportion of total global production.

Table 7 The world's 10 largest TNCs, 2013

Rank	Company	Revenue ($ billion)	Profits ($ billion)	HQ country
1	Royal Dutch Shell	481.7	26.6	Netherlands
2	Wal-Mart Stores	469.2	17.0	USA
3	Exxon Mobil	449.9	44.9	USA
4	Sinopec Group	428.2	8.2	China
5	China National Petroleum	408.6	18.2	China
6	BP	388.3	11.6	UK
7	State Grid	298.4	12.3	China
8	Toyota Motor	265.7	11.6	Japan
9	Volkswagen	247.6	27.9	Germany
10	Total	234.3	13.7	France

TNCs and nation states (countries) are the two main elements of the global economy. The governments of countries individually and collectively set the rules for the global economy, but the bulk of investment is through TNCs which are the main drivers of 'global shift'. Under this process manufacturing industry at first and more recently services have relocated in significant numbers from developed countries to selected developing countries as TNCs have taken advantage of lower labour costs and other ways to reduce costs. It is this process that has resulted in the emergence of an increasing number of newly industrialised countries since the 1960s. The development of successive generations of newly industrialised countries is the major success story of globalisation.

Twenty years ago the vast majority of the world's TNCs had their headquarters in North America, Western Europe and Japan. However, over the last two decades the emerging economies of the newly industrialised countries such as South Korea, China and India have been accounting for an increasing slice of the global economy. Much of this economic growth has been achieved through the expansion of their own most important companies, first domestically (as national corporations) and more recently on an international basis (as TNCs).

TNCs have a huge impact on the global economy in general and in the countries in which they choose to locate in particular. They play a major role in world trade in terms of what and where they buy and sell. A considerable proportion of world trade is intra-firm, taking place within TNCs. Table 8 considers the possible advantages and disadvantages of Nike to the USA (its headquarters country) and Vietnam (an outsourcing country).

Table 8 The potential advantages and disadvantages of TNCs – Nike to the USA and Vietnam

Country	Possible advantages	Possible disadvantages
USA: headquarters	Positive employment impact and stimulus to the development of high-level skills in design, marketing and development in Beaverton, Oregon; direct and indirect contribution to local and national tax base.	Another US firm that does not manufacture in its own country – indirect loss of jobs and the negative impact on balance of payments as footwear is imported; trade unions complain of an uneven playing field because of the big contrast in working conditions between developing and developed countries.
Vietnam: outsourcing	Creates substantial employment in Vietnam; pays higher wages than local companies; improves the skills base of the local population; the success of a global brand may attract other TNCs to Vietnam, setting off the process of cumulative causation; exports are a positive contribution to the balance of payments; sets new standards for indigenous companies; contribution to local tax base helps pay for improvements to infrastructure.	Concerns over the exploitation of cheap labour and poor working conditions; allegations of the use of child labour; company image and advertising may help to undermine national culture; concerns about the political influence of large TNCs; the knowledge that investment could be transferred quickly to lower-cost locations.

The spread of a global **consumer culture** has been important to the success of many TNCs. The **mass media** have been used very effectively to encourage consumers to 'want' more than they 'need'. The power of **brands** and their global marketing strategies cannot be underestimated. This is particularly so in food, beverages and fashion.

The role of technology

Advances in technology have affected all aspects of global economic activity. Major advances in transportation and telecommunications systems have significantly reduced the geographical barriers separating countries and peoples. **Transport systems** are the means by which materials, products and people are transferred from place to place. **Communications systems** are the ways in which information is transmitted from place to place in the form of ideas, instructions and images. As time has progressed, the **diffusion** of new ideas has speeded up so that a technical breakthrough in one part of the world has had an impact on other parts of the world much more quickly than ever before.

The **internet** has been essential to the development and speed of globalisation. It is the fastest-growing mode of communication ever. It took 38 years for radio to reach 50 million users, 13 years for television to reach this mark, but just 4 years for the internet. It has been estimated that the number of internet users around the world increased from 361 million in 2000 to 2.4 billion in 2012. This gives a global internet penetration rate (percentage of the population) of 34.4 per cent. By world region this varied from 78.6 per cent in North America to 15.6 per cent in Africa.

The internet has allowed TNCs to manage complex operations all over the world and to talk to its customers in large numbers directly. TNCs can react more quickly than ever before to changing consumer demand.

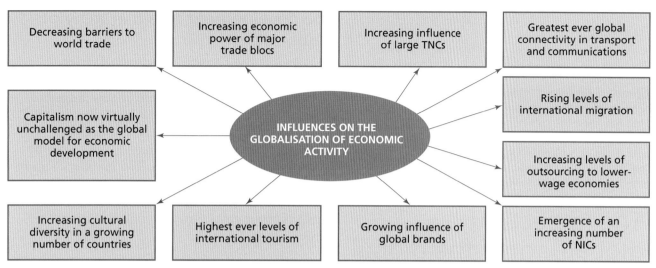

Figure 25 Influences on the globalisation of economic activity

Other factors responsible for economic globalisation

Figure 25 shows the main influences on the globalisation of economic activity. Until the post-1950 period, industrial production was mainly organised within individual countries. This has changed rapidly in the last 60 years or so with the emergence of a **new international division of labour** (NIDL). The NIDL divides production into different skills and tasks that are often spread across a number of countries. The following are some other factors responsible for economic globalisation:

- The increasing complexity of international trade flows as the NIDL has developed.
- Major advances in trade liberalisation under the World Trade Organization. The barriers to world trade (tariffs, quotas and regulations) are much lower today than in the past. This means that there is more incentive to trade.
- The emergence of fundamentalist free-market governments in the USA and the UK around 1980. The economic policies, such as privatisation, developed by these governments influenced policy-making in many other countries.
- The emergence of an increasing number of newly industrialised countries.
- The integration of the former Soviet Union and its Eastern European satellites into the capitalist system after the fall of communism in the late 1980s. Now, no significant group of countries stands outside the free market global system.
- The opening up of other economies, particularly those of China and India, as these countries wanted to benefit from the process of globalisation.

- The deregulation of world financial markets, allowing a much greater level of international competition in financial services.

Figure 26 The fall of the Berlin Wall – the beginning of the integration of eastern Europe into the free market system. One photo shows a remaining part of the wall, the other marks the position where the wall once was.

Activities

1 Define (a) globalisation (b) transnational corporation.
2 Why have TNCs been so important in the process of globalisation?
3 a How important are brands to TNCs?
  b Which brands impress you the most?
4 What has been the role of technology in globalisation?
5 Explain the new international division of labour.
6 State three factors responsible for economic globalisation.

The impacts of globalisation: the global scale

The changing world economic order

The rapid growth of newly industrialised countries has brought about major changes in the economic strength of countries. Figure 27 shows the extent to which the global share of GDP has changed since 1980. The decline in the share of the EU has been particularly sharp over this time period. In contrast the share of the BRIC countries (Brazil, Russia, India, China) has been very significant indeed. In 1990 the developed world controlled about 64 per cent of the global economy as measured by gross domestic product. This fell to 52 per cent by 2009 – one of the most rapid economic changes in history! Some countries have benefited much more than others in the changing world economic order.

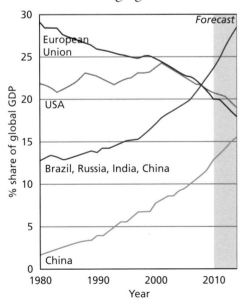

Figure 27 The global share of GDP

The development of a hierarchy of global cities

The emergence of a network of **global cities** has been an important part of the process of globalisation. A global city is one that is judged to be a significant nodal point in the global economic system. These are major financial and decision-making centres. New York, London and Tokyo are the world's major global cities. The number of global cities has increased as the process of globalisation has advanced, and so has the level of competition between major cities. Attracting more business creates jobs and wealth.

The international movement of workers

In recent decades the international movement of workers has spanned a much wider range of countries than ever before. This refers to both countries of origin and destination. There are now over 100 million migrant workers around the world. Migration of labour is a key feature of globalisation, but in some countries it can be a very controversial issue.

The global movement of commodities

People around the world have a greater choice of international commodities than ever before, although in many developing countries the prices of such commodities are out of the reach of many people. However, even though people may not have the money to purchase a commodity, they become aware of the lifestyle it attempts to portray because of advertising. Relatively cheap products from China and other NICs have helped to keep inflation low in many developed countries and allowed far more people to buy consumer goods.

The increasing uniformity of landscapes

Globalisation has undoubtedly had a significant impact on the increasing uniformity of landscapes. For example, Figure 28 illustrates the main ways in which common urban characteristics have diffused around the world. Many developments will of course also encompass local traits as well, but the strong global elements will be clear to see. Industrial, agricultural, tourist and transport landscapes have also become more uniform as similar processes worldwide have influenced their development. Some people find this disappointing as more and more places lose there 'uniqueness'.

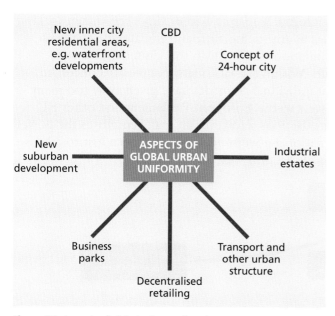

Figure 28 Aspects of global urban uniformity

Environmental degradation

In many parts of the world industrialisation, urbanisation, capital intensive farming and other major processes are having a devastating impact on the environment causing:

- air pollution
- deforestation
- land degradation and desertification
- salinisation and contamination of water supplies
- landscape change
- declines in biodiversity.

Much more decisive international action will be required to limit the environmental impact of economic activity.

Cultural diffusion

Cultural diffusion is the process of the spreading of cultural traits from one place to another. The mixing of cultures is a major dimension of globalisation. This has occurred through:

- migration, which circulates ideas, values and beliefs around the world
- the rapid spread of news, ideas and fashions through the mass media, trade and travel
- the growth of global brands such as Coca-Cola and McDonald's, which serve as common reference points
- the internet, which has allowed individual and mass communication on a scale never available before
- the transport revolution, which has facilitated the mass movement of people and products around the world.

The development of mass tourism

Recent decades have witnessed the globalisation of international tourism. International tourist arrivals reached just over 1 billion in 2012, over 40 per cent higher than in 2000! This is an average of almost 4 per cent a year, marking it out as a high-growth industry.

Global civil society

The development of **global civil society** (environmental groups, protest movements, charities, trade unions etc.) has been an important aspect of the diffusion of ideas around the world. It has spawned new networks of communication which are not government and company based. These networks monitor the actions of governments and companies, spreading their criticisms rapidly to all those who want to take an interest. Political protests now occur almost simultaneously in countries far apart because of the power and effectiveness of instant global communication.

Table 9 Examples of the impacts of globalisation at global, national and local scales

Global	National	Local
The growing power of TNCs and global brands	Concerns about loss of sovereignty to regional and international organisations	Small local businesses often find it difficult to compete with major global companies
The emergence of an increasing number of NICs	Increased cultural diversity from international migration	Closure of a TNC branch plant can cause high local unemployment
Development of a hierarchy of global cities	Higher levels of incoming and outgoing international tourism	The populations of many local communities have become more multicultural
The increasing complexity of the world economy	TNCs employing an increasing share of the workforce	Greater variety of international cuisine
The emergence of English as the working language of the 'global village'	Increasing incidences of trans-boundary pollution	Families now more likely to be spread over different countries due to increased international migration
The emergence of powerful trade blocs	The growth of anti-globalisation movements as people worry about how important decisions are made	Lower cost of international travel in real terms
Environmental degradation caused by increasing economic activity	TNCs avoiding paying tax in some countries through 'creative accounting' – a very controversial issue!	The development of 'ethnic villages' in large urban areas

The impact of globalisation: the national scale

People in many countries are concerned about the apparent **loss of sovereignty** of nation states. The loss of sovereignty results from the ceding of national autonomy to other organisations. The American sociologist Benjamin Barber views the changes taking place as anti-democratic and threatening the very foundations upon which the United States were built. Within the EU, many people are concerned that individual countries have given away too many powers to the European Parliament and other EU organisations. However, in many countries the power of national governments has also been lost to TNCs and global civil society (Figure 29).

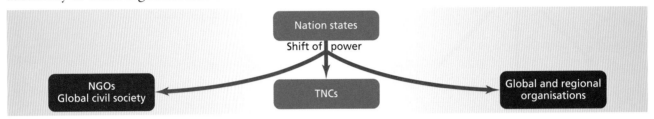

Figure 29 The shift of power from nation states

TNCs have increased their influence in many countries and become major employers. However, TNCs can close operations in one country and open up in another very quickly. This is a worry for governments as many jobs can be lost in major business closures. Some governments have found it difficult to collect what they think is a fair amount of tax from TNCs. This has become an increasingly controversial issue.

High levels of international migration have increased cultural diversity in many countries. This can bring advantages and disadvantages to host countries. Tensions can arise when economic conditions are difficult and there are not enough jobs to go round. For many countries the growth in international tourism has made it an important source of foreign currency and a more important source of employment. However, an increase in tourism can have costs as well as benefits.

As the scale of economic activity expands, increasing consumption of resources, particularly water, can cause tensions between neighbouring countries. Competition for energy and other resources drives prices upwards, making life particularly difficult for low-income countries and the poor in higher-income countries. As the scale of global economic activity has increased, concerns about trans-boundary pollution have grown. A major pollution incident in one country may have significant consequences in a neighbouring country. Anti-globalisation movements have developed in many countries to voice concerns over a range of issues associated with globalisation. Table 10 summarises some of the costs and benefits of globalisation to the UK.

Table 10 The costs and benefits of globalisation to the UK

Perspective	Benefits	Costs
Economic	As one of the world's most 'open' economies, the UK attracts a very high level of foreign direct investment, creating significant employment and contributing to GDP; a high level of investment abroad by UK companies also increases national income. Financial deregulation has enhanced the position of the City as one of the world's top three financial centres. Low-cost manufactured goods from China and elsewhere have helped keep inflation low.	High job losses in traditional industries due to global shift and deindustrialisation. TNCs can move investment away from the country as quickly as they can bring it in, causing loss of jobs and corporation tax. Speculative investment, causing economic uncertainty, has increased with financial deregulation. There is a widening gap between the highest and lowest-paid workers.
Social	Economic growth has facilitated high levels of spending on education and health in particular. Globalisation is a large factor in the increasingly cosmopolitan nature of UK society. The transport and communications revolution has transformed lifestyles.	A strong economy has attracted a very high level of immigration in recent years, with increasing concerns that this is unsustainable.
Political	Strong trading relationships with a large number of other countries brings political influence; as a member of the EU, the UK can extend its influence to areas where is was not previously well represented.	Voter apathy as many people see loss of political power to EU and major TNCs. International terrorism is a growing threat with increasing ethnic diversity, rapid transportation and more open borders.
Environmental	Deindustrialisation has improved environmental conditions in many areas; increasing international cooperation to solve cross-border environmental issues gives a better chance of such problems being addressed.	Population growth has an impact on the environment, with increasing demand for land, water and other resources. Rapid industrial growth in China and elsewhere has an impact on the global environment, including the UK.

The impact of globalisation: the local scale

Think of the area or region in which you live. What is the evidence that it has been affected by globalisation? It may be, for example:

- Small local businesses have found it hard to compete with major global companies. Many local areas have mounted 'support local shops' campaigns. The unique character of shopping areas can be damaged if too many local shops close and valuable jobs can be lost.

- The populations of many local communities have become more multicultural and in the process more culturally diverse. One example might be a greater variety of foreign restaurants and ethnic shops today compared with 20 years ago.
- In urban areas of a significant size, a number of ethnic 'villages' may be recognised.
- Increased international migration might mean that families in a region are now more likely to be spread over different countries. For younger people the motive is often to enhance career prospects and improve their quality of life, but retirement migration is also an important trend. People are more mobile than ever before.

Case study: The Tata Group and its global links

Indian companies, both private and government-owned organisations, are becoming increasingly transnational in their operations. India really emerged as a newly industrialised country in the 1990s when important economic reforms began to open up the country to foreign investment and made it easier for Indian companies to forge international links and to operate abroad. Other significant policy changes since 2000 have contributed to the recent rapid growth of Indian outward FDI.

Tata is perceived to be India's best-known global brand. Tata was founded in 1868 by Jamsetji Tata as a trading company. Tata Group is an Indian transnational conglomerate company which remains family-owned. With its headquarters in Mumbai, it encompasses seven business sectors:

- Communications and information technology
- Engineering
- Materials
- Services
- Energy
- Consumer products
- Chemicals.

Tata Group has over 100 companies with each of them operating independently. Some of the largest of these companies are Tata Steel, Tata Motors, Tata Consultancy Services, Tata Power, Tata Chemicals, and Tata Global Beverages. In recent decades Tata has expanded rapidly around the world. Tata Group now has operations in more than 80 countries and receives more than 58 per cent of its revenue from outside India. In 2012, the total number of employees worldwide was 456 000. Tata Group has steadily moved up the 'value chain' by producing more sophisticated and higher-value products.

For example, Tata has a considerable presence in the UK. Key acquisitions there have included:

- Tetley Group by Tata Tea for $430 million in 2000
- Corus Group by Tata Steel for $13 billion in 2007
- Jaguar and Land Rover by Tata Motors for $2.5 billion in 2008.

The objective has often been to buy world-renowned brands that are synonymous with high quality. If a brand name is well known it becomes much easier to increase sales in foreign markets if the company is being well managed.

In 2011 the Tata Group as a whole employed about 45 000 people in the UK with 19 400 in Tata Steel and 19 000 at Tata-owned Jaguar/Land Rover. Other Tata companies active in the UK are Tata Consultancy Services and Tata Global Beverages.

Tata Group has set great store by its reputation for social responsibility which began in India, but which has also spread abroad in more recent years. It was awarded the Carnegie Medal for Philanthropy in 2007.

Figure 30 Jaguar/Land Rover – a long-standing UK brand acquired by Tata in 2008

Case study analysis

1 Why do you think Tata Group has expanded abroad so rapidly in recent decades?
2 Describe Tata Group's presence in the UK.
3 Look at the website for the Tata Group (www.tata.com) to see the latest developments in this large group of companies. Produce a brief factfile of your findings.

Sheep farming in New Zealand – New Zealand is a major exporter of food

Key questions

- What are the main features of an agricultural system?
- What are the causes and effects of food shortages and the possible solutions to this problem?

● Agricultural systems

Individual farms and general types of farming can be seen to operate as a **system**. A farm requires a range of **inputs** such as labour and energy so that the **processes** that take place on the farm, such as ploughing and harvesting, can be carried out. The aim is to produce the best possible **outputs** such as milk, eggs, meat and crops. A profit will only be made if the income from selling the outputs is greater than expenditure on the inputs and processes. Figure 1 is an input-process-output diagram for a wheat farm.

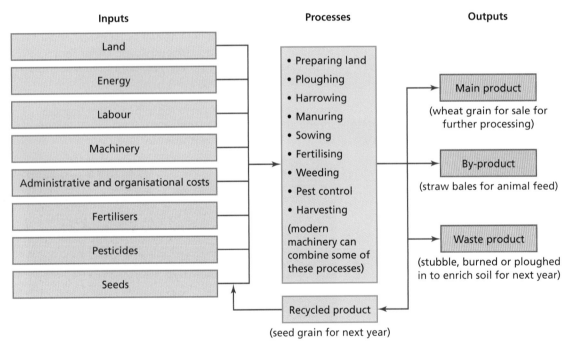

Inputs	Processes	Outputs
Land	• Preparing land	Main product
Energy	• Ploughing	(wheat grain for sale for further processing)
Labour	• Harrowing	
Machinery	• Manuring	By-product
Administrative and organisational costs	• Sowing	(straw bales for animal feed)
Fertilisers	• Fertilising	
Pesticides	• Weeding	Waste product
Seeds	• Pest control	(stubble, burned or ploughed in to enrich soil for next year)
	• Harvesting	
	(modern machinery can combine some of these processes)	

Recycled product
(seed grain for next year)

Figure 1 Systems diagram for a wheat farm

Different types of agricultural system can be found within individual countries and around the world. The most basic distinctions are between:

- arable, pastoral and mixed farming
- subsistence and commercial farming
- extensive and intensive farming
- organic and non-organic farming.

Arable, pastoral and mixed farming

Arable farms cultivate crops and are not involved with livestock. An arable farm may concentrate on one crop (monoculture) such as wheat, or may grow a range of different crops. The crops grown on an arable farm may change over time. For example, if the market price of potatoes increases, more farmers will be attracted to grow this crop. **Pastoral farming** involves keeping livestock such as dairy cattle, beef cattle, sheep and pigs. **Mixed farming** involves cultivating crops and keeping livestock together on a farm. Usually on a mixed farm at least part of the crop production will be used to feed the livestock.

Figure 2 Arable farming in the Nile valley with the pyramids in the background

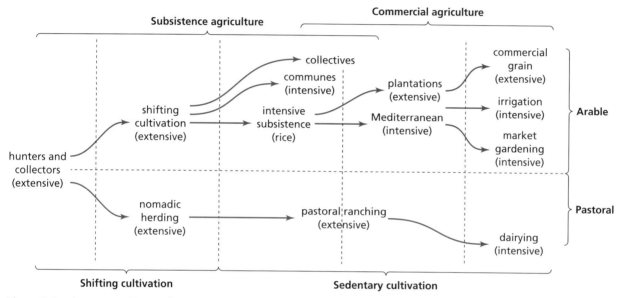

Figure 3 Farming types and levels of development

Subsistence and commercial farming

Subsistence farming is the most basic form of agriculture where the produce is consumed entirely or mainly by the family who work the land or tend the livestock. If a small surplus is produced it may be sold or traded. Examples of subsistence farming are shifting cultivation and nomadic pastoralism (Figure 3). Subsistence farming is generally small scale and labour intensive with little or no technological input.

In contrast, the objective of **commercial farming** is to sell everything that the farm produces. The aim is to maximise yields in order to achieve the highest profits possible. Commercial farming can vary from small scale to very large scale. The very largest farms are often owned by TNCs.

Figure 4 The output of local poultry production in a Moroccan market

Extensive and intensive farming

Extensive farming is where a relatively small amount of agricultural produce is obtained per hectare of land, so such farms tend to cover large areas of land.

Inputs per unit of land are low. Extensive farming can be both arable and pastoral in nature. Examples of extensive farming are sheep farming in Australia and wheat cultivation on the Canadian Prairies. In contrast, **intensive farming** is characterised by high inputs per unit of land to achieve high yields per hectare. Examples of intensive farming include market gardening, dairy farming and horticulture. Intensive farms tend to be relatively small in terms of land area.

Figure 5 Viticulture in the Rioja region of northern Spain

Organic farming

Organic farming does not use manufactured chemicals, so production is without chemical fertilisers, pesticides, insecticides and herbicides. Instead animal and green manures are used along with mineral fertilisers such as fish and bone meal. Organic farming therefore requires a higher input of labour than mainstream farming. Weeding is a major task in this type of farming. Organic farming is less likely to result in soil erosion and is less harmful to the environment in general. For example, there will be no nitrate runoff into streams and much less harm to wildlife.

Organic farming tends not to produce the 'perfect' potato, tomato or carrot. However, because of the increasing popularity of organic produce it commands a substantially higher price than mainstream farm produce.

Activities

1 Describe the inputs, processes and outputs for the wheat farm shown in Figure 1.
2 a Explain the difference between arable and pastoral farming.
  b What is mixed farming?
3 Examine the differences between (a) commercial and subsistence farming and (b) intensive and extensive farming.
4 Describe the characteristics of organic farming.

The influence of natural and human inputs on agricultural land use

A wide range of factors combine to influence agricultural land use and practices on farms. These can be placed under the general headings of physical, social/cultural, economic and political factors.

Physical factors

North America, for example, has many different physical environments. This allows a wide variety of crops to be grown and livestock kept. New technology and high levels of investment have steadily extended farming into more difficult environments. Irrigation has enabled farming to flourish in the dry south-west while new varieties of wheat have pushed production northwards in Canada. However, the physical environment remains a big influence on farming. There are certain things that technology and investment can do little to alter. So relief, climate and soils set broad limits as to what can be produced. This leaves the farmer with some choices, even in difficult environments. The farmer's decisions are then influenced by economic, social/cultural and political factors.

Temperature is a critical factor in crop growth as each type of crop requires a minimum growing temperature and a minimum growing season. Latitude, altitude and distance from the sea are the major influences on temperature. Precipitation is equally important. This is not just the annual total but the way it is distributed throughout the year. Long, steady periods of rainwater to infiltrate into the soil are best, making water available for crop growth. In contrast, short heavy downpours can result in surface runoff, leaving less water available for crop growth and soil erosion.

Soil type and fertility have a huge impact on agricultural productivity. Often, areas that have never been cleared for farming were ignored because soil fertility was poor or perceived to be poor. In some regions wind can have a serious impact on farming, for example causing bush fires in some US states such as California. Locally, aspect and the angle of slope may also be important factors in deciding how to use the land.

In Canada, farming is severely restricted by climate. Less than 8 per cent of the total area of the country is farmed. Seventy per cent of Canada lies north of the thermal limit for crop growth.

Water is vital for agriculture. **Irrigation** is an important factor in farming not just in North America, but in many other parts of the world as well. Table 1 compares the main types of irrigation. This is an example of the 'ladder' of agricultural technology, with surface irrigation being the most traditional method and subsurface (drip) irrigation the most advanced technique.

Table 1 Types of irrigation

	Efficiency (%)
Surface – used in over 80% of irrigated fields worldwide	
Furrow Traditional method; cheap to install; labour-intensive; high water losses; susceptible to erosion and salinisation	20–60
Basin Cheap to install and run; needs a lot of water; susceptible to salinisation and waterlogging	50–75
Aerial (using sprinklers) – used in 10–15% of irrigation worldwide	
Costly to install and run; low-pressure sprinklers preferable	60–80
Sub-surface ('drip') – used in 1% of irrigation worldwide	
High capital costs; sophisticated monitoring; very efficient	75–95

Figure 6 Goats feeding from a bowl (because the ground is frozen) in cold central Asia

Economic factors

Economic factors include transport, markets, capital and technology. The cost of growing different crops or keeping different livestock varies. The market prices for agricultural products will vary also and can change from year to year. The necessary investment in buildings and machinery can mean that some changes in farming activities are very expensive. These would be more difficult to achieve than other, cheaper changes. Thus it is not always easy for farmers to react quickly to changes in consumer demand.

In most countries there has been a trend towards fewer but larger farms. Large farms allow **economies of scale** to operate which reduce the unit costs of production. As more large farms are created, small farms find it increasingly difficult to compete and make a profit. Selling to a larger neighbouring farm may be the only economic solution. The EU is an example of a region where average farm size varies significantly. Those countries with a large average farm size generally have more efficient agricultural sectors than countries with a small average farm size.

Figure 7 A food market in Morocco

Agricultural technology is the application of techniques to control the growth and harvesting of animal and vegetable products. The development and application of agricultural technology requires investment and thus it is an economic factor. The status of a country's agricultural technology is vital for its food security and other aspects of its quality of life. An important form of aid is the transfer of agricultural technology from more advanced to less advanced countries.

Social/cultural factors

What a particular farm and neighbouring farms have produced in the past can be a significant influence on current farming practices. There is a tendency for farmers to stay with what they know best and often a sense of transgenerational responsibility to maintain family farming tradition. Tradition matters more in some farming regions than others.

Land tenure means the ways in which land is or can be owned. In the past inheritance laws have had a huge impact on the average size of farms. In some countries it has been the custom on the death of a farmer to divide the land equally between all his sons, but rarely between daughters. Also, dowry customs may include the giving of land with a daughter on marriage. The reduction in the size of farms by these processes often reduced them to operating at only a subsistence level.

In most societies women have very unequal access to, and control over, rural land and associated resources. It is now generally accepted that societies with well-recognised property rights are also the ones that thrive best economically and socially.

Political factors

The influence of government on farming has steadily increased in many countries. For example, in the USA the main parts of government farm policy over the past half-century have been:

- Price support loans: loans that tide farmers over until they sell their produce.
- Production controls: these limit how much a farmer can produce of surplus crops.
- Income supplements: these are cash payments to farmers for major crops in years when market prices fail to reach certain levels.

Thus the decisions made by individual farmers are heavily influenced by government policies such as those listed above. An agricultural policy can cover more than one country, as evidenced by the EU's Common Agricultural Policy.

Activities

1. List the main physical factors that can influence farming.
2. Summarise the information presented in Table 1.
3. Why has the size of farms steadily increased in many agricultural regions?
4. Briefly state the importance of advances in agricultural technology.
5. Give an example of how a social/cultural factor can have an impact on farming.
6. How can political factors influence farming?

Interesting note

In terms of agricultural exports the major countries by value of exports in 2011 were the USA, the Netherlands, Germany, France and Brazil.

Case study: An agricultural system – intensive rice production in the Lower Ganges Valley

Location

An important area of intensive subsistence rice cultivation is the lower Ganges valley (Figure 8) in India and Bangladesh. The Ganges basin is India's most extensive and productive agricultural area and its most densely populated. The delta region of the Ganges occupies a large part of Bangladesh, one of the most densely populated countries in the world. Rice contributes over 75 per cent of the diet in many parts of the region. The physical conditions in the lower Ganges valley and delta are very suitable for rice cultivation:

- temperatures of 21 °C and over throughout the year, allowing two crops to be grown annually (rice needs a growing season of only 100 days)
- monsoon rainfall over 2000 mm providing sufficient water for the fields to flood, which is necessary for wet rice cultivation
- rich alluvial soils built up through regular flooding over a long time period during the monsoon season
- an important dry period for harvesting the rice.

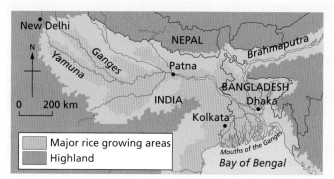

Figure 8 The Lower Ganges valley

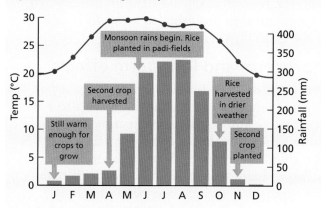

Figure 9 Climate graph for Kolkata

A water intensive staple crop

Rice is the staple or main food crop in many parts of Asia. This is not surprising considering its high nutritional value. Current rice production systems are extremely water intensive. Ninety per cent of agricultural water in Asia is used for rice production. The International Rice Research Institute estimates that it takes 5000 litres of water to produce one kilogram of rice. Much of Asia's rice production can be classed as intensive subsistence cultivation where the crop is grown on very small plots of land using a very high input of labour. Rice cultivation by small farmers is sometimes referred to as 'pre-modern intensive farming' because of the traditional techniques used, in contrast to intensive farming systems in developed countries such as market gardening which are very capital intensive.

'Wet' rice is grown in the fertile silt and flooded areas of the lowlands while 'dry' rice is cultivated on terraces on the hillsides. A **terrace** is a levelled section of a hilly cultivated area. Terracing is a method of soil conservation. It also prevents the rapid runoff of irrigated water. Dry rice is easier to grow but provides lower yields than wet rice.

The farming system

Padi-fields (flooded parcels of land) characterise lowland rice production. Water for irrigation is provided either when the Ganges floods or by means of gravity canals. At first, rice is grown in nurseries. It is then transplanted when the **monsoon rains** flood the padi-fields. The flooded fields may be stocked

with fish for an additional source of food. The main rice crop is harvested when the drier season begins in late October. The rice crop gives high yields per hectare. A second rice crop can then be planted in November, although water supply can be a problem in some areas for the second crop.

Figure 10 Rice padi-field scene in lower Ganges valley

Water buffalo are used for work. This is the only draught animal adapted for life in wetlands. The water buffalo provide an important source of manure in the fields. However, the manure is also used as domestic fuel. The labour-intensive nature of rice cultivation provides work for large numbers of people. This is important in areas of very dense population where there are limited alternative employment opportunities. The low incomes and lack of capital of these subsistence farmers means that hand labour still dominates in the region. It takes an average of 2000 hours a year to farm one hectare of land. A high labour input is needed to:

- build the embankments (bunds) that surround the fields – these are stabilised by tree crops such as coconut and banana
- construct irrigation canals where they are required for adequate water supply to the fields
- plant nursery rice, plough the padi-field, transplant the rice from the nursery to the padi-field, weed and harvest the mature rice crop
- cultivate other crops in the dry season and possibly tend a few chickens or other livestock.

Rice seeds are stored from one year to provide the next year's crop. During the dry season when there may be insufficient water for rice cultivation, other crops such as cereals and vegetables are grown. Farms are generally small, often no more than one hectare in size. Many farmers are tenants and pay for use of the land by giving a share of their crop to the landlord.

Case study analysis

1 Describe the location of the Lower Ganges valley.
2 Why is rice cultivation in the area considered to be an intensive form of agriculture?
3 Explain why the physical environment provides good conditions for rice cultivation.
4 Describe the inputs, processes and outputs of this type of agriculture.

● Causes and effects of food shortages

Causes

Food shortages can occur because of both natural and human problems. The natural problems that can lead to food shortages include:

- soil exhaustion
- drought
- floods
- tropical cyclones
- pests
- disease.

However, economic and political factors can also contribute to food shortages. Such factors include:

- low capital investment
- rapidly rising population
- poor distribution/transport difficulties
- conflict situations.

In late 2012, the UN warned of an imminent worldwide food crisis, highlighting three major problems:

- global grain reserves at critically low levels
- rising food prices creating unrest in many countries
- extreme weather resulting in the climate being 'no longer reliable'.

In the same year the Food and Agriculture Organisation estimated that around the world 870 million people were malnourished, with the food crisis growing in Africa and the Middle East.

The impact of such problems has been felt most intensely in developing countries, where adequate food stocks to cover emergencies affecting food supply usually do not exist. However, developed countries have not been without their problems. For example, in recent years both the USA and Australia have suffered severe drought conditions. So developed countries are not immune from the physical problems that can cause food shortages. However, they invariably have the human resources to cope with such problems, so actual food shortages do not generally occur.

Short-term and long-term effects

The effects of food shortages are both short-term and longer-term. Malnutrition can affect a considerable number of people, particularly children within a relatively short period when food supplies are significantly reduced. With malnutrition people are less resistant to disease and more likely to fall ill. Such diseases include beri beri (vitamin B1 deficiency), rickets (vitamin D deficiency) and kwashiorkor (protein deficiency). People who are continually starved of nutrients never fulfil their physical or intellectual potential. Malnutrition reduces people's capacity to work so that land may not be properly tended and other forms of income successfully pursued. This is threatening to lock parts of the developing world into an endless cycle of ill-health, low productivity and underdevelopment.

Case study: **A region suffering from food shortages – Sudan and South Sudan**

The countries of Sudan and South Sudan (Figure 11), which were the single country of Sudan until 2011, have suffered food shortages for decades. The long civil war and drought have been the main reasons for famine in Sudan, but there are many associated factors as well (Figure 12). The civil war, which lasted for over 20 years, was between the government in Khartoum and rebel forces in the western region of Darfur and in the south (now South Sudan). A Christian Aid document in 2004 described the Sudan as 'A country still gripped by a civil war that has been fuelled, prolonged and part-financed by oil'. One of the big issues between the two sides in the civil war was the sharing of oil wealth between the government-controlled north and the south of the country where much of the oil is found. The United Nations has estimated that up to 2 million people were displaced by the civil war and more than 70 000 people died from hunger and associated diseases. At times, the UN World Food Programme stopped deliveries of vital food supplies because the situation was considered too dangerous for the drivers and aid workers.

Figure 11 Sudan and South Sudan

Physical factors
- Long-term decline of rainfall in southern Sudan
- Increased rainfall variability
- Increased use of marginal land leading to degradation
- Flooding

Social factors
- High population growth (3%) linked to use of marginal land (overgrazing, erosion)
- High female illiteracy rates (65%)
- Poor infant health
- Increased threat of AIDS

Agricultural factors
- Highly variable per capita food production; long-term the trend is static
- Static (cereals and pulses) or falling (roots and tubers) crop yields
- Low and falling fertiliser use (compounded by falling export receipts)
- Lack of a food surplus for use in crisis

Economic/political factors
- High dependency on farming (70% of labour force; 37% of GDP)
- Dependency on food imports (13% of consumption 1998–2000) whilst exporting non-food goods, e.g. cotton
- Limited access to markets to buy food or infrastructure to distribute it
- Debt and debt repayments limit social and economic spending
- High military spending

Drought in southern Sudan compounds low food intake; any remaining surpluses quickly used

Shorter-term factors leading to increased Sudanese food insecurity and famine

Both reduce food availability in Sudan and inflate food prices

Conflict in Darfur reduces food production and distribution

Situation compounded by:
- Lack of government political will
- Slow donor response
- Limited access to famine areas
- Regional food shortages

Figure 12 Summary of causes of famine in the Sudan

Figure 13 The fertile banks of the river Nile in Sudan with desert beyond

The separation of Sudan into two countries has not occurred easily. There has been intermittent fighting in border regions. This, along with economic instability, has undermined agricultural production. In March 2013 the World Food Programme warned that more than 4.1 million people were likely to be short of food in South Sudan in that year. This is approximately 40 per cent of the new country's population.

Case study analysis

1 Describe the location of Sudan and South Sudan.
2 a With the help of Figure 12, explain the causes of food shortages in recent decades.
  b Suggest what needs to happen for the situation to improve.
3 How bad was the food shortage situation in South Sudan in 2013?

Possible solutions to food shortages

Food aid

In the short term and in some instances the medium term, food aid is absolutely vital to cope with food shortages. When disaster strikes there is no alternative to this strategy. According to the charity ActionAid there are three types of food aid:

- **relief food aid** which is delivered directly to people in times of crisis
- **programme food aid** which is provided directly to the government of a country for sale in local markets (this usually comes with conditions from the donor country)
- **project food aid** which is targeted at specific groups of people as part of longer-term development work.

Figure 14 Food aid being delivered in Somalia

The USA and the EU together provide about two-thirds of global food aid deliveries. At the international level the main organisations are the UN World Food Programme (WFP), the UN Food and Agriculture Organization (FAO) and the Food Aid Convention.

Food aid is vital to communities in many countries, particularly in Africa but also in parts of Asia and Latin America. However, it is not without controversy:

- The charity CARE has criticised the method of US food aid to Africa. CARE sees the selling of heavily subsidised US produced food in African countries as undermining the ability of African farmers to produce for local markets, making countries even more dependent on aid to avoid famine. CARE wants the USA to send money to buy food locally instead.
- Friends of the Earth say that a genetically modified rice, not allowed for human consumption and originating in the USA, has been found in food aid in West Africa.
- Food aid is very expensive, not least because of the high transport costs involved.

There have been recent concerns that food aid may be required for even more people in the future. In recent years, the term 'global food crisis' has been used more and more by the media. Steep increases in the price of food have caused big problems in a number of countries. Major protests about the price of food have taken place in countries including Haiti, Indonesia, the Philippines and Egypt. The World Bank warned that progress on development could be destroyed by rapidly rising food costs.

The Green Revolution

Figure 15 Green Revolution crops being harvested in Brazil

The package of agricultural improvements generally known as the Green Revolution was seen as the answer to the food problem in many parts of the developing world in the post-1960 period. India was one of the first countries to benefit when a high-yielding variety (HYV) seed programme started in 1966–67. In terms of production it was a turning point for Indian agriculture, which had virtually reached stagnation. The programme introduced new hybrid varieties of five cereals: wheat, rice, maize, sorghum and millet. All were drought-resistant with the exception of rice, were very responsive to the application of fertilisers, and had a shorter growing season than the traditional varieties they replaced. Although the benefits of the Green Revolution are clear, serious criticisms have also been made. The two sides of the story can be summarised as follows:

Advantages

- Yields are twice to four times greater than for traditional varieties.
- The shorter growing season has allowed the introduction of an extra crop in some areas.
- Farming incomes have increased, allowing the purchase of machinery, better seeds, fertilisers and pesticides.
- The diet of rural communities is now more varied.
- Local infrastructure has been upgraded to accommodate a stronger market approach.
- Employment has been created in industries supplying farms with inputs.
- Higher returns have justified a significant increase in irrigation.

Disadvantages

- High inputs of fertiliser and pesticide are required to optimise production. This is costly in both economic and environmental terms. In some areas rural indebtedness has risen sharply.
- HYVs require more weed control and are often more susceptible to pests and diseases.
- Middle and higher-income farmers have often benefited much more than the majority on low incomes, thus widening the income gap in rural communities. Increased rural-to-urban migration has often been the result.
- Mechanisation has increased rural unemployment.
- Some HYVs have an inferior taste.
- The problem of salinisation has increased along with the expansion of the irrigated area.

In recent years a much greater concern has arisen about Green Revolution agriculture. The problem is that the high-yielding varieties introduced during the Green Revolution are usually low in minerals and vitamins. Because the new crops have displaced the local fruits, vegetables and legumes that traditionally supplied important vitamins and minerals, the diet of many people in the developing world is now extremely low in zinc, iron, vitamin A and other micronutrients.

The Green Revolution has been a major factor enabling global food supply to keep pace with population growth, but with growing concerns about a new food crisis, new technological advances may well be required to improve the global food security situation.

UNEP's options for improving food security

The United Nations Environment Programme has argued that increasing food energy efficiency provides a critical path for significant growth in food supply without compromising environmental sustainability.

- Options with short-term effects are (a) price regulation on commodities and larger cereal stocks to decrease the risk of highly volatile prices and (b) reduce/remove subsidies on biofuels to cut the capture of cropland by biofuels.
- Options with mid-term effects are (a) reduce the use of cereals and food fish in animal feed, (b) support farmers in developing diversified

eco-agricultural systems that provide critical ecosystem services (for example water supply and regulation) as well as adequate food to meet local and consumer needs and (c) increased trade and improved market access by improving infrastructure and reducing trade barriers.

- Options with long-term effects are (a) limit global warming, including the promotion of climate-friendly agricultural production systems and land use policies at a scale to help mitigate climate change and (b) raise awareness of the pressures of increasing population growth and consumption patterns on sustainable ecosystem functioning.

Activities

1 Describe the different types of food aid.
2 Why is food aid sometimes controversial?
3 Discuss the advantages and disadvantages of Green Revolution farming.
4 Comment briefly on UNEP's options for improving food security.

Pulp and paper mill, British Columbia, Canada

Key questions
- What are the stages of an industrial system?
- What are the factors influencing the distribution and location of factories and industrial zones?

● Industrial systems and types

Manufacturing industry as a whole, or an individual factory, can be regarded as a system. Industrial systems, like agricultural systems, have inputs, processes and outputs (Figure 1).

- **Inputs** are the elements that are required for the processes to take place. Inputs include raw materials, labour, energy and capital.
- **Processes** are the industrial activities that take place in the factory to make the finished product. For example, in the car industry, processes include moulding sheet steel into the shaped panels that make up the car, welding and painting.
- **Outputs** comprise the finished product or products that are sold to customers. Sometimes **by-products** may be produced. A by-product is something that is left over from the main production process which has some value and therefore can be sold. All manufacturing industries produce **waste product** which has no value and must be disposed of. Costs will be incurred in the disposal of waste product.

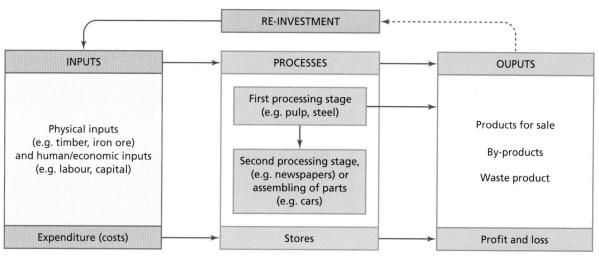

Figure 1 Industrial systems diagram

To 'manufacture' means to 'make'. Manufacturing industry is the general term used for the secondary sector of economic activity. Manufacturing is often described or classified by the use of opposing terms, such as 'heavy industry' and 'light industry' (Table 1).

In this case iron and steel would be an example of a heavy industry, using large amounts of bulky raw materials, processing on a huge scale and prod final products of a significant size. In con assembly of computers is a light indus

Processing and assembly industries

A significant distinction is between processing and assembly industries. Processing industries are based on the direct processing of raw materials. Again, the iron and steel industry would be an example, using large quantities of iron ore, coal and limestone. Processing industries are often located close to their raw materials. In contrast, assembly industries put together parts and components that have been made elsewhere. A large car assembly plant will use thousands of components to build a car. Assembly industries usually have a much wider choice of location than processing industries and thus they are often described as **footloose** industries.

High-technology industry

High-technology industry is the fastest growing manufacturing industry in the world. It all began in the 1960s in Silicon Valley (the Santa Clara valley), south of San Francisco. Since then it has spread across the world. Virtually all developed countries and NICs have at least one high-technology cluster (companies grouped together in one region). 'High-tech' companies use or make silicon chips, computers, software, robots, aerospace components and other very technically advanced products. These companies put a great deal of money into scientific research. Their aim is to develop newer, even more advanced products. Think of the latest products from companies such as Apple, Samsung, Sony and Nokia.

High-technology industries often cluster together in science parks, the idea for which was originally created in the USA. They are often found in close proximity to leading universities because of the need to employ well-qualified graduates in science and technology and to be aware of the latest research taking place in universities. The Cambridge Science Park is a major example in the UK. The clustering of high-technology industry means that companies can collaborate easily on joint projects, and highly skilled workers can move easily from one company to another.

Table 1 Classification of industry

Classification contrasts	Characteristics
Large scale and small scale	Depending on the size of plant and machinery, and the numbers employed.
Heavy and light	Depending on the nature of processes and products in terms of unit weight.
Market oriented and raw material oriented	Depending on the location of the industry or firm, which is drawn either towards the market or the raw materials required – usually because of transportation costs.
Processing and assembly	Processing involves the direct processing of raw materials; assembly is to do with putting together parts and components.
Capital intensive and labour intensive	Depending on the ratio of investment in plant and machinery to the number of employees.
Fordist and flexible	Fordist industries, named after the assembly-line methods used in the early automobile industry, mass produce on a large scale making standardised products. Flexible industries make a range of specialised products using high technology to respond quickly to changes in demand.
National and transnational	Many firms in the small- to medium-size range manufacture in only one country. Transnationals, which are usually extremely large companies, produce in at least two countries but may manufacture in dozens of nations.

Interesting note

The Boeing aircraft factory in Washington state, USA is the largest building in the world by volume. This is the assembly site for the company's largest aircraft.

Activities

1 Explain the industrial systems diagram shown in Figure 1.
2 With regard to manufacturing industry, explain the difference between (a) heavy and light, (b) processing and assembly and (c) capital intensive and labour intensive.
3 Suggest why high-technology industries often cluster together.

● Factors affecting the location of industry

Every day, decisions are made about where to locate industrial premises, ranging from small workshops to huge industrial complexes. In general, the larger the company the greater the number of real alternative locations available. For each possible location a wide range of factors can have an impact on total costs and thus influence the decision-making process. The factors affecting industrial location differ from industry to industry and their relative importance is subject to change over time. These factors can be broadly subdivided into physical and human (Table 2). They relate both to individual factories and to industrial zones.

Table 2 Physical and human factors influencing industrial location

Physical factors	Human factors
Site: The availability and cost of land is important. Large factories in particular will need flat, well-drained land on solid bedrock. An adjacent water supply may be essential for some industries. **Raw materials**: Industries requiring heavy and bulky raw materials which are expensive to transport will generally locate as close to these raw materials as possible. **Energy**: At times in the past, industry needed to be located near fast-flowing rivers or coal mines. Today, electricity can be transmitted to most locations. However, energy-hungry industries, such as metal smelting, may be drawn to countries with relatively cheap hydro-electricity, such as Norway. **Natural routeways and harbours**: These were essential factors in the past and are still important today as many modern roads and railways still follow natural routeways. Natural harbours provide good locations for ports and the industrial complexes often found at ports. **Climate**: Some industries such as aerospace and film benefit directly from a sunny climate. Indirect benefits such as lower heating bills and a more favourable quality of life may also be apparent.	**Capital (money)**: Business people, banks and governments are more likely to invest money in some areas than others. **Labour**: Increasingly it is the quality and cost of labour rather than the quantity that are the key factors here. The reputation, turnover and mobility of labour can also be important. **Transport and communications**: Transport costs are lower in real terms than ever before but remain important for heavy, bulky items. Accessibility to airports, ports, motorways and key railway terminals may be crucial factors for some industries. **Markets**: the location and size of markets is a major influence for some industries. **Government influence**: government policies and decisions can have a big direct and indirect impact on the location of industry. Governments can encourage industries to locate in certain areas and deny them planning permission in others. **Quality of life**: highly skilled personnel who have a choice about where they work will favour areas where the quality of life is high (leisure facilities, good housing, attractive scenery etc).

The combined influence of a range of factors will have an impact on the decision-making of a company in terms of the following:

- Location – companies decide on particular locations for a variety of reasons. Most will look to the location that is seen as the 'least-cost location' or the 'highest-profit' location. A poor choice of location can mean a company making a loss and eventually closing. An excellent location, resulting in considerable profits, may prompt a company to expand.
- Scale of production – the amount of a product a company plans to produce will be an important factor in deciding location. Companies can achieve economies of scale by manufacturing more of a product. However, if they decide on a larger scale of production they have to be sure that they (a) have a physical site large enough for the desired scale of production, (b) can recruit sufficient skilled labour in the region and (c) will have enough customers for their higher scale of production.
- Methods of organisation – companies can follow various methods of organisation from traditional to highly innovative. Location factors can influence such decisions. The most advanced companies in an industry tend to be very capital intensive while more traditional companies tend to be more labour intensive.
- The product or range of products manufactured – many large companies produce a range of products. Some locations may be more suited to the production of one product than another because of the cost factors involved.

● Industrial agglomeration

Industrial agglomeration is the clustering together of economic activities. Agglomeration can result in companies enjoying the benefits of external economies of scale. This means the lowering of a firm's costs due to external factors. The success of one company may attract other companies from the same industry group. External economies of scale can be subdivided into:

- urbanisation economies, which are the cost savings resulting from urban location due to factors such as the range of producer services available and the investment in infrastructure already in place
- localisation economies which occur when a firm locates close to suppliers (backward linkages) or firms that it supplies (forward linkages). This reduces transport costs, allows for faster delivery, and facilitates a high level of personal communication between firms.

Industrial estates

An **industrial estate** is an area zoned and planned for the purpose of industrial development. Industrial estates are also known as industrial parks and trading estates. A more 'lightweight' version is the business park or office park, which has offices and light industry, rather than larger-scale industry.

Industrial estates can be found in a range of locations, from inner cities to rural areas. Industrial estates are usually located close to transport infrastructure, especially where more than one form of transport meet. The logic behind industrial estates includes:

- concentrating dedicated infrastructure in a small area to reduce the per-business expense of that infrastructure
- attracting new business by providing an integrated infrastructure in one location
- separating industry from residential areas to try to reduce the environmental and social impact
- eligibility of industrial estates for grants and loans under regional economic development policies.

The changing location of manufacturing

Changes in the location of manufacturing industry can be recognised at a range of scales:

- The global shift in manufacturing industry from the developed world to NICs and developing countries has already been mentioned as part of the process of globalisation (see Topic 3.1).
- Within each country, rich or poor, there are areas where manufacturing is highly concentrated and other regions where it is largely absent. In the USA the north-east 'manufacturing belt', which covers only one-eighth of the country, has over 35 per cent of all manufacturing jobs although one hundred

years ago the figure was around 70 per cent. Over the last 60 years in particular, industry in the USA has been drawn towards the 'sunbelt' states of the south and west for a number of important locational reasons. Similar concentrations can be recognised in other countries as well as changes in location over time. Everywhere the most significant locational change has been from traditional manufacturing regions, more often than not on coalfields, to higher quality of life regions offering the hard and soft infrastructural requirements of modern industry.

- Within individual regions or countries, manufacturing has historically concentrated in and around the major urban areas. However, in recent decades there has been a significant shift of industry towards **greenfield** rural locations. This movement has been so great that it is generally recognised as the most important locational change of manufacturing in the developed world since 1950.
- At the urban scale the relative shift from inner city to suburbs increased as the twentieth century progressed. Although there has been much debate about the demise of the inner city in the developed world, many would agree that the loss of employment, much of it in manufacturing, was the initiating factor in the cycle of inner city decline.

Figure 2 In 2012, Vietnam was the fourth largest footwear producer in the world after China, India and Brazil

Activities

1 Describe and explain two physical factors and two human factors that affect the location of industry.
2 Why does industrial agglomeration occur?

3 What is an industrial estate?
4 Suggest reasons for the shift of manufacturing industry from urban to rural areas.

Case study: Bangalore – India's high-tech city

Bangalore (Figure 3) is the most important city in India for high-technology industry. Bangalore's pleasant climate, moderated by its location on the Deccan Plateau over 900 m above sea level, is a significant attraction to foreign and domestic companies alike. Known as the 'Garden City', Bangalore claims to have the highest quality of life in the country. Because of its dust-free environment, large public sector undertakings such as Hindustan Aeronautics Ltd and the Indian Space Research Organisation were established in Bangalore by the Indian government. In addition, the state government has a long history of support for science and technology. There are many colleges of higher education in this sector and there has been large-scale investment in science and technology parks. The city prides itself on a 'culture of learning' which gives it an innovative leadership within India.

In the 1980s Bangalore became the location for the first large-scale foreign investment in high technology in India when Texas Instruments selected the city above a number of other possibilities. Other TNCs soon followed as the reputation of the city grew. Important backward and forward linkages were steadily established over time. This was a classic example of the process of cumulative causation. Apart from ICT industries, Bangalore is also India's most important centre for aerospace and biotechnology.

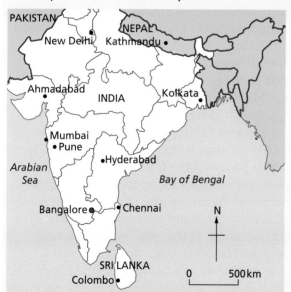

Figure 3 Location of Bangalore

India's ICT sector has benefited from the filtering down of business from the developed world. Many European and North American companies which previously outsourced

their ICT requirements to local companies are now using Indian companies. Outsourcing to India occurs because:

- labour costs are considerably lower
- a number of developed countries have significant ICT skills shortages
- India has a large and able English-speaking workforce (there are about 80 million English speakers in India).

Since 1981, Bangalore's population has grown from 2.4 million to over 9.6 million in 2011. The city has grown into a major international hub for ICT companies. It has the nickname of the 'Silicon Valley of India'. Bangalore has steadily built up a large pool of highly skilled labour that can undertake a wide range of complex tasks in high-technology industries. There has been very high investment into the city's infrastructure to accommodate such a high rate of expansion. The city's landscape has changed dramatically, with many new glass and steel skyscrapers and numerous cybercafes.

Bangalore

- Bangalore is the fourth largest technology cluster in the world after Silicon Valley, Boston and London.
- The number of ICT companies increased from 13 in 1991 to 2200 in 2013.
- The ICT industry is divided into three main clusters: Electronics City, International Technology Park and the Software Technology Park. New, smaller clusters have emerged in recent years.
- Major companies include Hewlett Packard, Siemens, Tata Consulting Services (TCS), Infosys Technologies, Wipro and Kshema Technologies.
- The city has attracted outsourcing right across the IT spectrum from software development to IT enabled services.
- The city boasts 21 engineering colleges.
- NASDAQ, the world's biggest stock exchange, opened its third international office in Bangalore in 2001.
- 80% of global ICT companies have based their India operations and R&D centres in Bangalore.
- Companies in Bangalore employ about 35% of India's pool of ICT professionals.
- Bangalore accounts for half of the 260 biotechnology companies in India.

Figure 4 Bangalore factfile

Case study analysis

1 Describe the location of Bangalore.
2 Explain the reasons for the development of Bangalore as a major international ICT hub.

3.4 Tourism

Expedition cruise ship off the coast of South Georgia, South Atlantic

Key questions

- What are the reasons for the growth of tourism in relation to the main attractions of the physical and human landscape?
- What are the benefits and disadvantages of tourism to receiving areas?
- Why is careful management of tourism required in order for the industry to be sustainable?

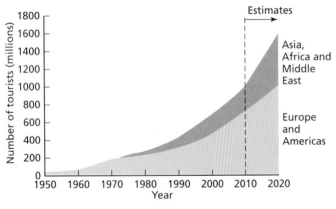

Figure 1 Growth in global tourism

Over the last 50 years tourism has developed into a major global industry which is still expanding rapidly (Figure 1). It is one of the major elements in the process of globalisation. Tourism is defined as travel away from the home environment (a) for leisure, recreation and holidays, (b) to visit friends and relations (VFR), and (c) and for business and professional reasons.

● The growth of tourism

Tourism has developed in response to the main attractions of the physical and human landscape. The medical profession was largely responsible for the growth of taking holidays away from home. During the seventeenth century doctors increasingly began to recommend the benefits of mineral waters, and by the end of the eighteenth century there were hundreds of spas in existence in Britain. Bath and Tunbridge Wells were among the most famous (Figure 2). The second stage in the development of holiday locations was the emergence of the seaside resort. Sea bathing is usually said to have begun at Scarborough in Britain about 1730. Those who could afford it were beginning to appreciate coastal landscapes in a new way.

Figure 2 The historical mineral waters in the spa town of Bath

The annual holiday, away from work, for the masses was a product of the Industrial Revolution, which brought big social and economic changes. However, until the latter part of the nineteenth century only the very rich could afford to take a holiday away from home.

The first **package tours** were arranged by Thomas Cook in 1841 in the UK. These took travellers from Leicester to Loughborough, 19 km away, to attend temperance (abstinence from alcoholic drink) meetings. At the time it was the newly laid railway network that provided the transport infrastructure for Cook to expand his tour operations. Of equal importance was the emergence of a significant middle-class with time and money to spare for extended recreation. It was not long before such activities spread to other countries. There was a growing appreciation of what human landscapes could offer, in particular the attractions of large cities such as Paris, Rome and London.

By far the greatest developments have occurred since the end of the Second World War, arising from the substantial growth in leisure time, affluence and mobility enjoyed in developed countries. However, it took the jet plane to herald the era of international mass tourism. In 1970 when Pan Am flew the first Boeing 747 from New York to London, scheduled planes carried 307 million passengers. By 2012 the number had reached 2.9 billion.

Reasons for the growth of global tourism

Table 1 shows the range of factors responsible for the growth of global tourism. More and more people have become aware of the attractions of the physical and human landscape in their own country and abroad, and rising living standards have allowed an increasing number of people to experience such attractions.

Table 1 Factors affecting global tourism

Economic
- Steadily rising real incomes
- The decreasing real costs of holidays
- Widening range of destinations within the middle-income range
- Heavy marketing of shorter foreign holidays aimed at those who have the time and disposable income to take an additional break
- Expansion of budget airlines
- 'Air miles' and other retail reward schemes aimed at travel and tourism
- 'Globalisation' has increased business travel considerably

Social
- Increase in the average number of days of paid leave
- Increasing desire to experience different cultures and landscapes
- Raised expectations of international travel with increasing media coverage of holidays, travel and nature
- High levels of international migration over the last decade or so, which means that more people have relatives and friends living abroad

Political
- Many governments have invested heavily to encourage tourism
- Government backing for major international events such as the Olympic Games and the World Cup

Recent data

In 2012 international tourist arrivals (overnight visitors) worldwide exceeded 1 billion for the first time ever, reaching a total of 1035 million tourists. In 1950 there were only 25 million international tourists. The World Tourism Organization forecasts an increase to 1.8 billion in 2030. International tourism receipts reached $1075 billion in 2012. Tourism accounts for 9 per cent of global GDP and one in eleven jobs. Europe remains the world region with the greatest number of both tourist arrivals and tourism receipts (Figure 3). People from developed countries still dominate global tourism but many emerging economies have shown very fast growth rates in recent years. When people can afford to travel they usually do. **Tourist generating countries** have a big impact on the flow of money around the world.

Fifty-two per cent of inbound tourism is for the purpose of leisure, recreation and holidays (Figure 4). The second most important reason is visiting friends and relatives. Inbound tourism by mode of transport in 2012 comprised:

- air 52 per cent
- road 40 per cent
- water 6 per cent
- rail 2 per cent.

Seasonality is the major problem with tourism as a source of employment, having a major impact on incomes and the quality of life during the less popular times of the year. Many popular tourist destinations try to extend the tourist season by staging music festivals and other events.

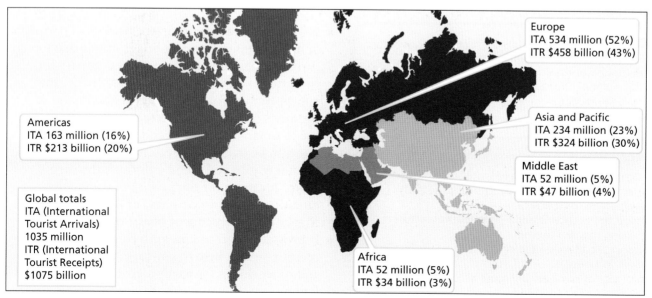

Europe
ITA 534 million (52%)
ITR $458 billion (43%)

Americas
ITA 163 million (16%)
ITR $213 billion (20%)

Asia and Pacific
ITA 234 million (23%)
ITR $324 billion (30%)

Middle East
ITA 52 million (5%)
ITR $47 billion (4%)

Global totals
ITA (International Tourist Arrivals)
1035 million
ITR (International Tourist Receipts)
$1075 billion

Africa
ITA 52 million (5%)
ITR $34 billion (3%)

Figure 3 International tourist arrivals and international tourism receipts, 2012

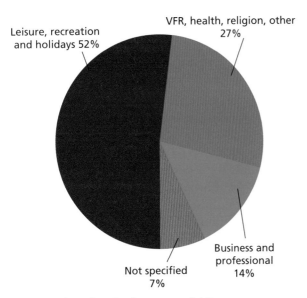

Figure 4 Inbound tourism by purpose of visit

Figure 5 The London Eye – one of the most popular tourist destinations in the UK

Interesting note

In 2012 the countries with the largest numbers of international tourist arrrivals were France (83 million), USA (67 million), China (58 million), Spain (58 million) and Italy (46 million).

Activities

1 Describe the growth of global tourism (Figure 1).
2 What were the factors responsible for the early development of tourism?
3 Discuss the economic, social and political factors affecting modern tourism.
4 Produce a bullet-point summary of the information shown in Figure 3.
5 Write a brief summary of Figure 4.

● The benefits and disadvantages of tourism to receiving areas

The economic impact

Many countries, both developed and developing, have put a high level of capital investment into tourism. This is money invested in hotels, attractions, airports, roads and other aspects of infrastructure that facilitate high-volume tourism. There has been considerable debate about the wisdom of such a strategy. Do the economic benefits outweigh the economic costs? The majority of people concerned with the tourist industry think they do, but critics of the impact of tourism have presented some strong arguments of their own. Figure 6 shows that tourism has many indirect as well as direct effects.

> What is thought of as the 'tourism industry' is only the tip of the iceberg
>
> **Tourism industry direct effect**
> Accommodation, recreation, catering, entertainment, transportation
>
> **Tourism economy (indirect effect)**
> Aircraft manufacturing, chemicals, computers, concrete, financial services, foods and beverages, furniture and fixtures, iron/steel, laundry services, metal products, mining, oil/gas suppliers, plastics, printing/publishing, rental car manufacturing, resort development, sanitation services, security, ship building, suppliers, textiles, utilities, wholesalers, wood

Figure 6 The direct and indirect economic impact of the tourist industry

Supporters of the development potential of tourism put forward the following arguments:

● It is an important factor in the balance of payments of many nations. Tourism brings in valuable foreign currency. This foreign currency is necessary for countries to pay for the goods and services they import from abroad. Many small developing countries have few other resources that they can use to obtain foreign currency.
● Tourism benefits other sectors of the economy, providing jobs and income through the supply chain. It can set off the process of cumulative causation whereby one phase of investment can trigger other subsequent phases of investment.
● It provides governments with considerable tax revenues which help to pay for education, health and other things for which a government has to find money.

- By providing employment in rural areas it can help to reduce rural-to-urban migration. Such migration is a major problem in many developing countries.
- A major tourism development can act as a **growth pole**, stimulating the economy of the larger region.
- It can create openings for small businesses such as taxi firms, beach facility hire companies and small cafés.
- It can support many jobs in the informal sector, which plays a major role in the economy of many developing countries.

Figure 7 Beach artist, Agadir, Morocco – an example of informal sector employment

However, critics argue that the value of tourism is often overrated because of the following:

- **Economic leakages** (Figure 8) from developing to developed countries run at a rate of between 60 and 75 per cent. Economic leakages are the part of the money a tourist pays for a foreign holiday that does not benefit the destination country because it goes elsewhere. With cheap package holidays, by far the greater part of the money paid stays in the country where the holiday was purchased.

Figure 8 Economic leakages

- Tourism is labour intensive, providing a range of jobs especially for women and young people. However, most local jobs created are menial, low-paid and seasonal. Overseas labour may be brought in to fill middle and senior management positions.
- Money borrowed to invest in the necessary infrastructure for tourism increases the national debt.
- At some destinations tourists spend most of their money in their hotels with minimum benefit to the wider community.
- Tourism might not be the best use for local resources which could in the future create a larger **multiplier effect** if used by a different economic sector.
- Locations can become over-dependent on tourism which causes big problems if visitor numbers fall.
- The tourist industry has a huge appetite for resources which often impinge heavily on the needs of local people. A long-term protest against tourism in Goa highlighted the fact that one five-star hotel consumed as much water as five local villages, and the average hotel resident used 28 times more electricity per day than a local person.
- International trade agreements such as the General Agreement on Trade in Services (GATS) allow the global hotel giants to set up in most countries. Even if governments favour local investors there is little they can do.

Figure 9 Cruise ship on the River Nile – tourism is Egypt's main source of foreign currency

The social and cultural impact

The traditional cultures of many communities in the developing world have suffered because of the development of tourism. The disadvantages include the following:

- the loss of locally owned land as tourism companies buy up large tracts of land in the most scenic and accessible locations
- the abandonment of traditional values and practices
- displacement of people to make way for tourist developments
- changing community structure – communities that were once very close socially and economically may be weakened considerably due to a major outside influence such as tourism

- abuse of human rights by large companies and governments in the quest to maximise profits
- alcoholism and drug abuse as drink and drugs become more available to satisfy the demands of foreign tourists
- crime and prostitution, sometimes involving children – 'sex tourism' is a big issue in certain locations such as Bangkok, but it is also present in some degree in most locations visited by large numbers of international tourists
- visitor congestion at key locations, hindering the movement of local people
- local people denied access to beaches to provide 'exclusivity' for visitors
- loss of housing for local people as more visitors buy second homes in popular tourist areas (Figure 10).

Figure 11 shows how the attitudes to tourism can change over time. An industry which is usually seen as very beneficial initially can eventually become the source of considerable irritation, particularly where there is a big clash of cultures.

1 **Euphoria**
  - Enthusiasm for tourist development
  - Mutual feeling of satisfaction
  - Opportunities for local participations
  - Flows of money and interesting contacts
2 **Apathy**
  - Industry expands
  - Tourists taken for granted
  - More interest in profit making
  - Personal contact becomes more formal
3 **Irritation**
  - Industry nearing saturation point
  - Expansion of facilities required
  - Encroachment into local way of life
4 **Antagonism**
  - Irritations become more overt
  - The tourist is seen as the harbinger of all that is bad
  - Mutual politeness gives way to antagonism
5 **Final level**
  - Environment has changed irreversibly
  - The resource base has changed and the type of tourist has also changed
  - If the destination is large enough to cope with mass tourism it will continue to thrive

Figure 11 Doxey's 'Index of Irritation' caused by tourism

However, tourism can also have positive social and cultural impacts:

- Tourism development can increase the range of social facilities for local people.
- It can lead to greater understanding between people of different cultures.
- Family ties may be strengthened by visits to relatives living in other regions and countries.
- Visiting ancient sites can develop a greater appreciation of the historical legacy of host countries.
- It can help develop foreign language skills in host communities.
- It may encourage migration to major tourist generating countries.
- A multitude of cultures congregating together for major international events such as the Olympic Games can have a very positive global impact.

The tourist industry and the various scales of government in host countries have become increasingly aware of the problems the industry creates. They are now using a range of management techniques in an attempt to mitigate such effects. Education is the most important element so that visitors are made aware of the most sensitive aspects of the host culture.

Figure 10 Entrance to a National Park in Andalucia, Spain. The graffiti refers to the number of foreigners buying up houses in the nearby village of Frigiliana.

Figure 12 Armed tourism police on a Nile cruise ship

Activities

1 Compare the direct and indirect effects of tourism.
2 Explain how economic leakages occur.
3 Explain the sequence of changes illustrated in Doxey's Index (Figure 11).
4 Research the social impact of international tourism in one destination.

The management and sustainability of tourism

Tourism has reached such a large scale in so many parts of the world that it can only continue with careful management. In most popular tourist destinations the stated objective is that tourism should be sustainable. However, sustainable tourism strategies have been much more successful in some areas than others. **Sustainable tourism** is tourism organised in such a way that its level can be sustained in the future without creating irreparable environmental, social and economic damage to the receiving area.

As the level of global tourism increases rapidly it is becoming more and more important for the industry to be responsibly planned, managed and monitored. Tourism operates in a world of finite resources where its impact is becoming of increasing concern to a growing number of people. At present, only about 5 per cent of the world's population have ever travelled by plane. However, this is undoubtedly

going to increase substantially, putting even greater pressure on tourist destinations.

Environmental groups are keen to make travellers aware of their **destination footprint**. This is the environmental impact caused by an individual tourist on holiday in a particular destination.

They are urging people to:

- 'fly less and stay longer'
- carbon-offset their flights
- consider 'slow travel'.

Tourists might consider the impact of their activities both for individual holidays but also in the longer term. For example, they may decide that every second holiday will be in their own country (not using air transport). It could also involve using locally run guesthouses and small hotels as opposed to hotels run by international chains. This enables more money to remain in local communities.

Figure 13 Sand dune restoration works, County Kerry, Ireland

Virtually every aspect of the industry now recognises that tourism must become more sustainable. **Ecotourism** is at the leading edge of this movement. Ecotourism is a specialised form of tourism where people experience relatively untouched natural environments such as coral reefs, tropical forests and remote mountain areas, and ensure that their presence does no further damage to these environments.

Figure 14 Combating severe informal footpath erosion on Mt Vesuvius, Italy

Protected areas

Over the course of the last 130 years or so, more and more of the world's most spectacular and ecologically sensitive areas have been designated for protection at various levels. The world's first National Park was established at Yellowstone in the USA in 1872. Now there are well over 1000 worldwide. Many countries have National Forests, Country Parks, Areas of Outstanding Natural Beauty, World Heritage Sites and other designated areas which merit special status and protection. Wilderness Areas with the greatest restrictions on access have the highest form of protection.

In many countries and regions there are often differences of opinion when the issue of special protection is raised. For example, in some areas jobs in mining, forestry and tourism may depend on developing presently unspoilt areas. So it is not surprising that values and attitudes can differ considerably when big decisions about the future of environmentally sensitive areas are being made. Often, a clear distinction has to be made between the objectives of **preservation** and **conservation**. Preservation is maintaining a location exactly as it is and not allowing development. Conservation is allowing for developments that do not damage the character of a location.

Tourist hubs

The concept of tourism hubs or clusters is a model that has been applied in a number of locations. The idea is to concentrate tourism and its impact in one particular area so that the majority of the region or country feels little of the negative impacts of the industry. Benidorm in Spain and Cancun in Mexico are examples where this model was adopted but both locations show how difficult it is to confine tourism within preconceived boundaries as the number of visitors increases and people want to travel beyond the tourist enclaves.

Quotas

Quotas seem to be one of the best remedies on offer. The UK Centre for Future Studies has suggested a lottery-based entrance system, an idea endorsed by Tourism Concern. Here, the number of visitors would not be allowed to exceed a sustainable level. This is an idea we are likely to hear much more about in the future.

Ecotourism in Ecuador

Ecuador's tourism strategy has been to avoid becoming a mass market destination but to market 'quality' and 'exclusivity' instead, in as eco-friendly a way as possible.

Ecotourism has helped to bring needed income to some of the poorest parts of the country. It has provided local people with a new alternative way of making a living. As such it has reduced human pressure on ecologically sensitive areas.

The main geographical focus of ecotourism has been in the Amazon rainforest around Tena, which has become the main access point. The ecotourism schemes in the region are usually run by small groups of indigenous Quichua Indians (Figure 15).

The Quichua people insist that all visitors must abide by certain rules and regulations

Exchanges of clothing or other personal items with community members are not allowed. Nor are community members allowed to accept gifts.

Avoid any displays of affection, even with close friends. In this community it is considered rude to hold hands or kiss in public.

When walking in the rainforest:
- do not touch any branches without looking carefully first. They may carry thorns, dangerous insects or even snakes
- do not pull on branches or vines – they may fall down on top of you.

If you need to go to the toilet, and facilities are not immediately available, go to the side of the rainforest track, never in or near a stream or lake.

Visitors should never go off alone. It's easy to get lost in the rainforest.

All rubbish, e.g. empty bottles and tubes, must be taken away by visitors.

Do not enter people's houses without being invited in. Do not make promises you may not be able to keep, e.g. to send back photos after the visit.

Always check first before touching plants or animals. They may cause a rash, or sting you. Do not collect plants, insects or animals unless you have permission.

Figure 15 Ecotourism in Ecuador's rainforest

Activities

1 Define 'sustainable tourism'.
2 What do you understand by the term 'destination footprint'?
3 Which environments in the region in which you live are protected and why?
4 What do you think of the idea of quotas for visitor numbers at certain locations?
5 What do you understand by the concept 'slow travel'?
6 Describe the operation of ecotourism in Ecuador's rainforest.

Case study: Jamaica – the benefits and disadvantages associated with the growth of tourism

Economic importance

Tourism has become an increasingly vital part of Jamaica's economy in recent decades. The contribution of tourism to total employment and GDP has risen substantially. It has brought considerable opportunities to its population, although it has not been without its problems. Jamaica has been determined to learn from the 'mistakes' of other countries and ensure that the population will gain real benefits from the growth of tourism.

Tourism's direct contribution to GDP in 2012 amounted to almost $1.3 billion or 8.4 per cent of total GDP. Adding all the indirect economic benefits increased the figure to almost $4.1 billion or 27.4 per cent of total GDP. Direct employment in the industry amounted to 90 000 but the overall figure which includes indirect employment is over three times as large. In the most popular tourist areas the level of reliance on the industry is extremely high.

Tourism is the largest source of foreign exchange for the country. The revenue from tourism plays a significant part in helping central and local government fund economic and social policies. Special industry taxes have gone directly into social development, healthcare and education, all of which are often referred to as 'soft infrastructure'. However, tourism has also spurred the development of 'hard infrastructure' such as roads, telecommunications and airports. Also, as attitudes within the industry itself are changing, larger hotels and other aspects of the industry have become more socially conscious. Classic examples are the funding of local social projects.

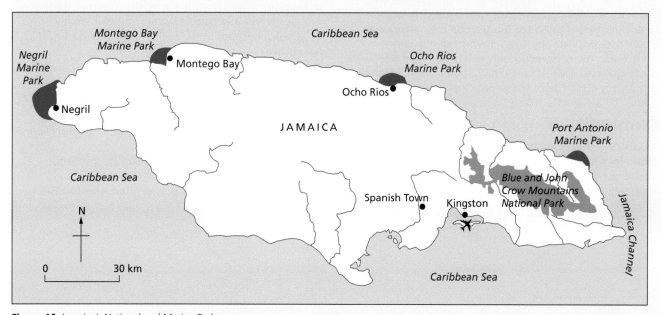

Figure 16 Jamaica's National and Marine Parks

National Parks and ecotourism

Figure 16 shows the location of Jamaica's National and Marine Parks. Further sites have been identified for future protection. The Jamaican government sees the designation of the parks as a positive environmental impact of tourism. Entry fees to the Parks pay for conservation. The desire of tourists to visit these areas and the need to conserve the environment to attract future tourism drives the designation and management process.

The marine parks are attempting to conserve the coral reef environments off the coast of Jamaica. They are at risk from damage from overfishing, industrial pollution and mass tourism. The Jamaica Conservation and Development Trust is responsible for the management of the National Parks, while the National Environmental Planning Agency has overseen the Government's sustainable development strategy since 2001.

Figure 17 A beach fringed with palm trees in Montego Bay Marine Park

Ecotourism is a developing sector of the industry with, for example, raft trips on the Rio Grande river increasing in popularity. Tourists are taken downstream in very small groups. The rafts, which rely solely on manpower, leave singly with a significant time gap between them to minimise any disturbance to the peace of the forest.

Community tourism

Considerable efforts are being made to promote **community tourism** so that more money filters down to the local population and small communities. The Sustainable Communities Foundation through Tourism (SCF) programme has been particularly active in central and south-west Jamaica. Community tourism is seen as an important aspect of **pro-poor tourism**. This is tourism that results in increased net benefits for poor people.

The Jamaica Tourist Board (JTB) is responsible for marketing the country abroad. Recently it used the fact that Jamaica was one of the host countries for the 2007 Cricket World Cup to good effect. The JTB also promotes the positive aspects of Jamaica culture and the Bob Marley Museum in Kingston has become a popular attraction. Such attractions are an important part of Jamaica's objective of reducing seasonality. The physical attractions of Jamaica almost sell themselves, so the government is putting much effort in trying to boost the island's human attractions.

The disadvantages of tourism

The high or 'winter' season runs from mid-December to mid-April when hotel prices are highest. The rainy season extends from May to November. It has been estimated that 25 per cent of hotel workers are laid off during the off-season. This has a major adverse impact on the standard of living of households reliant on the tourist industry. It also of course means that expensive tourism infrastructure is underused for part of the year.

Although seasonality is seen as the major problem associated with tourism in Jamaica, other negative aspects include:

- the environmental impact of tourism which includes traffic congestion and pollution at popular locations, and the destruction of the natural environment to make way for tourism infrastructure
- the heavy use of resources, particularly water, by hotels
- under-use of facilities in the off-season
- socio-cultural problems, illustrated by the behaviour of some tourists which clashes with the island's traditional morals – and some people have a negative image of Jamaica because of its level of violent crime and harassment.

Case study analysis

1   Explain the importance of tourism to the economy of Jamaica.
2   Describe the location of Jamaica's National and Marine Parks.
3   Define (a) 'community tourism' and (b) 'pro-poor tourism'.
4   Discuss the main problems associated with tourism.

3.5 Energy

Oil refinery, Milford Haven, UK

Key questions

- How important are non-renewable fossil fuels, renewable energy supplies, nuclear power and fuelwood; globally and in different countries at different levels of development?
- What are the benefits and disadvantages of nuclear power and renewable energy sources?

Non-renewable and renewable energy supplies

Non-renewable sources of energy are the **fossil fuels** and nuclear fuel. Fossil fuels consisting of hydrocarbons (coal, oil and natural gas) were formed by the decomposition of prehistoric organisms in past geological periods. These resources are finite so that as they are used up the supply that remains is reduced. Eventually, these non-renewable resources could become completely exhausted. The burning of fossil fuels creates considerable amounts of pollution and is the major source of greenhouse gas emissions. Climate change due to these emissions is by far the biggest environmental problem facing the planet.

Renewable energy can be used over and over again. These resources are mainly forces of nature that are **sustainable** and which usually cause little or no environmental pollution. Renewable energy includes hydro-electric, biofuels, wind, solar, geothermal, tidal and wave power.

At present, non-renewable resources dominate global energy. The challenge is to transform the global **energy mix** to achieve a better balance between renewables and non-renewables.

There is a huge gap in energy consumption between rich and poor countries. Wealth is the main factor explaining the energy gap. The use of energy can improve the quality of life in so many ways. That is why most people who can afford to buy cars, televisions and washing machines do so. However, there are other influencing factors, with climate at the top of the list.

The demand for energy has grown steadily over time. Figure 1 shows a global increase of over 60 per cent between 1987 and 2012. The fossil fuels dominate the global energy situation. Their relative contribution in 2012 was: oil 33 per cent, coal 30 per cent, natural gas 24 per cent. In contrast, hydro-electricity (HEP) accounted for 6.6 per cent and nuclear energy 4.5 per cent. Figure 1 includes commercially traded energy only. It excludes fuels such as wood, peat and animal waste which, though important in many countries, are unreliably documented in terms of consumption statistics.

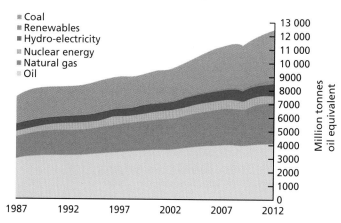

Figure 1 Changes in world energy consumption by type, 1987–2012

Figure 2 Aircraft refuelling at Gatwick airport, UK

Consumption by type of fuel varies widely by world region (Figure 3):

- **Oil** – only in Asia Pacific is the contribution of oil less than 30 per cent and it is the main source of energy in four of the six regions shown in Figure 3. In the Middle East it accounts for almost 50 per cent of consumption.
- **Coal** – only in the Asia Pacific region is coal the main source of energy. In contrast it accounts for less than 5 per cent of consumption in the Middle East and South and Central America. China was responsible for 50.2 per cent of global coal consumption in 2012.
- **Natural gas** – natural gas is the main source of energy in Europe and Eurasia and it is a close second to oil in the Middle East. Its lowest share of the energy mix is 11 per cent in Asia Pacific.
- **Hydro-electricity** – the relative importance of hydro-electricity is greatest in South and Central America (25 per cent). Elsewhere its contribution varies from 6 per cent in Africa to less than 1 per cent in the Middle East.
- **Nuclear energy** – nuclear energy is not presently available in the Middle East and it makes the smallest contribution of the five traditional energy sources in Asia Pacific, Africa and South and Central America. It is most important in Europe and Eurasia and North America.
- **Renewables** – consumption of renewable energy other than HEP is rising rapidly, but from a very low base. Renewables made the largest relative contribution to energy consumption in Europe and Eurasia.

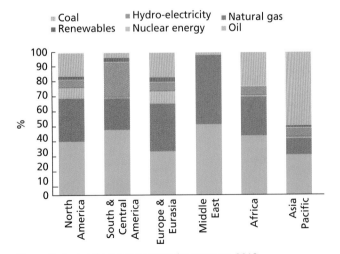

Figure 3 Regional energy consumption patterns, 2012

Figure 4 Fuel station on the river Amazon

Figure 5 shows per capita energy consumption around the world. The highest consumption countries, such as the USA and Canada, use more than 6 tonnes oil equivalent per person, while almost all of Africa and much of South America and Asia uses less than 1.5 tonnes oil equivalent per person. Figure 6 is a model showing the relationship between resource use in general and the level of economic development. This model applies well to energy usage, with the newly industrialised countries having the highest rates of growth.

In terms of usage by type of energy, some general points can be made:

- The most developed countries tend to use a wide mix of energy sources as they are able both to invest in domestic energy potential and to buy energy from abroad.
- The high investment required for nuclear electricity means that only a limited number of countries produce electricity this way. However, many countries that could afford the investment chose not to adopt this strategy.
- Richer nations have been able to invest more money in renewable sources of energy.

In the poorest countries fuelwood is an important source of energy, particularly where communities have no access to electricity.

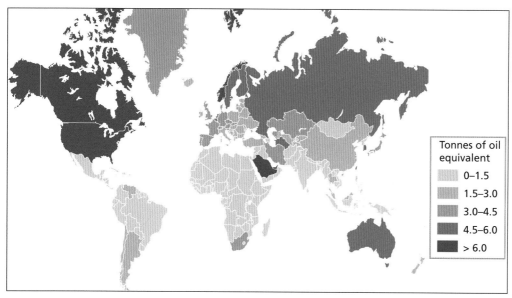

Figure 5 Energy consumption per capita, 2012

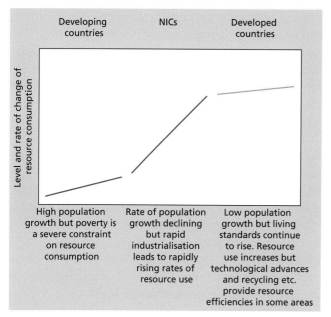

Figure 6 Model of the relationship between resource use and the level of economic development

Fuelwood in developing countries

In developing countries about 2.5 billion people rely on fuelwood, charcoal and animal dung for cooking. Fuelwood and charcoal are collectively called fuelwood, which accounts for just over half of global wood production. Fuelwood provides much of the energy needs for Sub-Saharan Africa. It is also the most important use of wood in Asia. In 2010, 1.2 billion people were still living without electricity. Figure 7 shows the countries with the greatest number of people lacking access to electricity.

Although at least one study claims that the global demand for fuelwood peaked in the mid-1990s, there can be no doubt that there are severe shortages in many countries. This is a major factor in limiting development.

In developing countries the concept of the 'energy ladder' is important. Here, a transition from fuelwood and animal dung to 'higher level' sources of energy occurs as part of the process of economic development. Income, regional electrification and household size are the main factors influencing the demand for fuelwood. Forest depletion is initially heavy near urban areas but slows down as cities become wealthier and change to other forms of energy. It is the more isolated rural areas that are most likely to lack connection to an electricity grid. In such areas the reliance on fuelwood is greatest. Wood is likely to remain the main source of fuel for the global poor in the foreseeable future.

The collection of fuelwood does not cause deforestation on the same scale as the clearance of land for agriculture, but it can seriously deplete wooded areas. The use of fuelwood is the main cause of indoor air pollution in developing countries.

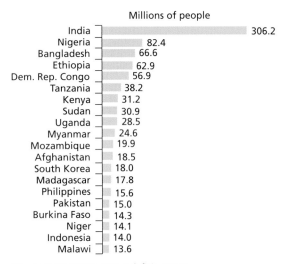

Millions of people

India	306.2
Nigeria	82.4
Bangladesh	66.6
Ethiopia	62.9
Dem. Rep. Congo	56.9
Tanzania	38.2
Kenya	31.2
Sudan	30.9
Uganda	28.5
Myanmar	24.6
Mozambique	19.9
Afghanistan	18.5
South Korea	18.0
Madagascar	17.8
Philippines	15.6
Pakistan	15.0
Burkina Faso	14.3
Niger	14.1
Indonesia	14.0
Malawi	13.6

Figure 7 Electricity access deficit, 2010

Interesting note

The individual countries consuming the most energy in 2012, as a percentage of the world total, were: China (21.9), USA (17.7), Russia (5.6), India (4.8) and Japan (3.8).

Activities

1 List the non-renewable sources of energy.
2 Describe the changes in world energy consumption shown in Figure 1.
3 To what extent do the types of energy consumption vary by world region?
4 Provide a bullet-point summary of Figure 5. Refer to all classes in the key.
5 a What is fuelwood?
  b Why is it such an important source of energy in the developing world?

● The benefits and disadvantages of nuclear power and renewable energy

Nuclear power

Until a few years ago the future of nuclear power looked bleak, with a number of countries apparently 'running down' their nuclear power stations and many other nations firmly set against the idea of introducing nuclear electricity. However, heightened fears about oil supplies, energy security and climate change have brought this controversial source of power back onto the global energy agenda.

No other source of energy creates such heated discussion as nuclear power. Concerns include the following:

● Power plant accidents – could release radiation into air, land and sea.
● Radioactive waste storage/disposal – most concern is over the small proportion of 'high-level waste'. This is so radioactive that it generates heat and corrodes all containers. It would cause death within a few days to anyone directly exposed to it. In the UK this amounts to about 0.3 per cent of the total volume of all nuclear waste. However, it accounts for about half the total radioactivity. No country has yet implemented a long-term solution to the nuclear waste problem. The USA and Finland have plans to build waste repositories deep underground in areas of known geological stability.
● Rogue state or terrorist use of nuclear fuel for weapons – as the number of countries with access to nuclear technology rises, such concerns are likely to increase. An interim report published in December 2008 by the US Congressional Commission on the Strategic Posture of the United States concluded, 'It appears that we are at a "tipping point" in nuclear proliferation. Part of the concern is that some countries which claim to be developing nuclear electricity only may well put themselves in a position to develop nuclear weapons.'
● High construction and decommissioning costs – recent estimates put an average price of about $6.3 billion on a new nuclear power plant. When a nuclear plant has come to the end of its useful life, the costs of decommissioning are high.
● Because of the genuine risks associated with nuclear power and the level of security secrecy required, it is seen by some people as less 'democratic' than other sources of power.
● The possible increase in certain types of cancer near nuclear plants – there has been much debate about this issue, but the evidence appears to be becoming more convincing.

At one time the rise of nuclear power looked unstoppable. However, a serious incident at the Three Mile Island nuclear power plant in Pennsylvania USA in 1979, and the much more serious Chernobyl disaster in the Ukraine in 1986, brought any growth in the industry to a virtual halt. No new nuclear power plants have been ordered in the USA since then, although public opinion has become more favourable in recent years as (a) Three Mile Island and Chernobyl recede into the past and (b) worries about polluting fossil fuels increase. Most of the recent nuclear power plants constructed have been in Asia.

The advantages of nuclear power are:

- Zero emissions of greenhouse gases – this has become more and more important as concern about climate change has risen. Along with hydropower, nuclear electricity is the major source of 'carbon-free' energy used today.
- Reduced reliance on imported fossil fuels – more countries have become concerned about energy security. Energy insecurity may lead to increased geopolitical tension and the potential for conflict as consumers attempt to secure supplies. This will be most likely within a 'business as usual' framework of reliance on fossil fuels. Nuclear power is seen by a number of governments as a tried and tested way of reducing reliance on energy imports. France is a classic example of how this has been done.
- Nuclear power is not as vulnerable to fuel price fluctuations as oil and gas – uranium, the fuel for nuclear plants, is relatively plentiful. Most of the main uranium mines are in politically stable countries.
- In recent years nuclear plants have demonstrated a very high level of reliability and efficiency as technology has advanced and experience has been built up.
- Nuclear technology has spin-offs in fields such as medicine and agriculture.

This decade will be crucial to the future of nuclear energy, with many countries making final decisions to extend or begin their nuclear electricity capability. The nuclear energy issue is likely to be a major political battleground in some countries.

Activities

1 State three advantages and three disadvantages of nuclear power.
2 When did the nuclear accidents at Three Mile Island and Chernobyl occur?
3 Why might nuclear electricity become more important in the future?

Renewable energy supplies

Countries are eager to harness renewable energy resources to:

- reduce their reliance on often dwindling domestic fossil fuel resources
- lower their reliance on costly fossil fuel imports
- improve their energy security with higher domestic energy production
- cut greenhouse gas emissions for a cleaner environment and to satisfy international obligations.

Hydro-electricity

Of the traditional five major sources of energy, HEP is the only one that is renewable. It is by far the most important source of renewable energy. The 'big four' HEP nations of China, Brazil, Canada and the USA account for almost 53 per cent of the global total (Table 1).

Table 1 HEP consumption, 2012

Rank	Country	Million tonnes oil equivalent	% share of world total
1	China	194.8	23.4
2	Brazil	94.5	11.4
3	Canada	86.0	10.4
4	USA	63.2	7.6
5	Russia	37.8	4.5
6	Norway	32.3	3.9

Most of the best HEP locations are already in use, so the scope for more large-scale development is limited. However, in many countries there is scope for small-scale HEP plants to supply local communities.

Although HEP is generally seen as a clean form of energy, it is not without its problems which include:

- large dams and power plants can have a huge negative visual impact on the environment
- the obstruction of the river for aquatic life
- deterioration in water quality
- large areas of land may need to be flooded to form the reservoir behind the dam
- submerging large forests without prior clearance can release significant quantities of methane, a greenhouse gas.

Newer alternative energy sources

The first major wave of interest in new alternative energy sources resulted from the energy crisis of the early 1970s. However, the relatively low price of oil in the 1980s, 1990s and the opening years of the present century dampened down interest in these energy sources. Then renewed concerns about energy in recent years and corresponding price increases kick-started the alternative energy industry again.

The main drawback to the new alternative energy sources is that they invariably produce higher cost electricity than traditional sources. However, the cost gap with non-renewable energy is narrowing. Figure 8 shows the sharp increase in the consumption of renewable energy (other than HEP) in the last decade. In 2012, this accounted for 1.9 per cent of global primary energy consumption. The newer sources of renewable energy making the largest contribution to global energy supply are wind power and biofuels.

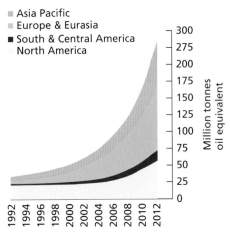

Figure 8 Renewable energy consumption by world region, 1992–2012

Wind power

The worldwide capacity of wind energy reached almost 237 GW by the end of 2011, up from 10 GW in 1998! The leaders in global wind energy are China (62 GW), the USA (47 GW), Germany (29 GW) and Spain (22 GW). Together these countries account for over 67 per cent of the world total. In recent years wind energy has reached the 'take-off' stage both as a source of energy and a manufacturing industry. The main advantages of wind energy are that compared with most other forms of renewable energy it can generate significant amounts of electricity, and it can be harnessed to a reasonable degree in most parts of the world.

Figure 9 A wind farm in California

Apart from establishing new wind energy sites, **repowering** is also beginning to play an important role. This means replacing first generation wind turbines with modern multi-megawatt turbines which give a much better performance. As wind turbines have been erected in more areas of more countries, the opposition to this form of renewable energy has increased. For example:

- People are concerned that huge turbines located nearby could blight their homes and have a significant impact on property values.
- There are concerns about the hum of turbines disturbing both people and wildlife.
- Skylines in scenically beautiful areas might be spoiled forever.
- Turbines can kill birds. Migratory flocks tend to follow strong winds but wind companies argue that they don't place turbines near migratory routes.
- Suitable areas for wind farms are often near the coast where land is both scenically beautiful and expensive.

The development of large offshore wind farms, for example in UK waters, has become an increasingly debatable issue, mainly due to the visual impact of such large installations. Safety concerns with regard to the movement of shipping are also an issue.

There has also been increasing debate about how much electricity wind turbines in many areas actually produce. There can be a big difference between the technical capacity of a wind turbine and the amount of electricity it actually produces.

Biofuels

Biofuels are fossil fuel substitutes that can be made from a range of crops including oilseeds, wheat and sugar. They can be blended with petrol and diesel. The biggest producers of biofuels are the USA, Brazil and China. By increasing biofuel production these countries have reduced the amount of oil they need to consume, which is the main reason behind biofuel production. Advocates of biofuels also argue that biofuels come from a renewable resource (crops); they can be produced wherever there is sufficient crop growth, helping energy security; and often produce cleaner emissions than petroleum-based fuels.

However, there are clear disadvantages in biofuel production. Increasing amounts of cropland have been used to produce biofuels, adding to the 'global food crisis'. Large amounts of land, water and fertilisers are needed for large-scale crop production. The manufacture of biofuels also uses significant amounts of energy, creating greenhouse gas emissions. In addition, biofuels have a lower energy output than traditional fuels. Initially, environmental groups such as Friends of the Earth and Greenpeace were very much in favour of biofuels, but as the damaging environmental consequences have become clear, such environmental organisations were the first to demand a rethink of this energy strategy.

Geothermal electricity

Geothermal energy is the natural heat found in the Earth's crust in the form of steam, hot water and hot rock (Figure 10). Rainwater may percolate several kilometres down in permeable rocks where it is heated due to the Earth's geothermal gradient. This is the rate at which temperature rises as depth below the surface increases. The average rise in temperature is about 30 °C per km, but the gradient can reach 80 °C near plate boundaries. This source of energy can be used directly for industry, agriculture, bathing and cleansing. For example, in Iceland hot springs supply water at 86 °C to 95 per cent of the buildings in and around Reykjavik.

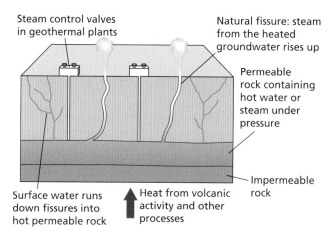

Figure 10 Geothermal power

The USA is the world leader in geothermal electricity. However, total production accounts for less than 0.4 per cent of the electricity used in the USA. Other leading countries using geothermal electricity are the Philippines, Italy, Mexico, Indonesia, Japan, New Zealand and Iceland. At present virtually all the geothermal power plants in the world operate on steam resources, and have an extremely low environmental impact.

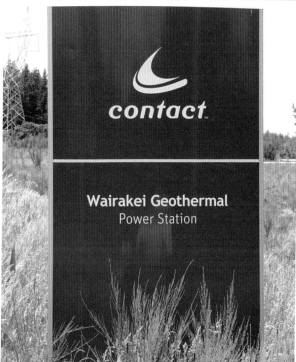

Figure 11 Geothermal power, Wairakei, New Zealand

The advantages of geothermal power for those countries that have access to this form of energy are:

- extremely low environmental impact
- geothermal plants occupy relatively small land areas
- not dependent on weather conditions (like wind and solar power)
- relatively low maintenance costs.

The limitations of this form of energy are:

- there are few locations worldwide where significant amounts of energy can be generated
- total global generation remains very small
- some of these locations are far from where the energy could be used
- installation costs of plant and piping are relatively high.

Solar power

Figure 12 Solar electricity being generated by photovoltaic panels in Spain

From a relatively small base the installed capacity of solar electricity is growing rapidly. In 2012, global solar power capacity passed 100 000 MW. This amounts to about 0.4 per cent of all global electricity generation. Experts say that solar power has huge potential for technological improvement which could make it a major source of global electricity in years to come. Germany, Italy, the USA, China and Japan currently lead the global market for solar power.

Solar electricity is currently produced in two ways:

- Photovoltaic systems – these are solar panels that convert sunlight directly into electricity.
- Concentrating Solar Power (CSP) systems – they use mirrors or lenses and tracking systems to focus a large area of sunlight into a small beam. This concentrated light is then used as a heat source for a conventional thermal power plant.

Table 2 Advantages and disadvantages of solar power

Advantages	Disadvantages
A completely renewable resource	Initial high cost of solar plants
No noise or direct pollution during electricity generation	Solar power cannot be harnessed during storms, on cloudy days or at night
Very limited maintenance required to keep solar plants running	Of limited use in countries with low annual hours of sunshine
Solar power technology is improving consistently over time and reducing costs	Large areas of land required to capture the sun's energy in order to generate significant amounts of power
Can be used in remote areas where it is too expensive to extend the electricity grid	
A generally positive public perception	

Tidal and wave power

Although currently in its infancy, a study by the Electric Power Research Institute estimated that as much as 10 per cent of US electricity could eventually be supplied by tidal energy. This potential could be equalled in the UK and surpassed in Canada. So for some countries, potential energy production from this source could be very high.

Tidal power plants act like underwater windmills, transforming sea currents into electrical current. Tidal power is more predictable than solar or wind power, and the infrastructure is less obtrusive. However, start-up costs are high. Thus, the 240 MW Rance facility in north-western France is the only utility-scale tidal power system in the world. However, the greatest potential is Canada's Bay of Fundy in Nova Scotia, but there are environmental concerns here. The main concerns are potential effects on fish populations and other marine life, levels of sedimentation building up behind facilities, and the possible impact on tides along the coast.

Because predicted building and maintenance costs are expensive, the return on investment takes a long time. Also, while generally predictable, tidal energy is not as dependable as fossil-fired or nuclear generation.

Wave energy is where generators are placed on the ocean's surface and energy levels are determined by the strength of the waves. The first experimental wave farm was opened in Portugal in 2008 at the Agucadoura Wave Park. However, due to technical problems the facility was shut down two months after opening. A number of research projects are in operation including one off the shores of Oregon in the USA. The costs and benefits of wave energy are broadly similar to those of tidal power.

Activities

1 Explain the difference between renewable and non-renewable sources of energy.
2 Give two advantages and two disadvantages for each of the following forms of renewable energy: wind, biofuels, solar, geothermal, tidal.
3 Apart from hydro-electricity, why does renewable energy contribute so little to global energy supply?
4 For the country in which you live, find out which forms of renewable energy are used and how much they contribute to total energy production.

Case study: Energy supply in China

China overtook the USA in total energy usage in 2009. The demand for energy in China continues to increase significantly as the country expands its industrial base. In 2012, China's energy consumption breakdown by energy sources was:

- coal 68.4 per cent
- oil 17.6 per cent
- hydro-electricity 7.1 per cent
- natural gas 4.7 per cent
- nuclear energy 0.8 per cent
- renewables 1.2 per cent.

China's energy policy has evolved over time. As the economy expanded rapidly in the 1980s and 1990s, much emphasis was placed on China's main energy resource, coal. China was also an exporter of oil until the early 1990s although it is now a very significant importer. Chinese investment in energy resources abroad has risen rapidly. Long-term energy security is viewed as essential if the country is to maintain the pace of its industrial revolution.

In recent years China has tried to take a more balanced approach to energy supply and at the same time reduce its environmental impact through:

- energy conservation
- placing a strong emphasis on domestic resources
- diversified energy development
- environmental protection
- mutually beneficial international cooperation.

The development of clean coal technology is an important aspect of China's energy policy. China is constructing clean coal plants at a rapid rate and gradually retiring older, more polluting power plants. China has recently built a small experimental facility near Beijing to remove carbon dioxide from power station emissions and use it to provide carbonation for beverages.

The further development of nuclear and hydropower is another important strand of Chinese policy. The country also aims to stabilise and increase the production of oil while augmenting that of natural gas and improving the national oil and gas network.

China's strategic petroleum reserve

Priority was also given to building up the national oil reserve. In 2007 China announced an expansion of its crude reserves into a two-part system. Chinese reserves would consist of a government-controlled strategic reserve complemented by mandated commercial reserves. The government-controlled reserves are being completed in three phases. This will protect China to a certain extent from fluctuations in the global oil price, which can arise for a variety of reasons.

Renewable energy policy

Total renewable energy capacity in China reached 226 GW in 2009. This included:

- 197 GW of hydro-electricity
- 25.8 GW of wind energy
- 3.2 GW of biomass
- 0.4 GW of grid-connected solar PV.

China's wind power capacity grew thirty-fold between 2005 and 2009 to become the second largest in the world behind the USA. China's wind turbine manufacturing industry is now the largest in the world. China is now also the largest manufacturer of solar PV.

The Three Gorges Dam

The Three Gorges Dam across the Yangtze river is the world's largest electricity generating plant of any kind. This is a major part of China's policy in reducing its reliance on coal. The dam is over 2 km long and 100 m high. The lake impounded behind it is over 600 km long. All of the originally planned components were completed in late 2008. There are 38 main generators giving the scheme a massive 22 500 MW generating capacity. The dam supplies Shanghai and Chongqing in particular with electricity. This is a multipurpose scheme that also increases the river's navigational capacity and reduces the potential for floods downstream. However, there was considerable opposition to the dam for a number of reasons (see p. 112).

Figure 13 The Three Gorges Dam

Case study analysis

1 When did China overtake the USA in total energy usage?
2 Describe China's energy consumption by source.
3 What are the main principles of China's current energy policy?
4 What name is given to China's stock of oil kept aside in case of an emergency?
5 Why has the Three Gorges Dam been so important to energy development in China?

Reservoir with water tower in Wales

Key questions
- What are the methods of water supply?
- How does the use of water vary between countries at different levels of economic development?
- What are the reasons for water shortages in some areas?
- Why is careful management required to ensure future supplies of water?

● The global water crisis

The longest a person can survive without water is about ten days. All life and virtually every human activity needs water. It is the world's most essential resource and a pivotal element in poverty reduction. But for about 80 countries, with 40 per cent of the world's population, lack of water is a constant threat. And the situation is getting worse, with demand for water doubling every 20 years. In those parts of the world where there is enough water, it is being wasted, mismanaged and polluted on a grand scale. In the poorest nations it is not just a question of lack of water; the paltry supplies available are often polluted.

● Methods of water supply

Water supply is the provision of water by public utilities, commercial organisations or by community endeavours. The objective in all cases is to supply water from its source to the point of usage. In 2010, about 85 per cent of the global population had access to a piped water supply through house connections or an improved water source through other means

than house supply, including standpipes, water kiosks, spring supplies and protected wells. This left almost 900 million people who did not have access to an improved water source and had to use unprotected wells or springs, canals, lakes or rivers for their water needs.

Dams and reservoirs

In the twentieth century, global water consumption grew sixfold, twice the rate of population growth. Much of this increased consumption was made possible by significant investment in water infrastructure, particularly dams and reservoirs affecting nearly 60 per cent of the world's major river basins. Figure 1 shows water supply and management methods in the large Canadian province of Alberta where water supply is a concern in many parts of the region.

A dam is a barrier that holds back water. Dams are mainly used to save, manage and prevent the flow of excess water into specific regions. They may also be used to generate hydro-electricity and provide road bridges across valleys. A reservoir is an artificial lake primarily used for storing water. Not all reservoirs are held behind dams, but the really large ones usually are. These are 'on channel' reservoirs where a dam has been built across an existing river (Figure 1). In contrast, 'off channel' reservoirs usually use depressions in the existing landscape or human-dug depressions to store water. They may be in close proximity to rivers so that water can be moved from one to the other, depending on whether storage or supply is the immediate objective.

The world's major dams are really massive structures capable of holding huge amounts of water in the reservoirs behind them. The volume of water in Lake Kariba, held behind the Kariba Dam in Zimbabwe, is a staggering 180.6 km^3! This water can be released gradually as and when required by the settlements downstream of the dam. Reservoir storage needs have increased as world population has grown. There are approximately 80 000 dams of varying sizes in the USA alone. Globally the construction of dams has declined since the height of the era in the 1960s and 1970s. This is because most of the best sites for dams are already in use or such sites are strongly protected by environmental legislation and therefore off-limits for construction.

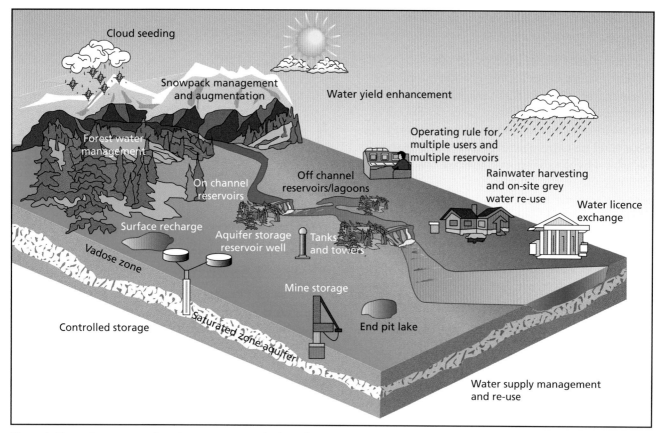

Figure 1 Alternative water supply and management methods in Alberta, Canada

An alternative to building new dams and reservoirs is to increase the capacity of existing reservoirs by extending the height of the dam. For example, the San Vicente Dam Raise Project in southern California is adding 36 m to the existing 67 m structure. At a cost of $530 million it will more than double the current capacity of the reservoir.

Wells and boreholes

A well or borehole is a means of tapping into various types of aquifers (water-bearing rocks), gaining access to groundwater. They are sunk directly down to the water table. The water table is the highest level of underground water. For many communities groundwater is the only water supply source. Aquifers provide approximately half of the world's drinking water, 40 per cent of the water used by industry and up to 30 per cent of irrigation water.

Typically a borehole is drilled by machine and is relatively small in diameter. Wells are relatively large in diameter and are often sunk by hand, although machinery may be used.

Water from groundwater sources can be used directly or stored (Figure 1) to build up a considerable surface supply. Figure 2 shows two types of artesian well. These are wells sunk through impermeable strata into strata receiving water from an area at a higher altitude than that of the well, so that there is sufficient pressure to force water to flow upwards. The well on the left in Figure 2 will require pumping or manually drawn buckets to bring water from the level to which it rises to the surface. The artesian well on the right is below the water table, so water will flow to the surface unaided (a flowing artesian well). About 35 per cent of all public water supply in England and Wales comes from groundwater. Groundwater is even more important in arid and semi-arid areas. This is the main source of water in oasis settlements such as those in the Sahara desert in north Africa.

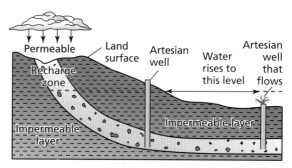

Figure 2 An aquifer and artesian wells

Desalination: the answer to water shortages?

Desalination plants are in widespread use in the Middle East where other forms of water supply are extremely scarce. Most of these plants distill water by boiling, generally using waste gases produced by oil wells. Without the availability of waste energy the process would be extremely expensive. This is the main reason why desalination plants are few and far between outside of the Middle East.

However, another method of desalination does exist. Originally developed in California in the mid-1960s for industrial use, the 'reverse osmosis' technique is now being applied to drinking water. Recent advances have substantially reduced the cost of reverse-osmosis systems. Large-scale systems using this new technology have been built in Singapore and Florida.

The sea water will still have to undergo conventional filter treatment to rid it of impurities such as microbes pumped into the sea from sewage plants. Thus it is likely that even when the technology has been highly refined, desalinated water will always be more expensive than water from conventional sources. However, desalination does have other advantages:

- It does not affect water level in rivers.
- It could mean that controversial plans for new reservoirs could be shelved.

However, desalination plants are expensive and do not offer a viable solution to the poorest countries unless costs can be drastically reduced.

Cloud seeding

Cloud seeding is a technique used to increase rainfall (or snowfall) in an area. It can be used directly over an agricultural area where rainfall is required immediately, or mountain or 'orogenic' cloud seeding can be used for snowpack augmentation, particularly in snowmelt-dominated basins like those originating in the Rocky Mountains in the USA and Canada (Figures 1 and 3). The more snow that falls in winter, the more water from snowmelt in spring. Cloud seeding is also sometimes used in major ski resorts to increase snowfall.

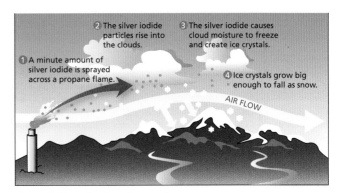

Figure 3 How cloud seeding works

Other methods of water supply

Other methods of water supply, shown on Figure 1, are largely self-explanatory:

- Forest water management can be very important in many areas. Land management activities can affect water flow and degrade the quality of water. Many countries rely on 'protection forests' to preserve the quality of drinking water supplies, alleviate flooding and to guard against erosion, landslides and the loss of soil.
- Water can be stored underground as well as on the surface, thus reducing losses from evaporation. Underground storage usually uses existing chambers such as abandoned mines.
- Water pricing and the granting of licences to use certain amounts of water from the public supply are now commonplace measures for large-scale users of water. If a licence holder does not need to use all the water it is entitled to use under the licence, the surplus water can be sold (water licence exchange).
- Households may be encouraged to use water butts and to trap rainwater by other methods, thus taking less from the piped public supply. They may also be encouraged to use 'grey water' to water gardens, for example. Grey water is water that has already been used, such as bath water.

● How water use varies

Figure 4 shows the contrasts in water use in developed and developing countries. In the latter, agriculture accounts for over 80 per cent of total water use, with industry using more of the remainder than domestic allocation. In the developed world agriculture accounts for slightly more than 40 per cent of total water use. This is lower than the amount allocated to industry. As in the developing world, domestic use is in third place.

As developing countries industrialise and urban-industrial complexes expand, the demand for water grows rapidly in the industrial and domestic sectors. As a result the competition with agriculture for water has intensified in many countries and regions. This is a scenario that has already played itself out in many developed countries where more and more difficult decisions are having to be made about how to allocate water.

There can also be large variations in water allocation within countries. For example, irrigation accounts for over 80 per cent of water demand in the west of the USA, but only about 6 per cent in the east. In general, precipitation declines from east to west in the USA.

The amount of water used by a population depends not only on water availability but also on levels of urbanisation and economic development. As global urbanisation continues, the demand for **potable water** (drinking water) in cities and towns will rise rapidly. In many cases demand will outstrip supply. In some countries water is delivered on a daily basis to urban areas that are not yet connected to the mains supply (Figure 5).

In terms of agriculture, more than 80 per cent of crop **evapotranspiration** comes directly from rainfall with the remainder from irrigation water diverted from rivers and groundwater. However, this varies considerably by region. In the Middle East and North Africa, where rainfall is low and unreliable, more than 60 per cent of crop evapotranspiration originates from irrigation.

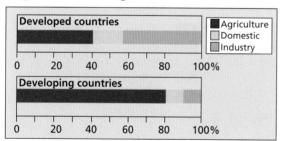

Figure 4 Water used for agriculture, industry and domestic purposes in the developed and developing worlds

Figure 5 Water collection and distribution in central Asia

Figure 6 A public drinking water point in Vienna, Austria

● Water shortages

110 000 km³ of precipitation falls onto the Earth's land surface each year. This would be more than adequate for the global population's needs, but much of it cannot be captured and the rest is very unevenly distributed. For example:

- Over 60 per cent of the world's population live in areas receiving only 25 per cent of global annual precipitation.
- The arid regions of the world cover 40 per cent of the world's land area, but receive only 2 per cent of global precipitation.
- The Congo river and its tributaries account for 30 per cent of Africa's annual runoff in an area containing 10 per cent of Africa's population.

Water scarcity is to do with the availability of potable water. **Physical water scarcity** is when physical access to water is limited. This is when demand outstrips a region's ability to provide the water needed by the population. It is the arid and semi-arid regions of the world that are most associated with physical water scarcity. Here temperatures and evapotranspiration rates are very high and precipitation low. In the worst affected areas, points of access to safe drinking water are few and far between. Egypt is a clear example of physical water scarcity, having to import more than half of its food because it does not have enough water to grow it domestically.

Figure 7 The dried-up bed of the Rio Oja, northern Spain

Economic water scarcity exists when a population does not have the necessary monetary means to utilise an adequate source of water. The unequal distribution of resources is central to economic water scarcity where the crux of the problem is lack of investment. This occurs for a number of reasons including political and ethnic conflict. Figure 8 shows that much of Sub-Saharan Africa is affected by this type of water scarcity.

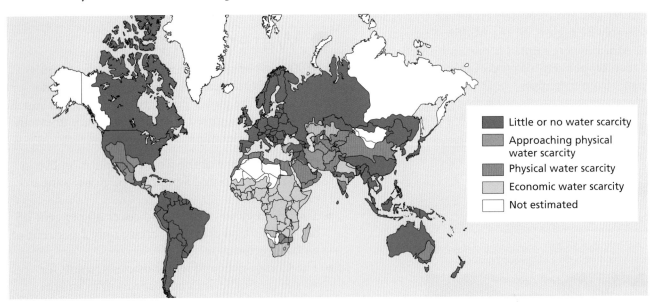

Little or no water scarcity

Approaching physical water scarcity

Physical water scarcity

Economic water scarcity

Not estimated

Figure 8 Physical water scarcity and economic water scarcity worldwide

Securing access to clean water is a vital aspect of development. The lack of clean, safe drinking water is estimated to kill over 4000 children per day. While deaths associated with dirty water have been virtually eliminated from developed countries, in developing countries most deaths still result from water-borne disease.

Water scarcity has been presented as the 'sleeping tiger' of the world's environmental problems, threatening to put world food supplies in jeopardy, limit economic and social development, and create serious conflicts between neighbouring drainage basin countries. The UN estimates that two-thirds of the world's population will be affected by 'severe water stress' by 2025. The situation will be particularly severe in Africa, the Middle East and South Asia. The UN notes that already a number of the world's great rivers such as the Colorado in the USA are running dry, and that groundwater is also being drained faster than it can be replenished. Many major aquifers have been seriously depleted, which will present serious consequences in the future. In China, the over-exploitation of aquifers has been a major factor in the decline in rice production in some areas.

The Middle East and North Africa face the most serious problems. Since 1972 the Middle East has withdrawn more water from its rivers and aquifers each year than is being replenished. Yemen and Jordan are withdrawing 30 per cent more from groundwater resources annually than is being naturally replenished. Israel's annual demand exceeds its renewable supply by 15 per cent. In Africa, 206 million people live in water stressed or water scarce areas.

A country is judged to experience **water stress** when water supply is below 1700 cubic metres per person per year. When water supply falls below 1000 cubic metres per person a year, a country faces **water scarcity** for all or part of the year.

The Water Project, a leading NGO, has recently stated the following with regard to water:

- At any one time, half of the world's hospital beds are occupied by patients suffering from water-borne diseases.
- Over one-third of the world's population has no access to sanitation facilities.
- In developing countries, about 80 per cent of illnesses are linked to poor water and sanitation conditions.
- One out of every four deaths of children under the age of 5 worldwide is due to a water-related disease.
- In developing countries, it is common for water collectors, usually women and girls, to have to walk several kilometres every day to fetch water. Once filled, pots and jerry-cans can weigh as much as 20 kg.

The link between poverty and water resources is very clear, with those living on less than $1.25 a day roughly equal to the number without access to safe drinking water. Access to safe water is vital in the prevention of diarrhoeal diseases which result in 1.5 million deaths a year, mostly among children

under 5. Improving access to safe water can be among the most cost-effective means of reducing illness and mortality. The *UN World Water Development Report* stated: 'The real tragedy is the effect it has on the everyday lives of poor people, who are blighted by the burden of water-related disease, living in degraded and often dangerous environments, struggling to get an education for their children and to earn a living, and to get enough to eat. The brutal truth is that the really poor suffer a combination of most, and sometimes all, of the problems in the water sector.'

The future

Scientists expect water scarcity to become more severe largely because:

- the world's population continues to increase significantly
- increasing affluence is inflating per capita demand for water
- of the increasing demands of biofuel production – biofuel crops are heavy users of water
- climate change is increasing aridity and reducing supply in many regions
- many water sources are threatened by various forms of pollution.

● Water management

The general opinion in the global water industry is that in the past the cost of water in the developed world has been too low to encourage users to save water. Higher prices would make individuals and organisations, both public and private, think more carefully about how much water they use. Higher prices would:

- encourage the systematic re-use of used or 'grey' water
- spur investment in recycling and reclamation systems
- lead to greater investment in the reduction of water losses.

However, many consumers still see water as a 'free' or very low-cost resource, and campaign groups are concerned that higher prices would have an unfair impact on people on low incomes. Water pricing for both domestic and commercial users is a sensitive issue. It has also become much more of a political issue as more and more countries have privatised their water resources.

Conserving irrigation water would have more impact than any other measure. Most irrigation is extremely inefficient, wasting half or more of the water used. A 10 per cent increase in irrigation efficiency would free up more water than is evaporated by all other users. The most modern drip irrigation systems are up to 95 per cent efficient, but require significant investment.

Although some industries have significantly reduced their use of water per unit of production, most water analysts believe that much more can be done. For example, production of 1 kg of aluminium can require up to 1500 litres of water. Other industries, such as paper production, are also very water intensive. Some countries such as Japan and Germany have made considerable improvements in industrial water use. For example, Japanese industry recycles more than 75 per cent of process water.

As water scarcity becomes more of a problem, the investment required to tackle this global challenge will rise. Table 1 shows the estimated investment needed by regions of the world for the period 2005–30. There are very large contrasts between the different regions. Delivering water to the points where it is required is a costly business in terms of both constructing and maintaining infrastructure. Overall, the sums of money illustrated in Table 1 are huge and money may need to be diverted from other sectors of national government funding. However, investment in water as a proportion of GDP has fallen by half in most countries since the late 1990s.

Urban sanitation services are very heavy users of water. Demand could be reduced considerably by adopting dry, or low-water use, systems such as dry composting toilets with urine separation systems. A number of pilot projects are in operation, such as the Gebers Housing Project in Stockholm.

Table 1 Water investment needs by area, 2005–30

	$ (trillions)
Asia/Oceania	9.0
South/Central America	5.0
Europe	4.5
USA/Canada	3.6
Africa	0.2
Middle East	0.3

Case study: The water problem in south-western USA

The USA is a huge user of water. Over the country as a whole there would not seem to be a water problem. However, the western states of the USA, covering 60 per cent of the land area with 40 per cent of the total population, receive only 25 per cent of the country's mean annual precipitation. Yet each day the west uses as much water as the east. It is the south-west in particular where the water problem is most intense. This is the area of the USA most vulnerable to water shortages.

Figure 9 Desert region in south-western USA

The south-west has prospered due to a huge investment in water transfer schemes. This has benefited agriculture, industry and settlement. Hundreds of aqueducts take water from areas of surplus to areas of shortage. The federal government has paid most of the bill but now the demand for water is greater than the supply. If the west is to continue to expand, a solution to the water problem must be found.

Although much of the south-west is desert or semi-desert, large areas of dry land have been transformed into fertile farms and sprawling cities. It all began with the Reclamation Act of 1902 which allowed the building of canals, dams and hydro-electric power systems in the states that lie, all or in part, west of the 100th meridian. Water supply was to be the key to economic development in general, benefiting not only the west but the USA as a whole.

California has benefited most from this investment in water supply. A great imbalance exists between the distributions of precipitation and population in the state. Seventy per cent of runoff originates in the northern one-third of the state but 80 per cent of the demand for water is in the southern two-thirds. While irrigation is the prime water user, the sprawling urban areas have also greatly increased demand. The 3.5 million hectares of irrigated land in California are situated mainly in the Imperial, Coachella, San Joaquin and lower Sacramento valleys. Figure 10 shows the major component parts of water transfer and storage in the state.

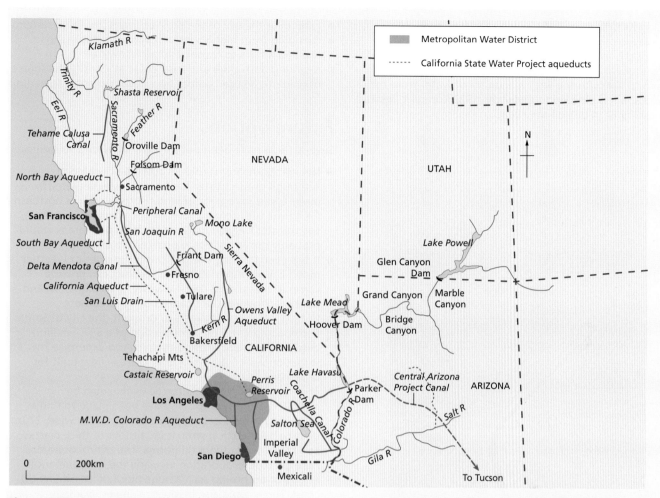

Figure 10 Water management schemes in California

Agriculture uses more than 80 per cent of the state's water, though it accounts for less than a tenth of the economy. Water development, largely financed by the federal government, has been a huge subsidy to California in general and to big water users in particular. However, recently there has been a move to bringing the price mechanism to bear on water resources.

The Colorado: a river under pressure

The 2333 km long Colorado river is an important source of water in the south-west. Over 30 million people in the region depend on water from the river. The river rises 4250 m up in the Rocky Mountains of northern Colorado and flows generally south-west through Colorado, Utah, Arizona and between Nevada and Arizona, and Arizona and California before crossing the border into Mexico. The river drains an area of about 632 000 km².

The Colorado was the first river system in which the concept of multiple use of water was attempted by the US Bureau of Reclamation. In 1922 the Colorado River Compact divided the seven states of the basin into two groups: Upper Basin and Lower Basin. Each group was

allocated 9.25 trillion litres of water annually, while a 1944 treaty guaranteed a further 1.85 trillion litres to Mexico. Completed in 1936, the Hoover Dam and Lake Mead marked the beginning of the era of artificial control of the Colorado. Despite the interstate and international agreements (between the USA and Mexico), major problems over the river's resources have arisen because population has increased along with rising demand from agriculture and industry.

The $4 billion Central Arizona Project (CAP) is the latest, and probably the last, big money scheme to divert water from this great river (Figure 11). Before CAP, Arizona had taken much less than its legal entitlement from the Colorado. It could not afford to build a water transfer system from the Colorado to its main cities and at the time the federal government did not feel that national funding was justified. Most of the state's water came from aquifers but it was overdrawing this supply by about two million acre-feet a year. If thirsty Phoenix and Tucson were to remain prosperous, something had to be done. The answer was CAP, which the federal government agreed to part-fund. Since CAP was completed in 1992, 1.85 trillion litres of water a year has been distributed to farms, Indian reservations, industries and fast-growing towns and cities along its 570 km route between Lake Havasu and Tucson. However, providing more water for Arizona has meant that less is available for California. In 1997 the federal government told California that the state would have to learn to live with the 5427 million m³ of water from the Colorado it is entitled to under the 1922 Compact, instead of taking 6414 million m³ a year.

Figure 11 Part of the Central Arizona Project

Resource management strategies

Implementation of the following strategies would conserve considerable quantities of water in the south-west of the USA:

- measures to reduce leakage and evaporation losses – up to 25 per cent of all water moved is currently lost in these ways
- recycling water in industry where, for example, it takes 225 000 litres to make one tonne of steel
- recycling municipal sewage for watering lawns, gardens and golf courses, which could be implemented or extended, as Los Angeles has already shown
- introducing more efficient toilet systems
- charging more realistic prices for irrigation water – many farmers pay well below the true cost of water pumped to them, while the rest is subsidised by the federal government

- extending the use of the most efficient irrigation systems
- changing from highly water-dependent crops such as rice and alfalfa to those needing less water
- requiring both cities and rural areas to identify the source of water to be used before new developments can begin.

Future options

Several ideas have been put forward for future strategies:

- New groundwater resources could be developed. Although groundwater has been heavily depleted in many areas, in regions of water surplus such as northern California they remain virtually untapped. However, the transfer of even more water from such areas would probably prove politically unacceptable.
- It has been claimed that various techniques of weather modification, especially cloud seeding, can provide water at reasonable cost. However, environmental and political considerations cannot be ignored here.
- In 1991, after several years of drought, the city of Santa Barbara approved the construction of a $37.4 million desalination plant. Although much too expensive for irrigation water, it is likely that more will be built for urban use.
- The frozen reserves of Antarctic water could be exploited. Serious proposals have been made to find a 100 million tonne iceberg (1.5 km long, 300 m wide, 270 m deep) off Antarctica, wrap it in sailcloth or thick plastic, and tow it to southern California. The critical questions here are cost, evaporation loss, and the environmental effects of anchoring such a huge block of ice off an arid coast. There could also be political implications.
- Offshore aqueducts might be constructed that would run under the ocean from the Columbia river in the north-west of the USA to California.

There is now general agreement that planning for the future water supply of the south-west should embrace all practicable options. Sensible management of this vital resource should rule out no feasible strategy if this important region is to sustain its economic viability and growing population.

Interesting note
The water level in the Colorado-fed Lake Mead, the USA's largest reservoir, has dropped by more than 30 m since the beginning of the twenty-first century. The basin is now only just over half full!

Case study analysis
1 Describe the imbalance in population and precipitation between the eastern and western parts of the USA.
2 Discuss the main uses of water in California.
3 Why is the Colorado river under so much pressure?
4 Explain the resource management strategies that can be used to try to improve the balance between supply and demand.

Environmental risks of economic development

3.7

Polluting factories and smog over Ulaanbaatar – flights in and out of the airport are regularly delayed due to smog in one of the world's most polluted cities

Key questions

- How can economic activities pose threats to the natural environment, locally and globally?
- How important are sustainable development and management to economic development?
- How important is resource conservation?

● The threat of economic activities to the natural environment

As the scale of global economic activity has increased, bringing considerable benefits to many people, the strain on the natural environment has become more obvious at all scales from local to global. The planet is experiencing a range of serious environmental challenges, many of which are interlinked. At the largest scale enhanced global warming is having an impact, or will have an impact, on the whole of the world.

Pollution

Pollution is contamination of the environment. It can take many forms – air, water, soil, noise, visual and others. Pollution has a major impact on people and the environment. Figure 1 shows how people are exposed to chemicals and how exposure to these chemicals can affect human health. The methods of exposure to pollutants are:

- breathing in chemical vapours and dust (inhalation)
- drinking or eating the chemical (ingestion)
- absorbing the chemical through the skin (absorption).

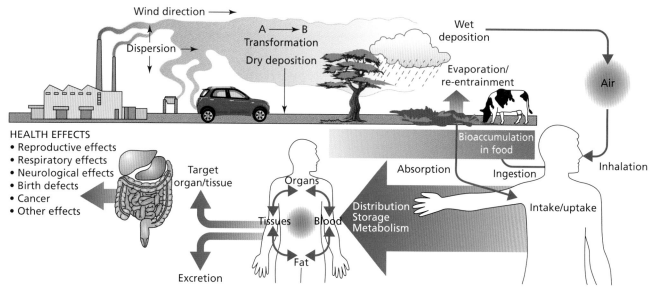

Figure 1 How exposure to pollution can affect human health

Air pollution

Of all types of pollution, air pollution has the most widespread effects on human health and the environment. Air pollution affects people and the environment at a range of scales, from local to global. In many parts of the developing world indoor air pollution is more severe than that experienced outdoors. This is the result of the use of biomass fuels for cooking and heating. In many poor countries this is the only option available.

Virtually every substance is **toxic** at a certain dosage. The most serious polluters are the large-scale processing industries which tend to form agglomerations as they have similar locational requirements (Table 1). The impact of a large industrial agglomeration may spread well beyond the locality and region to cross international borders.

For example, **prevailing winds** in Europe generally carry pollution from west to east. Thus the problems caused by acid rain in Scandinavia have been due partly to industrial activity in the UK. Dry and wet deposition can be carried for considerable distances. For example, pollution found in Alaska in the 1970s was traced back to the Ruhr industrial area in Germany.

Table 1 The most polluting industries

Industrial sector	Examples
Fuel and power	Power stations, oil refineries
Mineral industries	Cement, glass, ceramics
Waste disposal	Incineration, chemical recovery
Chemicals	Pesticides, pharmaceuticals, organic and inorganic chemicals
Metal industries	Iron and steel, smelting, non-ferrous metals
Others	Paper manufacture, timber preparation, uranium processing

Table 2 Major air pollutants

Pollutant	Sources	Effects
Ozone: a gas that can be found in two places. Near the ground (the troposphere) it is a major part of smog. This should not be confused with the protective layer of ozone in the upper atmosphere (stratosphere), which screens out harmful ultraviolet rays.	Ozone is not created directly, but is formed when nitrogen oxides and volatile organic compounds mix in sunlight. Nitrogen oxides come from burning gasoline, coal and other fossil fuels. The many types of volatile organic compounds come from sources ranging from factories to trees.	Ozone near the ground can lead to more frequent asthma attacks in people who have asthma, and can cause sore throats, coughs and breathing difficulty, even leading to premature death. It can also damage plants and crops.
Carbon monoxide: a gas that comes from the burning of fossil fuels, mostly in cars. It cannot be seen or smelled.	Carbon monoxide is released when engines burn fossil fuels. Emissions are higher when engines are not tuned properly, and when fuel is not completely burnt. Cars emit a lot of the carbon monoxide found outdoors. Furnaces and heaters in the home can emit high concentrations of carbon monoxide if they are not properly maintained.	Carbon monoxide makes it hard for body parts to get the oxygen they need. Exposure to carbon monoxide makes people feel dizzy and tired and gives them headaches. In high concentrations it can be fatal.
Nitrogen dioxide: a reddish-brown gas that comes from the burning of fossil fuels. It has a strong smell at high concentrations.	Nitrogen dioxide mostly comes from power plants and cars. It is formed in two ways: when nitrogen in fuel is burnt, and when nitrogen in the air reacts with oxygen at very high temperatures. It can also react in the atmosphere to form ozone, acid rain and particles.	High concentrations of nitrogen dioxide can give people coughs and make them short of breath. People exposed to nitrogen dioxide for a long time are more vulnerable to respiratory infections. Nitrogen dioxide reacts in the atmosphere to form acid rain, which can harm plants and animals.
Particulate matter: solid or liquid matter that is suspended in the air. To remain in the air, particles must usually be between 0.00005 mm and 0.1 mm.	Particulate matter can be divided into two types: coarse and fine. Coarse particles are formed from sources such as road dust, sea spray and construction. Fine particles are formed when fuel is burnt in vehicles and power plants.	Particulate matter that is small enough can enter the lungs and cause health problems, for example more frequent asthma attacks, respiratory problems and premature death.
Sulfur dioxide: a corrosive gas that cannot be seen or smelled at low levels, but has a 'rotten egg' smell at higher concentrations.	Sulfur dioxide mostly comes from burning coal or oil in power plants. Also from factories making chemicals, paper or fuel. It reacts in the atmosphere to form acid rain and particles.	Sulfur dioxide exposure can affect people who have asthma or emphysema by making it more difficult for them to breathe. It can also irritate eyes, nose and throat. It can harm trees and crops, damage buildings, and reduce visibility.
Lead: a blue-grey metal that is very toxic; found in a variety of forms and places.	Outside, lead comes from cars using fuel that is not unleaded. Also from power plants and other industrial sources. Inside, lead paint can still be found in older houses; old lead pipes can be a source of lead in drinking water.	Large amounts of lead can affect young children's brain development and cause kidney problems. For adults exposure to lead can increase the chance of heart attack and stroke.

Toxic air pollutants: includes a number of chemicals, for example arsenic, asbestos, benzene, dioxin.	Each pollutant comes from a different source, but many are created in chemical plants when fossil fuels are burnt. Some pollutants can be found in building materials or in food and water supplies.	Toxic air pollutants can cause cancer, and some can cause birth defects, skin and eye irritation, and breathing problems.
Stratospheric ozone depleters: chemicals that can destroy the ozone in the stratosphere. These include chlorofluorocarbons (CFCs), halons, chlorine and bromine.	CFCs are used in air conditioners and refrigerators, as they work well in coolants. Also in aerosol cans, fire extinguishers and solvents.	If stratospheric ozone is destroyed, people are exposed to more ultraviolet radiation from the sun, which can lead to skin cancer and eye problems. It can also harm plants and animals.
Greenhouse gases: gases that stay in the air for a long time and warm the planet by trapping sunlight. This is called the 'greenhouse effect' because the gases act like glass in a greenhouse. Greenhouse gases include carbon dioxide, methane and nitrous oxide.	Carbon dioxide is the most important greenhouse gas. It comes from burning fossil fuels in vehicles, power plants, houses and industry. Methane is released during the processing of fossil fuels; also from natural sources such as cows and padi-fields. Nitrous oxide comes from industrial sources and decaying plants.	The greenhouse effect can lead to changes in the climate, for example more extreme temperatures, higher sea levels, changes in forest composition, which in turn cause health problems for people.

Table 2 summarises the sources and effects on people and the environment of the major air pollutants. Air pollution is a massive environmental problem leading to, among other things, global warming, acid rain and the deterioration of the ozone layer.

Pollution is the major externality of industrial and urban areas. **Externalities** are the side-effects, positive and negative, of an economic activity that are experienced beyond its site. Pollution is at its most intense at the focus of pollution-causing activities, declining with distance from such concentrations (Figure 2). For some sources of pollution it is possible to map the externality gradient and field. In general, health risk and environmental damage are greatest immediately around the source of pollution and the risk decreases with distance from the source. However, atmospheric conditions and other factors can complicate this pattern.

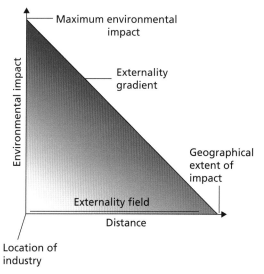

Figure 2 Externality gradient and field

> **Interesting note**
> The countries with the largest total carbon dioxide emissions are China, the USA, India, Russia and Japan. In terms of emissions per capita, the USA and Russia are the largest polluters.

Water pollution

Each year about 450 km³ of wastewater are discharged into rivers, streams and lakes around the world. While rivers in more affluent countries have become steadily cleaner in recent decades, the reverse has been true in much of the developing world. It has been estimated that 90 per cent of sewage in developing countries is discharged into rivers, lakes and seas without any treatment. The UN estimates that almost half the population in many developing world cities do not have access to safe drinking water. For example, the Yamuna river, which flows through Delhi, has 200 million litres of sewage drained into it each day. For many people the only alternative to using this water for drinking and cooking is to turn to water vendors who sell tap water at greatly inflated prices.

Although most people in developed countries think that their water supplies are clean and healthy there is growing concern in some quarters about traces of potentially dangerous medicines that may be contaminating tap water and putting unborn babies at risk, according to a report published in the UK in September 2008. One newspaper headline read 'Is our water being poisoned with a cocktail of drugs?'. Scientists are worried that powerful and toxic anti-cancer drugs are passing unhindered through sewage works and making their way back into the water supply.

Figure 3 A polluted river in Christchurch, New Zealand

China's rapid economic growth has led to widespread environmental problems. Pollution problems are so severe in some areas that the term 'cancer village' has become commonplace. In the village of Xiditou, south-east of Beijing, the cancer rate is 30 times the national average. This has been blamed on water and air contaminated by chemical factories. Tests on tap water have found traces of highly carcinogenic benzene that were 50 per cent above national safe limits. In the rush for economic growth, local governments eagerly built factories, but they had very limited experience of environmental controls.

- The Chinese government admits that 300 million people drink polluted water.
- This comes from polluted rivers and groundwater.
- 30 000 children die of diarrhoea or other water-borne illnesses each year.
- The river Liao is the most polluted, followed by waterways around Tianjin and the River Huai.

Noise and light pollution

Not all pollution involves inhalation, ingestion and absorption. Noise and light pollution are increasing hazards in developed societies. Noise pollution is disturbing or excessive noise that may harm the activity or balance of human or animal life. Most outdoor noise is caused by machines and modes of transport. Outdoor noise is generally referred to as 'environmental noise'.

The increase in air traffic is one of the major contributors to noise pollution (and air pollution). A large area is currently affected by aircraft noise from Heathrow airport, near London, stretching from the southern outskirts of Maidenhead in the west to the edge of Camberwell in the east (Figure 4). In this area, 600 000 people are affected by noise levels of 55 decibels or over. People living close to the airport are affected by noise levels of 75 decibels. Significant annoyance from aircraft noise begins at 50 decibels.

The proposed third runway at Heathrow will increase the number of flights from 420 000 a year to 700 000 and will bring far more people within the area affected by aircraft noise. A recent study has highlighted the link between exposure to noise and ill health, noting in particular exposure to night-time aircraft noise and high blood pressure. The latter can lead to heart attacks and strokes.

Light pollution is excessive or obtrusive artificial light. It is an externality of a developed society. Its sources include the interior and external lighting of all sorts of buildings, advertising and street lighting. It is most severe in highly industrialised and densely populated areas. It can have a serious impact on human health, causing fatigue, loss of sleep, headaches and loss of amenity.

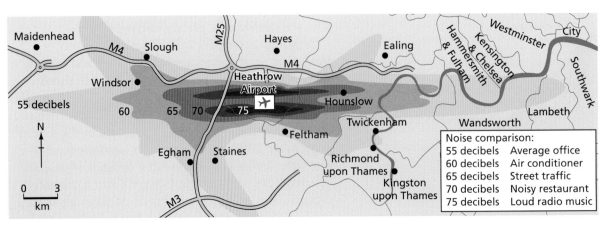

Figure 4 Heathrow airport, with surrounding noise levels

The relative risks of incidental and sustained pollution

It is important to consider the different impact between **incidental pollution** (one-off pollution incidents) and **sustained pollution** (longer-term pollution). The former is mainly linked to major accidents caused by technological failures and human error. Causes of the latter include ozone depletion and global warming. Some of the worst examples of incidental pollution are shown in Table 3.

Major examples of incidental pollution such as Chernobyl and Bhopal can have extremely long-lasting consequences which are often difficult to determine in the earlier stages. The effects of both accidents are still being felt more than two decades after they occurred.

Table 3 Major examples of incidental pollution

Location	Causes and consequences
Seveso, Italy	In July 1976 a reactor at a chemical factory near Seveso in northern Italy exploded, sending a toxic cloud into the atmosphere. An 18 km² area of land was contaminated with the dioxin TCDD. The immediate after-effects – a small number of people with skin inflammation – were relatively mild. However, the long-term impact has been much worse. The population is suffering increased numbers of premature deaths from cancer, cardiovascular disease and diabetes.
Bhopal, India	A chemical factory owned by Union Carbide leaked deadly methyl isocyanate gas during the night of 3 December 1984. The plant was operated by a separate Indian subsidiary which worked to much lower safety standards than those required in the USA. It has been estimated that 8000 people died within two weeks and a further 8000 have since died from gas-related diseases. The NGO Greenpeace puts the total fatality figure at over 20 000. Bhopal is recognised as the world's worst industrial disaster.
Chernobyl, Ukraine	The world's worst nuclear power plant accident occurred at Chernobyl, Ukraine in April 1986. Reactor number four exploded, sending a plume of highly radioactive fallout into the atmosphere which drifted over extensive parts of Europe and eastern North America. Two people died in the initial explosion and over 336 000 people were evacuated and resettled. In total, 56 direct deaths and an estimated 4000 extra cancer deaths have been attributed to Chernobyl. The estimated cost of $200 billion makes Chernobyl the most expensive disaster in modern history.
Harbin, China	An explosion at a large petrochemical plant in the north-east Chinese city of Harbin in 2005 released toxic pollutants into a major river. Benzene levels were 108 times above national safety levels. Benzene is a highly poisonous toxin which is also carcinogenic. Water supplies to the city were suspended. Five people were killed in the blast and more than 60 injured. Ten thousand residents were temporarily evacuated.

It is usually the poorest people in a society who are exposed to the risks from both incidental and sustained pollution. In the USA the geographical distribution of both ethnic minorities and people on the lowest incomes has been found to be strongly linked to the distribution of the worst kinds of pollution.

Activities

1 Define pollution.
2 What are the means of human exposure to pollutants?
3 Which industries are the largest polluters?
4 Describe the sources and effects of two major air pollutants.
5 Briefly explain Figure 2.
6 Why is water pollution much more of a problem in the developing world than in the developed world?
7 What are the main sources of noise pollution?
8 Define light pollution.
9 What is the difference between incidental pollution and sustained pollution?

● Enhanced global warming

Economic activities generating greenhouse gases

There is no doubt amongst geographers and scientists that the Earth's climate is changing and that human economic activities are at the very least a significant cause of these changes. The Earth-atmosphere system has a natural greenhouse effect (Figure 5) that is essential to all life on Earth, but large-scale pollution of the atmosphere by economic activities has created an **enhanced greenhouse effect**. This is causing temperatures to increase beyond the limits of the natural greenhouse effect. Many parts of the world are experiencing unexpected changes in their weather. Some of these changes could have disastrous consequences for the populations of the areas affected if they continue to get more severe. Human activity has significantly increased the amount of greenhouse gases in the atmosphere and this has caused temperature to rise more rapidly than ever before. As the economies of China, India and other NICs expand even further, greenhouse gas emissions will continue to increase.

- The present rate of change is greater than anything that has happened in the past. In the twentieth century, average global temperatures rose by 0.6 °C. Most of this increase took place in the second half of the century.
- The predictions are for a further global average temperature increase of between 1.6 °C and 4.2 °C by 2100.

The main greenhouse gases being created by human activity are:

- carbon dioxide
- methane
- nitrous oxides
- chlorofluorocarbons
- ozone.

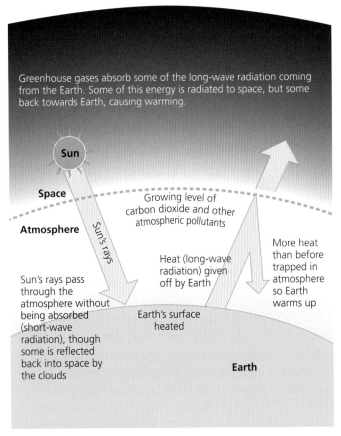

Greenhouse gases absorb some of the long-wave radiation coming from the Earth. Some of this energy is radiated to space, but some back towards Earth, causing warming.

Sun

Space

Growing level of carbon dioxide and other atmospheric pollutants

Atmosphere

Sun's rays

Sun's rays pass through the atmosphere without being absorbed (short-wave radiation), though some is reflected back into space by the clouds

Heat (long-wave radiation) given off by Earth

Earth's surface heated

More heat than before trapped in atmosphere so Earth warms up

Earth

Figure 5 The greenhouse effect

The consequences of enhanced global warming

There are many potential consequences of enhanced global warming including the following:

- **Global temperature variations and heatwaves**: in general, higher latitudes and continental regions will experience temperature increases significantly greater than the global average. There will be a rising probability of heatwaves with more extreme heat days and fewer very cold days.
- **Rising sea levels**: sea levels will respond more slowly than temperatures to changing greenhouse gas concentrations. Sea levels are currently rising at around 3 mm per year and the rise has been accelerating. Rising sea levels are due to a combination of **thermal expansion** and the melting of ice sheets and glaciers. Thermal expansion is the increase in water volume due to temperature increase alone. A global average sea level rise of 0.4 m from this cause has been predicted by the end of this century.
- **Increasing acidity in oceans**: as carbon dioxide levels rise in the atmosphere, more of the gas is dissolved in surface waters, creating carbonic acid. Since the start of the industrial revolution the acidity activity of the oceans has increased by 30 per cent. This is having a significant impact on coral reefs and shellfish.

Figure 6 Great Barrier Reef – increasing acidity in oceans is having an impact on coral reefs

- **Melting of ice caps and glaciers**: satellite photographs show ice melting at its fastest rate ever. The area of sea ice in the Arctic Ocean has decreased by 15 per cent since 1960, while the thickness of the ice has fallen by 40 per cent. In 2007, the sea ice around Antarctica had melted back to a record low. At the same time, the movement of glaciers towards the sea has speeded up. A satellite survey between 1996 and 2006 found that the net loss of ice rose by 75 per cent. Temperatures in western Antarctica have increased sharply in recent years, melting ice shelves and changing plant and animal life on the Antarctic Peninsula. Ice melting could cause sea levels to rise by a further 5 m (on top of thermal expansion). Hundreds of millions of people live in coastal areas within this range.

Figure 7 Ice melting at the edge of Antarctica

- **The warm water of the Gulf Stream**: the Gulf Stream originates in the Gulf of Mexico, and its continuation the North Atlantic Drift has a major influence on the climates of the east coast of North America and western Europe. Climatologists are concerned that melting Arctic ice could disrupt these warm ocean currents resulting in temperatures in western Europe falling by at least 5 °C. Winters in the UK and other neighbouring countries would be much colder than they are now.
- **Thawing peat bogs**: an area of permafrost spanning one million km² has started to melt for the first time since it formed 11 000 years ago at the end of the last ice age. The area, which covers the entire sub-Arctic region of western Siberia, is the world's largest frozen peat bog and scientists fear that as it thaws, it will release billions of tonnes of methane, a greenhouse gas 20 times more potent than carbon dioxide, into the atmosphere.
- **El Niño**: this phenomenon is a change in the pattern of wind and ocean currents in the Pacific Ocean. This causes short-term changes in weather for countries bordering the Pacific, such as flooding in Peru and drought in Australia. El Niño events tend to occur every two to seven years. There is concern that rising temperatures could increase their frequency and/or intensity.
- **Growth of the tropical belt**: a study published in 2007 warned that the Earth's tropical belt was expanding north and south. A further 22 million km² of the Earth are experiencing a tropical climate compared with 1980. The poleward movement of subtropical dry belts could affect agriculture and water supplies over large areas of the Mediterranean, the south-western USA, northern Mexico, southern Australia, southern Africa and parts of South America. The extension of the tropical belt will put more people at risk from tropical diseases.

Figure 8 Giant climate change banner, Brussels

- **Changing patterns of rainfall**: the amount and distribution of rainfall in many parts of the world could change considerably. Generally, regions that get plenty of rainfall now are likely to receive even more. And regions with low rainfall are likely to get less. The latter will include the poor arid and semi-arid countries of Africa. In 2009, the heaviest rain in 53 years battered Dhaka, the capital of Bangladesh.
- **Declining crop yields**: higher temperatures have already had an impact on global yields of wheat, corn and barley. A recent study revealed that crop yields fall between 3 and 5 per cent for every 0.5 °C increase in average temperature. Food shortages could begin conflicts between different countries.
- **Impact on wildlife**: many species of wildlife may be wiped out because they will not have a chance to adapt to rapid changes in their environment. For example, the loss of Arctic ice will have a huge effect on polar bears and other species that live and hunt among the ice floes.

Activities

1 a Describe the natural greenhouse effect.
  b How has this effect been 'enhanced' by economic activities?
2 List the main greenhouse gases.
3 Discuss three potential consequences of enhanced global warming.

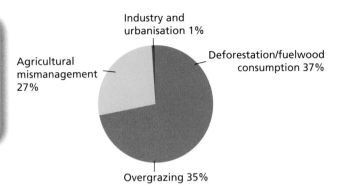

Figure 9 Causes of land degradation

Soil erosion and desertification

The extent and causes of soil erosion

Soil degradation is a global process. It involves both the physical loss (erosion) and the reduction in quality of topsoil associated with nutrient decline and contamination. It has a significant impact on agriculture and also has implications for the urban environment, pollution and flooding. The loss of the upper soil horizons containing organic matter and nutrients and the thinning of soil profiles reduces crop yields on degraded soils. Soil degradation can cancel out gains from improved crop yields. The statistics on soil degradation make worrying reading:

- Globally it is estimated that 2 billion hectares of soil resources have been degraded. This is equivalent to about 15 per cent of the Earth's land area.
- During the past 40 years nearly one-third of the world's cropland has been abandoned because of soil erosion and degradation.
- In Sub-Saharan Africa, nearly 2.6 million km² of cropland has shown a 'consistent significant decline' according to a March 2008 report. Some scientists consider this to be a 'slow-motion disaster'.
- It takes natural processes about 500 years to replace 25 mm of topsoil lost to erosion. The minimum soil depth for agricultural production is 150 mm.

In temperate areas much soil degradation is a result of market forces and the attitudes adopted by commercial farmers and governments. In contrast, in the tropics much degradation results from high population pressure, land shortages and lack of awareness. The greater climate extremes and poorer soil structures in tropical areas give greater potential for degradation in such areas compared with temperate latitudes. This difference has been a significant factor in development, or the lack of it.

The main cause of soil degradation is the removal of the natural vegetation cover, leaving the surface exposed to the elements. Figure 9 shows the human causes of degradation, with deforestation and overgrazing being the two main problems. The resulting loss of vegetation cover is a leading cause of wind and water erosion.

Deforestation occurs for a number of reasons including the clearing of land for agricultural use, for timber, and for other activities such as mining. Such activities tend to happen quickly whereas the loss of vegetation for fuelwood, a massive problem in many developing countries, is generally a more gradual process. **Overgrazing** is the grazing of natural pastures at stocking intensities above the livestock carrying capacity. Population pressure in many areas and poor agricultural practices have resulted in serious overgrazing. This is a major problem in many parts of the world, particularly in marginal ecosystems.

Agricultural mismanagement is also a major problem due to a combination of a lack of knowledge and the pursuit of short-term gain against consideration of longer-term damage. Such activities include shifting cultivation without adequate fallow periods, absence of soil conservation measures, cultivation of fragile or marginal lands, unbalanced fertiliser use and the use of poor irrigation techniques.

Local soil degradation

Figure 10 illustrates how a combination of causes and processes can operate in an area to result in soil degradation. The diagram shows a range of different economic activities which have an impact on the soil. Can you think of other economic activities that you could reasonably expect to find in such an area? What impact would these activities have on the soil? Notice how the diagram shows an increase in the area characterised by sealing as the urban area expands at the expense of farmland.

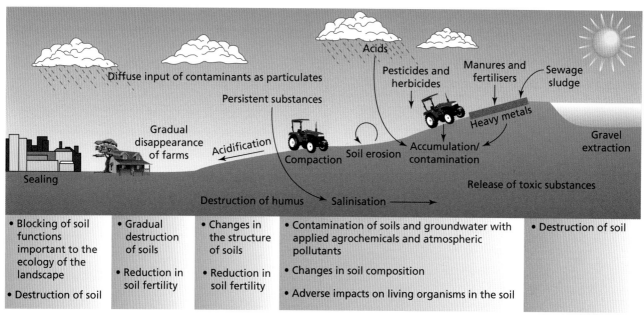

Figure 10 The causes and processes of local soil degradation

Environmental and socio-economic consequences of soil erosion

The environmental and socio-economic consequences of soil degradation are considerable. Such consequences can occur with little warning as damage to soil is often not perceived until it is far advanced. Socio-economic consequences are both on-site (local) and off-site (beyond the locality), while environmental consequences are primarily off-site.

Desertification is the gradual transformation of habitable land into desert. It is arguably the most serious environmental consequence of soil degradation. Desertification is usually caused by climate change and/or by destructive use of the land. Natural causes of desertification include temporary drought periods of high magnitude and long-term climate change towards aridity. The main human causes are:

- overgrazing
- overcultivation
- deforestation.

Desertification occurs when already fragile land in arid and semi-arid areas is over-exploited. It is a considerable problem in many parts of the world, for example on the margins of the Sahara desert in north Africa and the Kalahari desert in southern Africa. In semi-arid areas such as the edge of the Kalahari desert in southern Africa, a combination of low and variable precipitation, nutrient-deficient soils and heavy dependence on subsistence farming makes soil degradation a significant threat. At present, 25 per cent of the global land territory and nearly 16 per cent of the world's population are threatened by desertification. Table 4 summarises the consequences of desertification.

Table 4 The consequences of desertification

Environmental	Economic	Social and cultural
• Loss of soil nutrients through wind and water erosion • Changes in composition of vegetation and loss of biodiversity as vegetation is removed • Reduction in land available for cropping and pasture • Increased sedimentation of streams because of soil erosion and sediment accumulations in reservoirs • Expansion of area under sand dunes	• Reduced income from traditional economy (pastoralism and cultivation of food crops) • Decreased availability of fuelwood, necessitating purchase of oil/kerosene • Increased dependence on food aid • Increased rural poverty	• Loss of traditional knowledge and skills • Forced migration due to food scarcity • Social tensions in reception areas for migrants

Dust storms, which can seriously damage crops, may also be a problem in such areas. Dust storms occur naturally wherever dry soils and strong winds combine, but human activity can increase their severity significantly. These human activities are removal of vegetation, overgrazing, overcultivation and surface disturbance by vehicles. All these practices can add to the severity of the problem. In the Sahel the increase in dust-storm frequency has been shown to coincide with periods of severe drought.

Figure 11 Darkhan, Mongolia: (a) on a clear day and (b) during one of the regular dust storms that are contributing to soil degradation. The number and strength of dust storms has increased in recent years.

Soil degradation: a threat to food security?

The increasing world population and the rapidly changing diet of hundreds of millions of people as they become more affluent is placing more and more pressure on land resources. Some soil and agricultural experts say that a decline in long-term soil productivity is already seriously limiting food production in the developing world.

Activities

1 What is soil erosion?
2 To what extent is soil erosion a major global problem?
3 Discuss the causes and processes of local soil degradation shown in Figure 10.
4 a Define desertification.
  b Explain the causes of desertification.
5 Compare the photographs in Figure 11.

● Examples of environments under threat

Agricultural change in Argentina's pampas

Traditionally, cattle rearing has dominated farming in the pampas of Argentina. The pampas is one of the world's great grasslands. It is a flat prairie with deep, fertile topsoil. However, rapid change is underway as crop production replaces cattle rearing over significant areas of the pampas. There are undoubted benefits to this process as farmers are responding to changing patterns of global demand. But there are also risks involved in such a radical change in land use.

The change from pastoral to arable farming has considerably increased chemical input onto the land. This is having a significant impact on the ecosystem. The World Wide Fund for Nature (WWF) is concerned that the pampas is now being over-farmed. This is endangering wildlife, including South American ostriches, pumas and wildcats. The WWF is also concerned about the widespread destruction of native grass.

Figure 12 The pampas – cropland

Oil production in the Niger delta

The Niger delta covers an area of 70 000 km², making up 7.5 per cent of Nigeria's land area. It contains over 75 per cent of Africa's remaining mangrove. A report published in 2006 estimated that up to 1.5 million tonnes of oil has been spilt in the delta over the past 50 years. The report compiled by WWF says that the delta is one of the five most polluted spots on Earth. Pollution is destroying the livelihoods of many of the 20 million people who live there. The pollution damages crops and fishing grounds and is a major contributor to the upsurge in violence in the region. People in the region are dissatisfied with bearing the considerable costs of the oil industry but seeing very little in terms of the benefits. The report accused the oil companies of not using the advanced technologies to combat pollution that are evident in other world regions. However, Shell claims that 95 per cent of oil discharges in the last five years have been caused by sabotage.

The flaring (burning) of unwanted natural gas found with the oil is a major regional and global environmental problem. The gas found here is not useful because there is no gas pipeline infrastructure to take it to consumer markets. It is estimated that 70 million m³ are flared off each day. This is equivalent to 40 per cent of Africa's natural gas consumption. Gas flaring in the Niger delta is the world's single largest source of greenhouse gas emissions.

Figure 13 Environmental problems in the Niger delta

The threat to Australia's Great Barrier Reef

The Great Barrier Reef (Figure 14) is one of the great tourist attractions in Australia. It includes over 2900 reefs, around 940 islands and cays, and stretches 2300 km along the coast of Queensland. The Great Barrier Reef Marine Park covers an area of 345 000 km². This is an ecosystem of immense diversity. However, there are several significant problems relating to the Great Barrier Reef:

- The impact of land-based pollution from agriculture, industry, residential areas and tourism is causing significant damage to the reef ecosystem.
- Overfishing – the use of dragnets in particular – can damage the coral.
- Coral bleaching is exacerbated by increasing sea temperatures due to global warming. This causes coral polyps to die, leaving only the white 'skeleton' of the coral, and the range of colours is lost.
- Tourists visiting the reef cause damage by dropping and dragging anchors, walking and diving on the reef and by pollution from the tourist boats visiting the area.

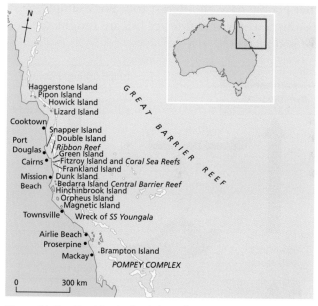

Figure 14 The Great Barrier Reef, Australia

● Sustainable development and management

The two key terms that have become increasingly important in terms of economic activity are **resource management** and **sustainable development**. Resource management is the control of the exploitation and use of resources in relation to environmental and economic costs. Sustainable development is a carefully calculated system of resource management which ensures that the current level of exploitation does not compromise the ability of future generations to meet their own needs.

Resource management in the European Union

Figure 15 shows what has happened in so many of the world's fishing grounds. Without careful resource management fish stocks could be totally depleted in some areas. Yet it is often difficult to get countries to agree on what to do. The European Union's Common Fisheries Policy is perhaps the most advanced international attempt to manage the fishing grounds belonging to this group of countries. While the fishing industry in the EU frequently complains that the amount of fish it is allowed to catch (the total allowable catch) is too low, environmental groups argue that the total allowable catch is much too high and that fishing in EU waters cannot be sustainable in the long term. Other people have an interest too. For example, consumers worry that if fewer fish are caught the price will increase.

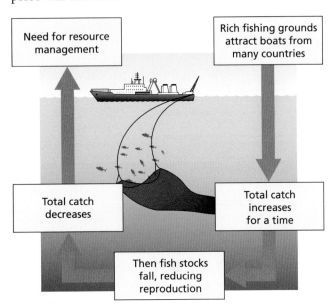

Figure 15 Fishing and resource management

The European Union also tries to manage agriculture in its member countries through its Common Agricultural Policy (CAP). In the early years the CAP's generous incentives for farmers encouraged high levels of production and the farming of marginal lands. It didn't seem that there was much thought about the environmental and other adverse consequences of maximising production. However, as the disadvantages became more obvious the CAP was reformed to take greater account of the environment. For example, farmers can now receive payments for taking their land out of agricultural production (set-aside).

It can be argued that the CAP is still a long way from truly sustainable agriculture, but there is no doubt it is moving in the right direction. The development of sustainable policies often occurs in stages.

Environmental impact statements and pollution control

Most countries now require some form of **environmental impact statement** for major projects such as a new road, an airport or a large factory. The objective is to identify all the environmental consequences and to try to minimise these as far as possible.

Industry has spent increasing amounts on research and development to reduce pollution – the so-called 'greening of industry'. In general, after a certain stage of economic development the level of pollution will decline (Figure 16). This is because countries have become more aware of their environmental problems with higher levels of economic activity and they have also created the wealth to invest in improving the environment.

The 1990s witnessed the first signs of 'product stewardship'. This is a system of environmental responsibility whereby producers take back a product, recycling it as far as possible, after the customer has finished with it. For example, in Germany the 1990 'take-back' law required car manufacturers to take responsibility for their vehicles at the end of their useful lives.

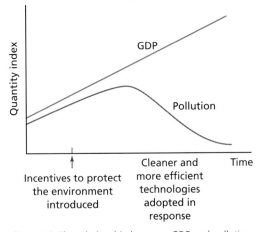

Figure 16 The relationship between GDP and pollution

International action

Increasingly, successful policies developed in one country are being followed elsewhere. A good example is the role of ecotourism in rainforest conservation (Figure 17). International organisations are the only hope of getting to grips with the really big problems, such as climate change. The success of international cooperation in tackling the hole in the ozone layer gives us reasonable hope for the future.

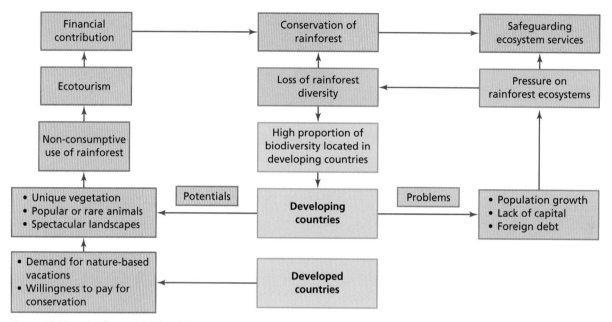

Figure 17 The role of ecotourism in rainforest conservation

Activities

1 What is meant by the terms (a) resource management and (b) sustainable development?
2 Explain the sequence of events shown in Figure 15.
3 Look at Figure 17. Describe and explain how ecotourism can enhance rainforest conservation.

● The importance of resource conservation

The **conservation of resources** is the management of the human use of natural resources to provide the maximum benefit to current generations while maintaining capacity to meet the needs of future generations. Conservation includes both the protection and rational use of natural resources. Conservation and the use of natural resources are social processes since they rely on people's behaviour, values and decisions. Both the demand for, and supply of, resources need to be planned and managed to achieve a sustainable system. In addition, there is growing pressure from individual governments and international organisations for a greater degree of equity in the use of the world's resources. They argue that there is a need for environmental policies and laws that contribute to more equitable sharing of the costs and benefits of conservation.

Conservation involves actions to use these resources most efficiently, thus extending their life as long as possible. For example, by **recycling** aluminium, the same piece of material is re-used in a series of products, reducing the amount of aluminum ore that must be freshly mined. Similarly, energy-efficient products help to conserve fossil fuels since the same energy services, such as lighting or transportation, can be attained with smaller amounts of fuel.

Figure 18 Various categories of recycling awaiting collection outside a suburban house in London, UK

Conservation also involves the **re-use** of resources. Plastic bags are an obvious example, but there are many others. Returning wire clothes hangers to dry cleaners, donating clothes and other items you no longer want to charities rather than dumping them, and repairing household items where possible rather than buying anew are all common examples in developed countries. In the recent recession, many shoe repairers increased their trade as people looked to extend the life of their shoes rather than have the expense of buying new shoes.

In many developing countries reusing plastic bottles, tins and other containers is a common form of conservation. There has long been a culture of re-use and recycling because people could not afford to buy new replacements. It is a common sight to see waste collectors roaming streets in search of re-usable items. The re-use of rubber tyres is an example.

Renewable resources, in contrast, can be seriously depleted if they are subjected to excessive harvest or otherwise degraded. No substitutes may be available for food products such as fish or agricultural crops. When the demise of biological resources causes the complete extinction of a species or the loss of a particular habitat, there can be no substitute for that diversity of life.

Various strategies can be used in the attempt to conserve resources. The agreement of quotas is an increasingly frequent resource management technique, illustrated by the EU's Common Fisheries Policy. **Quotas** involve agreement between countries to take only a predetermined amount of a resource. Quotas may change on an annual or longer time period basis. Much further along the line is **rationing**. This is very much a last-resort management strategy when demand is massively out of proportion to supply. For example, individuals might only be allowed a very small amount of fuel and food per week. Some senior citizens will remember that this happened during the Second World War.

At various times the use of **subsidies** has been criticised by environmentalists. It has been argued that reducing or abandoning some subsidies would aid conservation.

Figure 19 Green waste recycling in the UK

Waste reduction and recycling

Recycling involves the concentration of used or waste materials, their reprocessing, and their subsequent use in place of new materials. If organised efficiently, recycling can reduce demand considerably on fresh deposits of a resource. Recycling also involves the recovery of waste. New technology makes it possible to recover mineral content from the waste of earlier mining operations. However, the proportion of a material recycled is strongly related to the cost in proportion to the price of the original raw material, although governments are doing more and more to weaken this relationship.

Supermarkets have made significant efforts in recent years to reduce the number of plastic bags used. Some now charge for bags while others just encourage re-use of bags. Supermarkets like Tesco in the UK claim to have drastically reduced the number of new plastic bags used by customers each year. However, environmental groups are putting pressure on supermarkets to do more, claiming that many products have far too much packaging and the supermarkets themselves are high users of energy.

Recycling not only conserves valuable resources, it is also fundamental in the reduction of **landfill**. Landfill is undesirable for a number of reasons:

- Leachate pollution: leachate is solution formed when water percolates through a permeable medium. The leachate may be toxic or carry bacteria when derived from solid waste.

- Biodegradable waste rotting in landfill creates methane which is 21 times more potent as a greenhouse gas than carbon dioxide.
- Increasingly large areas of land are required for such sites.

Figure 20 Collecting and sorting plastic bottles in Ulaanbaatar, Mongolia

Product stewardship

Product stewardship is an approach to environmental protection in which manufacturers, retailers and consumers are encouraged or required to assume responsibility for reducing a product's impact on the environment. Also called 'extended producer responsibility' it is a growing aspect of recycling. In many cases this is a system of environmental responsibility whereby producers take back a product, recycling it as far as possible, after the customer has finished with it. For manufacturers, this includes planning for – and if necessary paying for – the recycling or disposal of the product at the end of its useful life. This may be achieved, in part, by redesigning products to use fewer harmful substances, to be more durable, re-useable and recyclable, and to make products from recycled materials.

Substitution

Substitution is the use of common and thus less valuable resources in place of rare, more expensive resources. An example is the replacement of copper by aluminium in the manufacture of a variety of products. Historically, when non-renewable resources have been depleted, new technologies

have been developed that effectively substitute for the depleted resources. New technologies have often reduced pressure on these resources even before they are fully depleted. For example, fibre optic cables has been substituted for copper ones in many electrical applications.

Activities

1 Explain the term 'conservation of resources'.
2 Define (a) re-use (b) recycling.
3 What is product stewardship?
4 How can substitution help to conserve important resources?

Energy efficiency

Meeting future energy needs in developing, emergent and developed economies while avoiding serious environmental degradation will require increased emphasis on radical new approaches which include:

- much greater investment in renewable energy
- conservation
- recycling
- carbon credits
- 'green' taxation.

Managing energy supply is often about balancing socio-economic and environmental needs. We have all become increasingly aware that this requires detailed planning and management. **Carbon credits** and **carbon trading** are an important part of the EU's environment and energy policies. Under the EU's emissions trading scheme, heavy industrial plants have to buy permits to emit greenhouse gases over the limit they are allowed (carbon credits) by government. However, this could be extended to other organisations such as banks and supermarkets. From 2008 the UK government has been offering the free provision of visual display electricity meters so that people can see exactly how much energy they are using at any time. Many countries are looking increasingly at the concept of **community energy**. Much energy is lost in transmission if the source of supply is a long way away. Energy produced locally is much more efficient. This will invariably involve **microgeneration**.

Table 5 summarises some of the measures governments and individuals can undertake to reduce the demand for energy and thus move towards a more sustainable situation.

Figure 21 Microgeneration – a small wind turbine on the roof of a school

Table 5 Examples of energy conservation measures

Government	Individuals
• Improve public transport to encourage higher levels of usage. • Set a high level of tax on petrol, aviation fuel etc. • Ensure that public utility vehicles are energy efficient. • Set minimum fuel consumption requirements for cars and commercial vehicles. • Impose congestion charging to deter non-essential car use in city centres. • Offer subsidies/grants to households to improve energy efficiency. • Encourage business to monitor and reduce its energy usage. • Encourage recycling. • Promote investment in renewable forms of energy. • Pass laws to compel manufacturers to produce more efficient electrical products.	**Transport** • Walk rather than drive for short local journeys. • Use a bicycle for short to moderate distance journeys. • Buy low fuel consumption/low emission cars. • Reduce car usage by planning more 'multi-purpose' trips. • Use public rather than private transport. • Join with others in a car pool. **In the home** • Use low-energy light bulbs. • Install cavity wall insulation. • Improve loft insulation. • Turn boiler and radiator settings down slightly. • Wash clothes at lower temperatures. • Purchase high energy efficiency appliances. • Don't leave appliances on standby.

● China's Pearl river delta

The Chinese economy has attained such a size and is continuing to grow so rapidly that it is now being called the 'new workshop of the world', a phrase first applied to Britain during the height of its industrial revolution in the nineteenth century. However, in the main industrial areas the environment has been put under a huge strain, leaving China with some of the worst pollution problems on the planet. One of China's main industrial regions is the Pearl river delta. It faces the challenge of continuing to grow economically while trying to protect its environment.

Case study: Environmental problems in the Pearl river delta

The Pearl river delta region, an area the size of Belgium, in south-east China (Figure 22) is the focal point of a massive wave of foreign investment into China. The Pearl river drains into the South China Sea. Hong Kong is located at the eastern extent of the delta, with Macau situated at the western entrance. The region's manufacturing industries already employ 30 million people but this number will undoubtedly increase in the future. Major industrial centres include Shunde, Shenzhen, Dongguan, Zhongshan, Zhuhai and Guangzhou.

In 2011, the Pearl river delta accounted for:

● 4.2 per cent of China's population
● 9.2 per cent of China's GDP
● 26.7 per cent of China's total exports.

The region is gradually growing into a single colossal megalopolis as individual cities expand outwards and move closer together.

Figure 22 The Pearl river delta

The three major environmental problems in the Pearl river delta are air pollution, water pollution and deforestation. In 2007, eight out of every ten rainfalls in Guangzhou were classified as acid rain. The high concentration of factories and power stations is the source of this problem, along with the growing number of cars in the province. The city has the worst acid rain problem in the province of Guangdong. The province's environmental protection bureau has reported that two-thirds of Guangdong's 21 cities were affected by acid rain in 2007. Overall, 45 per cent of the province's rainfall in 2007 was classified as acid rain.

Almost all the urban areas have overexploited their neighbouring uplands, causing a considerable reduction in vegetation cover. This has resulted in serious erosion. Half of the wastewater in Guangdong's urban areas is not treated before being dumped into rivers, compared with the national average of 40 per cent. Chemical oxygen demand (COD) is a key measurement of water pollution. Guangdong's government is working to reduce COD and also to cut sulfur dioxide emissions.

The environmental protection bureau classifies the environmental situation as 'severe' and says the government is committed to taking the 'necessary measures' to reduce pollution. Among the measures used to tackle the problems are (a) higher sewage treatment charges, (b) stricter pollution regulations on factories and (c) tougher national regulations on vehicle emissions.

Analyses of pollution problems in the Pearl river delta and elsewhere in China have focused strongly on institutional factors relating to the incentive structure for local government officials (to achieve a high level of economic growth) and the limited powers and independence of the Ministry of Environmental Protection.

Case study analysis

1   Describe the location of the Pearl river delta.
2   What are the causes of pollution in the Pearl river delta?
3   List the main measures used to tackle pollution in the Pearl river delta.

● End of theme questions

Topic 3.1: Development

Where a Billion People Still Live Without Electricity

According to a new report from the World Bank, 1.6 billion people gained access to electricity between 1990 and 2000, 70 percent of them in urban areas. But, as of 2010, 1.2 billion people were still living without it—173 million of them in urban areas. Because urban populations have been swelling even as access to electricity has grown, the global urban electrification rate actually hasn't changed much in 30 years, sitting at around 95 percent:

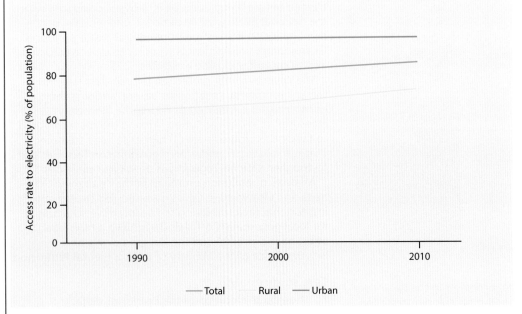

Figure 1 Global trends in the electrification rate, 1990–2010

The numbers in rapidly urbanising countries like India are particularly stark. Here are the populations, in millions of people, without access to electricity in the 10 countries with the farthest to go, as of 2010:

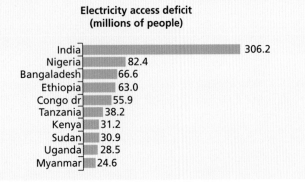

Electricity access deficit
(millions of people)

India 306.2
Nigeria 82.4
Bangladesh 66.6
Ethiopia 63.0
Congo dr 55.9
Tanzania 38.2
Kenya 31.2
Sudan 30.9
Uganda 28.5
Myanmar 24.6

Figure 2 Electricity access deficit for the 10 major countries

1 a How many people worldwide still lack access to electricity?
  b Describe the trends illustrated in Figure 1.
2 a Discuss the reasons for the different access rates in urban and rural areas.
  b Why has there been little or no change in the overall access to electricity?

3 a Comment on location and characteristics of the 10 countries with the greatest electricity access deficit (Figure 2).
  b How does lack of access to electricity hinder the development process?

Topic 3.2: Food production

Table 1 Adverse influences on global food production and distribution

Nature of adverse influence	Effect of adverse influence
Economic	• Demand for cereal grains has outstripped supply in recent years • Rising energy prices and agricultural production and transport costs have pushed up costs along the farm-to-market chain • Serious underinvestment in agricultural production and technology in LEDCs has resulted in poor productivity and underdeveloped rural infrastructure • The production of food for local markets has declined in many LEDCs as more food has been produced for export
Ecological	• Significant periods of poor weather and a number of severe weather events have had a major impact on harvests in key food-exporting countries • Increasing problems of soil degradation in both MEDCs and LEDCs • Declining biodiversity may impact on food production in the future
Socio-political	• The global agricultural production and trading system, built on import tariffs and subsidies, creates great distortions, favouring production in MEDCs and disadvantaging producers in LEDCs • An inadequate international system of monitoring and deploying food relief • Disagreements over the use of trans-boundary resources such as river systems and aquifers

1 a Why has the demand for cereal grains increased in recent years?
   b Suggest why the demand for cereal grains has outstripped supply.
   c Why has the production of food for local markets declined in many developing countries (LEDCs)?
2 a Discuss the ways in which severe weather conditions can reduce food production.
   b What is soil degradation and how does it affect food production?
3 a What is food aid (food relief)?
   b Why are people often critical of the way is which food aid is organised?

Topic 3.3: Industry

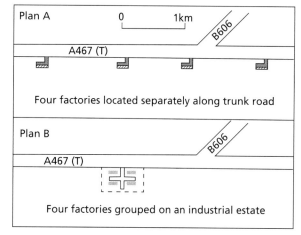

Figure 3 Groupings of factories

1 a How do Plans A and B differ?
   b Describe the characteristics of an industrial estate.
   c Suggest two ways in which the factories might save money by locating together on an industrial estate.
2 a Why might the local council and people living in the general area prefer Plan B to Plan A?
   b Discuss one possible disadvantage to the local community of Plan B to Plan A.
3 Suggest why high technology industries often cluster together.

Topic 3.4: Tourism

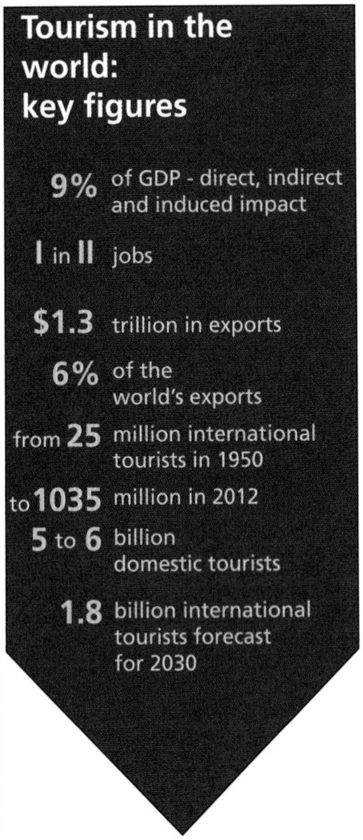

Tourism in the world: key figures

9% of GDP - direct, indirect and induced impact

1 in **11** jobs

$1.3 trillion in exports

6% of the world's exports

from **25** million international tourists in 1950

to **1035** million in 2012

5 to **6** billion domestic tourists

1.8 billion international tourists forecast for 2030

Tourism key to development, prosperity and well-being

- An ever increasing number of denstinations have opened up and invested in tourism, turning tourism into a key driver of socio-economic progress through export revenues, the creation of jobs and enterprises, and infrastructure development.

- Over the past six decades, tourism experienced continued expansion and diversification, becoming one of the largest and fastest-growing economic sectors in the world. Many new destinations have emerged apart from the traditional favourites of Europe and North America.

- Despite occasional shocks, international tourist arrivals have shown virtually uninterrupted growth – from 25 million in 1950, to 278 million in 1980, 528 million in 1995, and 1035 million in 2012.

Long-term outlook

- International tourist arrivals worldwide will increase by 3.3% a year from 2010 to 2030 to reach 1.8 billion by 2030 according to UNWTO long term forecast, Tourism Towards 2030.

- Between 2010 and 2030, arrivals in emerging destinations (+4.4% a year) are expected to increase at double the pace of those in advanced ecconomies (+2.2% a year).

- The market share of emerging economies increased from 30% in 1980 to 47% in 2012, and is expected to reach 57% by 2030, equivalent to over one billion international tourist arrivals.

Figure 4 Tourism in the world – key figures

1 a Define tourism.
  b Describe the increase in international tourism between 1950 and 2012 and the forecast for 2030.
  c Discuss three reasons for the growth of international tourism.
2 a What proportion of global employment is related to tourism?
  b Explain the difference between direct and indirect employment in the tourist industry.
3 a What proportion of world trade is related to tourism?
  b Why is tourism more important to the economies of some countries than others?
4 Discuss the disadvantages of tourism to developing countries.

Topic 3.5: Energy

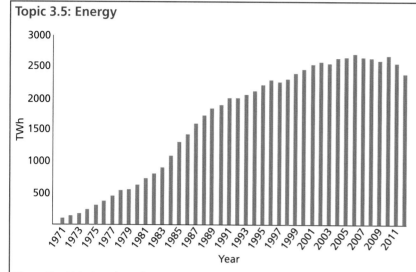

There are now over 430 commercial nuclear power reactors operating in 31 countries. About 70 more reactors are under construction. Sixteen countries depend on nuclear power for at least a quarter of their electricity. In 2011 and 2012, output declined due to cutbacks in Japan and Germany following the Fukushima accident.

Figure 5   Global nuclear electricity production 1971-2012

1   a   How many countries operate nuclear power reactors?
    b   Name two countries that produce nuclear electricity.
2   a   Describe the changes in global nuclear electricity production between 1971 and 2012 (Figure 5).
    b   Suggest reasons for the changes you identify.

3   a   Some countries that could build nuclear power plants have decided not to. Explain three concerns about the production of nuclear electricity.
    b   Give three advantages of nuclear power according to people who support the use of this form of electricity production.

Topic 3.6: Water

BILLIONS DAILY AFFECTED BY WATER CRISIS

Without water, life would not exit. It is a prerequisite for all human and economic development.

Yet today, 780 million people – about one in nine–lack access to clean water. More than twice that many, 2.5 billion people, don't have access to a toilet.

In most developed nations, we take access to safe water for ground But this wasn't always the case. A little more than 100 years ago, New york, London and Paris were centers of infectious disease. Child death rates were as high then as they are now in much of sub-saharan Africa. It was sweeping reforms in water and sanitation that enabled human progress to leap forward. It should come as no surprise that in 2007, a poll by the British Medical Journal found that clean water and sanitation comprised the most important medical advancement since 1840.

The health and economic impacts of today's global water crisis are staggering.

• More than 3.4 million people die each year from water, sanitation and hygiene-related casuses. Nearly all deaths, 99 percent, occur in the developing world.

• 2.5 billion people lack access to improved sanitation; 1.1 billion still practise open defecation.

• Lack of access to clean water and sanitation kills children at a rate equivalent of a jumbo jet crashing every four hours.

• 443 million school days are lost each year due to water-related illness.

• Women and children bear the primary responsibility for water collection in the majority of households. This is time not spent working at an income-generating job, caring for family members or attending school.

Figure 6   The global water crisis

1   a   How many people worldwide lack access to clean water and proper sanitation?
    b   In which parts of the world are these problems most severe?
    c   How do water and sanitation problems impact on i) health ii) education and iii) employment?
2   a   Define potable water.
    b   Explain the difference between physical water scarcity and economic water scarcity.
3   How can the problems of lack of access to clean water and proper sanitation in poor countries be tackled?

Topic 3.7: Environmental risks of economic development

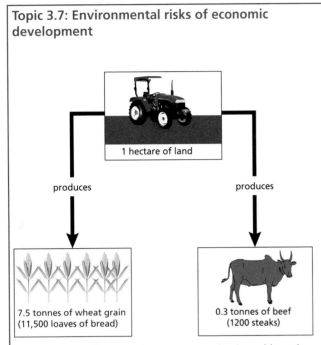

Figure 7 The efficiency of energy conversion in arable and pastoral farming

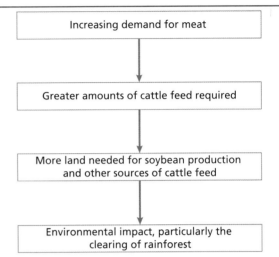

Figure 8 The environmental impact of the increasing demand for meat

1 a Compare the production of wheat grain and beef from one hectare of land.
   b Which type of farming will feed the most people?
2 a To what extent and why does the consumption of meat vary between developed and developing countries?
   b Why is the global demand for meat increasing at a significant rate?
3 Discuss the impact of this increasing demand for meat on the environment.

Geographical skills

● Scale

Most Ordnance Survey maps that we use are either at a 1:50 000 or a 1:25 000 scale. On a 1:50 000 map, 1 cm on the map relates to 50 000 cm on the ground. On a 1:25 000 map every 1 cm on the map relates to 25 000 cm on the ground. In every kilometre there are 100 000 centimetres (1000 m × 100 cm). So:

● on a 1:50 000 map every 2 cm corresponds to a kilometre
● on a 1:25 000 map every 4 cm corresponds to a kilometre.

A 1:25 000 map is more detailed than a 1:50 000 map and is therefore an excellent source for geographical enquiries. 1:50 000 maps provide a more general overview of a larger area. You may come across other scales, for example 1:10 000 and 1:2500.

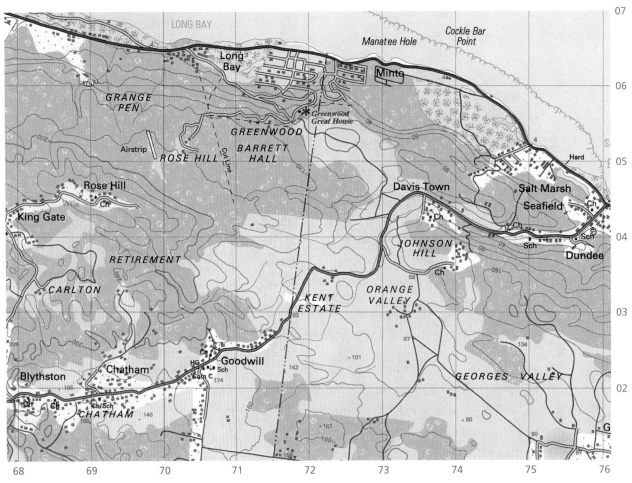

Figure 1 Part of the 1:50 000 map of Jamaica

Measurement on maps is made easier by grid lines. These are the regular horizontal and vertical lines you can see on an Ordnance Survey map.

The horizontal lines are called **northings** and the vertical lines are called **eastings**. They help to pinpoint the exact location of features on a map.

● Grid and square references

Grid references are the six-figure references that locate precise positions on a map. The first three figures are the eastings and these tell us how far a position is across the map. The last three figures are the northings and these tell us how far up the map a position is. An easy way to remember which way round the numbers go is 'along the corridor and up the stairs'.

In Figure 1, the church at Rose Hill is located at 691045 and the church at Davis Town is found at 737043.

Sometimes a feature covers an area rather than a point, for example all of the villages and the areas of woodland in Figure 1. Here a grid reference is inappropriate, so we use four-figure square references.

- The first two numbers refer to the eastings.
- The last two numbers refer to the northings.

The point where the two grid lines meet is at the bottom left-hand corner of the square. So in Figure 1, most of the village of Seafield is found in 7504. Some features may occur in two or more squares, for example Long Bay is found in squares 7006 and 7106.

● Direction

Directions can be expressed in two ways:

- compass points, for example south-west
- compass bearings or angular directions, for example 45°.

Sixteen compass points are commonly used. Some of these are shown in Figure 2.

Compass bearings are more accurate than compass points but they can be quite confusing. Compass bearings show variations from magnetic north. This is slightly different from the grid north on the Ordnance Survey map (which is the way in which the northings go). True north is different again – this is the direction of the North Pole.

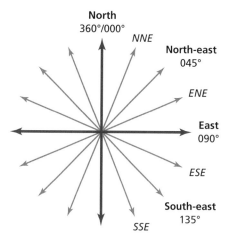

Figure 2 Compass points

● Relief and gradient

Contour lines

A **contour line** is an imaginary line that joins places of equal height.

- When the contour lines are spaced far apart the land is quite flat.
- When the contour lines are very close together the land is very steep (when the land is too steep for contour lines a symbol for a cliff is used).
- When contour lines are close together at the top, and then get further apart, this suggests a concave slope.
- When contour lines are close together at the bottom and far apart at the top, this suggests a convex slope.

Figure 3 Cable car from Lago di Fadala to Marmaloda Glacier

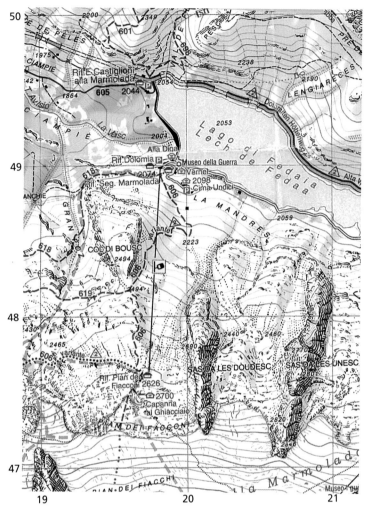

Figure 4 Part of the 1:25 000 map showing Marmaloda Glacier and Lago di Fadala

Gradients

The **gradient** of a slope is its steepness. We can get a rough idea of the gradient by looking at the contour pattern. If the contour lines are close together the slope is steep, and if they are far apart the land is quite flat. However, these are not very accurate descriptions. To measure gradient accurately we need two measurements:

- the vertical difference between two points (this can be worked out using the contour lines or spot heights)
- the horizontal distance between two places – this may or may not be a straight line (for example, a meandering stream would not be straight).

Working out gradient

Make sure that you use the same units for both vertical and horizontal measurements.

Divide the difference in horizontal distance (D) by the height (H). If the answer is, for example, '10' or '5' express it as '1:10' ('one in ten') or '1:5' ('one in five'). This means that for every 10 metres along you rise (or drop) one metre, or for every five metres in length the land rises (or drops) one metre. Alternatively, divide the height (H) by the difference in horizontal distance (D) and multiply by 100 per cent (H/D × 100%). This expresses the gradient as a percentage.

Activities

Study the OS map shown in Figure 1.

1. How far is it:
   a in a straight line
   b by road
   from the school in Goodwill to the school in the middle of Dundee?
2. What is the length of the coastline (to the nearest kilometre) as shown on the map extract?
3. Approximately how long is the air strip?
4. How wide is (a) the coral in Long Bay and (b) the mangrove forest between Minto and Salt Marsh?
5. What is the six-figure grid reference for:
   a the two schools at Dundee
   b Greenwood Great House?
6. What is found at 705023?

7. Give the four-figure grid reference for Chatham and for Davis Town.
8. Suggest reasons why there is an airstrip in grid square 6905.
9. In what direction is:
   a Long Bay from Davis Town
   b Goodwill from Rose Hill?

Study Figures 3 and 4.

10. In what direction is the cable car moving (towards the glacier)?
11. Approximately what is the altitude of (a) the lake, and (b) the Marmaloda glacier? (Note that contours are shown at 25 m intervals.)
12. Using map evidence, suggest why a hydro-electric power station was built at the head of Lago di Fadala.

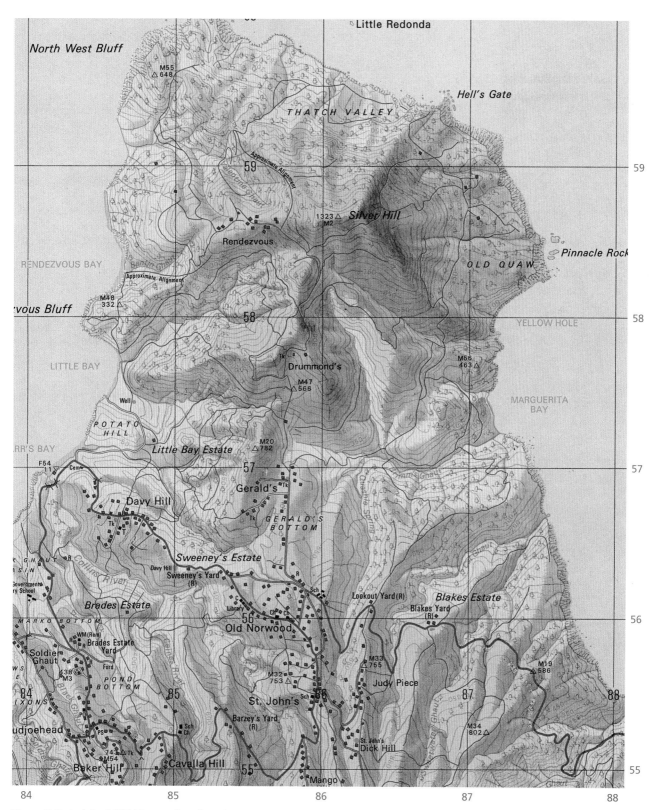

Figure 5 Part of the 1:25 000 map of northern Montserrat

Activities

Study Figure 5.

1 What is the height of (a) Silver Hill (8658) and (b) Baker Hill (8455)? (Note that the contours on this map are drawn at 50 feet intervals; assume that 3 feet equals 1 m.)
2 In what direction does Little Bay (8457) face?
3 How steep is the slope between Silver Hill and the coastline at Thatch Valley (8659)? Measure from the peak of Silver Hill to the nearest point of the coast in Little Redonda. Express your answer as a 'one in x slope'.

4 Describe the relief (height and gradient) of squares 8658 (Silver Hill), 8457 (Potato Hill) and 8655 (Judy Piece).
5 Following an eruption of the Soufrière volcano in 1997, much of the southern third of the island was evacuated. Plans were made to develop the northern part of Montserrat. Study the map:
  a Comment on the problems of developing the northern part of the island.
  b Which, in your opinion, is the best location to develop housing, services and economic activity. Give reasons for your answer.

● Cross-sections

A **cross-section** is a view of the landscape as it would appear if sliced open, or if you were to walk along it. It shows variations in gradient and the location of important physical and human features. Here's how to draw a cross-section:

1 Place the straight edge of a piece of paper between the two end points (Figure 6).
  a Mark off every contour line (in areas where the contours are very close together you could measure every second contour or significant contours, for example every 100 m).
  b Mark off important geographical features.
2 Align the straight edge of the piece of paper against a horizontal line on graph paper, which is exactly the same length as the line of the section. Use a vertical scale of 1 cm:50 m, or 1 cm:100 m; if you use a smaller scale (for example 1 cm:5 m) you will end up with a slope that looks Himalayan!
  a Mark off with a small dot each of the contours and the geographic features.
  b Join up the dots with a freehand curve.
  c Label the features.
  d Remember to label the horizontal and vertical scales, the title, and the grid references for the starting and finishing points.

Figure 5 shows a cross-section of an extinct volcano in France. See how steep the volcano looks when a scale of 1 cm:50 m is used compared with 1 cm:100 m.

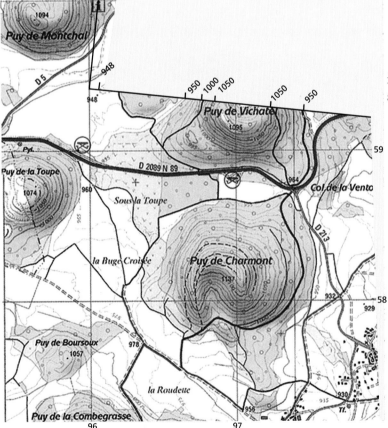

Figure 6 Drawing a cross-section

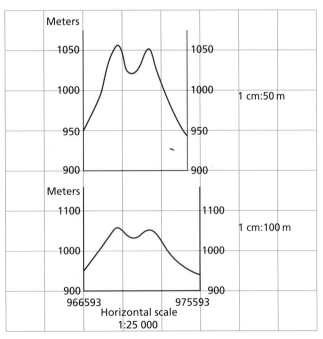

Figure 7 Cross-sections of Puy de Vichatel

Describing river landscapes

The long profile of a river can be shown on a line graph when the height of a river above base level is plotted against distance from its source. As rivers evolve through time and over distance, streams pass through a series of distinct changes. Figure 8 shows the long profile of a river and illustrates these stages.

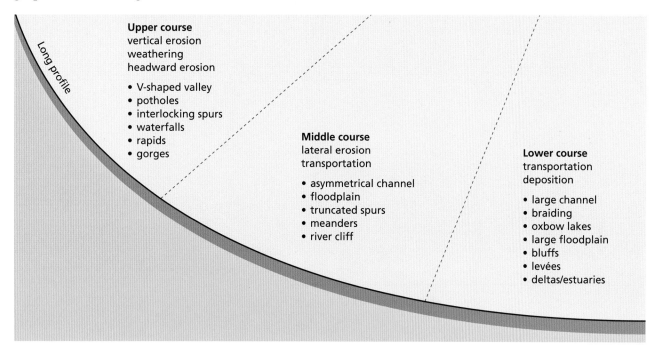

Upper course
vertical erosion
weathering
headward erosion

- V-shaped valley
- potholes
- interlocking spurs
- waterfalls
- rapids
- gorges

Middle course
lateral erosion
transportation

- asymmetrical channel
- floodplain
- truncated spurs
- meanders
- river cliff

Lower course
transportation
deposition

- large channel
- braiding
- oxbow lakes
- large floodplain
- bluffs
- levées
- deltas/estuaries

Long profile

Figure 8 Features associated with different stages of a river

Describing the stage of the river

- Is the river in its upper, middle or lower course?
- Use the contour lines to describe the shape of the river valley – a V-shaped valley with close contour lines suggests the upper course; more gentle slopes with a broad, flat floodplain suggests the lower course.
- Look at the size and shape of the river channel.
- Is the channel constrained by relief – that is, does it flow around interlocking spurs?
- Does the river meander across a flat floodplain?
- What are the features of the river? Can you identify any of the features listed in Figure 8 from map evidence?

Rivers have had a profound effect on both the site and situation of human settlements. Human activities have also had an increasing impact on drainage basins and river channels. Map evidence can be used to identify the relationship between rivers and human activities.

Describing a river's influence on site and situation

- Is the river navigable? (Is it straight and wide enough to allow boats to pass up it?)
- Does the river valley provide the only flat land in an area of rough terrain?
- Does the valley provide a natural routeway for roads and railway lines?
- Do settlements avoid the river's floodplain and locate on higher dry-point sites?
- Are settlements located at crossing points on a river? (The name endings of the settlement, such as 'ford' and 'bridge', are evidence of this.)

The human impact on river systems

- Is there evidence of forest clearance and wetland reclamation for agriculture?
- Does the map show any of the following land use changes which can affect a river and its drainage basin: mining activity, industrialisation, urbanisation, land drainage schemes?
- Has there been any direct interference with rivers through reservoir construction, channel straightening, dams, new channels?
- Are there any obvious sources of pollution (industry, sewage works) on the map?

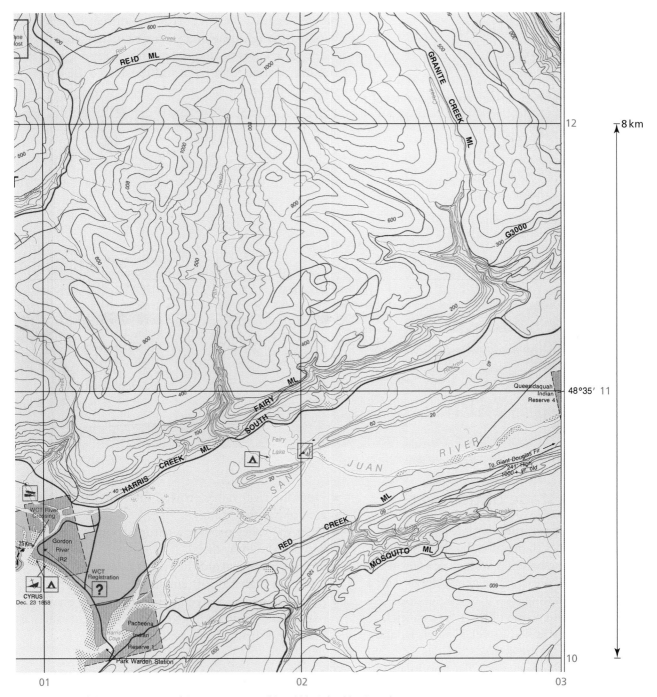

Figure 9 Part of the 1:14 000 map of the West Coast Trail in British Columbia, Canada

Activities

Study Figure 9.

1 Describe the relief (height and gradient) of the map extract.
2 In which direction does the Fairy Creek flow?
3 a Describe the valley of the San Juan river.
  b How does this compare with the southern part of Fairy Creek?

4 a What is the altitude (height) of the source (start) of Fairy Creek?
  b What is its altitude when it reaches Fairy Lake?
  c What is the distance from the source of Fairy Creek to Fairy Lake?
  d What is the gradient of Fairy Creek between its source and Fairy Lake? (Express the answer as a 1:x gradient, where gradient = vertical difference/horizontal distance.)

Coastal landforms

Describing coastal scenery

- Does the coastline have steep slopes and cliffs suggesting a coastline of erosion? Or are there wide expanses of sand and mud suggesting deposition?
- Are there many headlands and bays indicating local changes in processes?
- Is the coastline broken by river mouths or estuaries?

- What is the direction of the coastline?
- Is there any evidence of longshore drift, for example spits, bars, tombolos?
- Are any of the features named? Give names and grid references.
- Is there any map evidence of human attempts to protect the coastline, for example groynes, sea walls, breakwaters?
- Does the map tell you whether the stretch of coastline is protected or open?

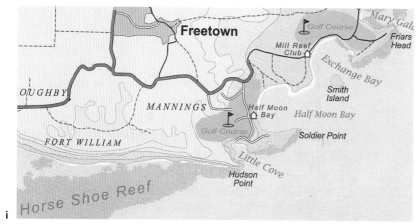

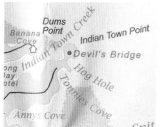

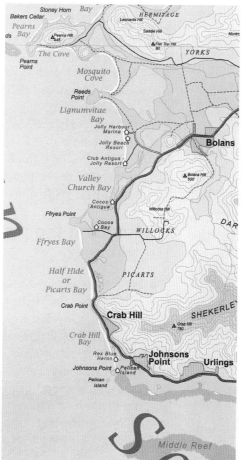

Figure 10 Extracts from a map of Antigua

Figure 11 Some coastal landforms in Antigua

Activities

1 Match each of the map extracts in Figure 10 with the three photographs in Figure 11.
2 Try to identify the cliff that is shown on Figure 11c. What is the map evidence to support your answer?
3 What type of feature is shown in Figure 11? Try to find the named example of the feature on one of the maps.
4 What features are shown in Figure 11b? Find located examples of these features on the map extracts.
5 What is the map evidence to suggest that tourism and recreation are important to this area?
6 Using map evidence, suggest how easy or difficult it may be to develop tourism further in the area.

Activity

Using Figure 12, give reasons why this area may be volcanic in origin.

● Rural settlement

Geographers should be able to find various types of information when studying rural settlements on a map. They should be able to comment on the following key terms:

- **Site**: the immediate location of a settlement – that is, the land upon which it is built, for example on a floodplain, close to a river, on a south-facing slope, on a crossroads, at a wet point or dry point.
- **Situation**: the relative location of a settlement to a larger area.
- **Function**: any service or employment opportunities that a settlement offers, for example commercial, recreational, industrial, agricultural, etc.
- **Shape**: the appearance of the settlement, for example linear, compact, T-shaped.

Describing and explaining the site and importance of a settlement

- Describe its location in relation to the relief of the area, for example valley floor, near or away from a river, direction of slope, dry point, wet point, exposed, sheltered, etc.
- Describe how many routes there are and which forms of transport are available. How important are the routes that meet at the settlement? How does the relief affect these routes?

Describing the form or shape of a settlement

- Is it nucleated or dispersed?
- Is it a linear or cruciform settlement?
- Is any of it modern? (Modern settlement may be recognised by a regular geometric street pattern and more widely spaced houses.)
- Have physical features influenced its shape? For example, a steep hillside or a floodplain may limit the growth of a town so that it becomes elongated, following the direction of the level or dry land.

● Describing volcanic landscapes

When describing volcanic landscapes consider the following questions:

- Is the land gently sloping or steep?
- Is there a crater present at the top of the volcano?
- Is there a central cone or are there lots of secondary cones?
- Are there any lava flows present?
- Is there vegetation present?
- Is there any other map evidence (quarrying, mining, tourism) that can be used to infer volcanic activity?

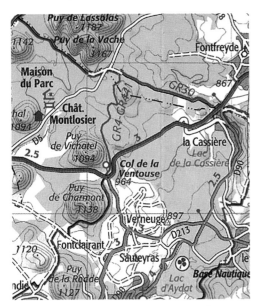

Figure 12 Extract from the 1:25 000 map of Lac d'Aydat

Describing the size and function of a settlement

- What is the size of the settlement? How many grid squares does it cover? (Each grid square represents 1 km².)
- Are the houses tightly packed together or dispersed?
- What functions are evident? For example, is it residential (housing), commercial (post office, administrative buildings), schools, industrial (works, quarries, railway sidings) and/or tourist-related (tourist information centre, viewpoints)?

Describing the situation of a settlement

- Using the whole map area, describe the site of the settlement in relation to large urban centres, motorways, roads, large rivers or other large-scale physical features such as hills or valleys.

- Is the settlement on a rail link?
- How accessible is the settlement to motorways, railways and large urban areas? It might be situated close to a motorway but have no direct access to it.

Describing and accounting for the general distribution of settlement

- Locate areas with little or no settlement. Account for the lack of settlement in terms of natural disadvantage, for example exposed position, steep gradient, flat land in danger of flooding. How is this land used?
- Locate areas of fairly close settlement. Account for this in terms of natural advantages for land use and occupation, for example farming, water, soil or south-facing slope.

Figure 13 Masca, Tenerife

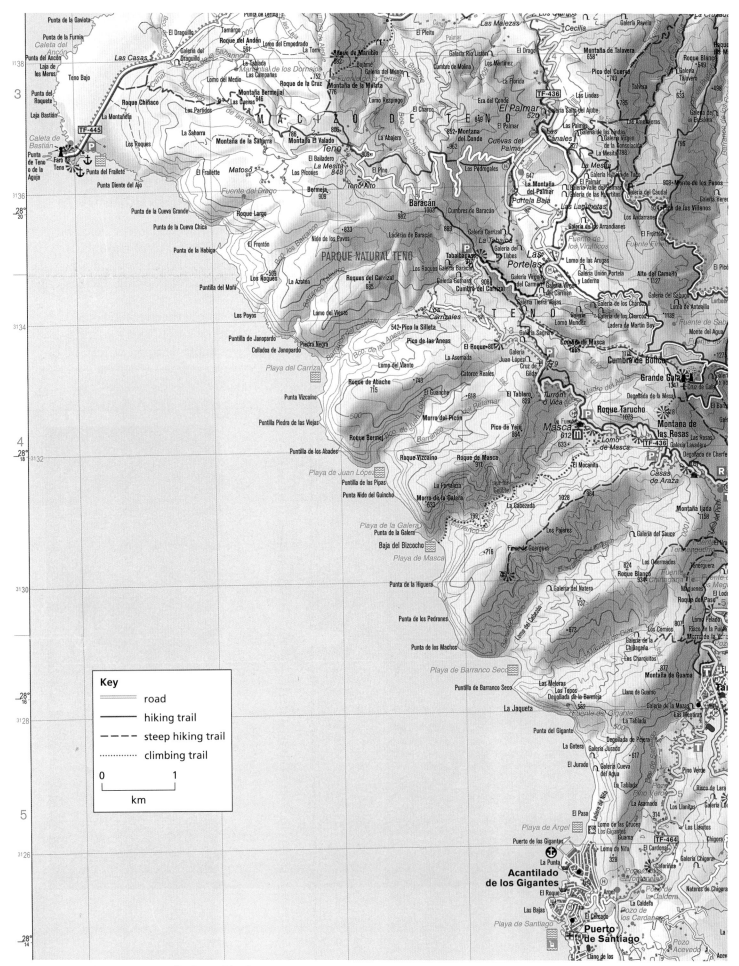

Figure 14 Part of south-west Tenerife, Canary Islands

Activities

Study Figure 14.

1 Describe the pattern of settlement (grey squares) as shown on the map extract.
2 Describe the relief and gradient of the area west of Masca.
3 Describe the road network (yellow lines) on the map extract.
4 What opportunities does this environment offer? Suggest reasons and use the map evidence and Figure 13 to help you.
5 What difficulties does this environment create? Again give evidence and suggest reasons.

Note: The red line is a hiking trail, the dashed red line is a steep hiking trail, and the red dots are climbing trails.

● Urban settlement

Urban landscapes

It is important to identify different land use types when describing an urban landscape.

- Identify the lines of communication, for example roads, railways, airports.
- Identify the types of land use, for example residential, industrial, commercial. What kind of residential land use is it? Is it terraced, semi-detached, working class, middle class? Is there industry present? If so, what kind? Can you see a CBD or grouping of shops?
- If the map is of a whole settlement, can you identify old and new areas?
- Use descriptive words when discussing the area: high-density/low-density housing; regular/haphazard roads; derelict/high-tech industry.
- Can you identify any other land uses, for example green spaces, derelict land, churches, schools, hospitals?

Industrial location

The factors determining industrial location are changing. In the early part of the twentieth century, heavy industries like iron and steel and car manufacturing were located close to raw materials and/or markets. Today, manufacturing industry is drawn to out-of-town or edge-of-town sites. One of the factors is space and another is cost. There is a more, cheaper land available away from the built-up areas of urban areas. Another factor is accessibility: edge-of-town sites are closer to communications (motorways and railways) and the residential areas where workers live. For many heavy industries, location by a deepwater channel is important for the import of raw material and the export of finished goods.

Activities

Study Figure 15.

1 Describe the site of St Catharines.
2 Suggest contrasting reasons for the lack of settlement in parts of squares 4081 and 4679.
3 a What industries are located in 4681 and 4781?
   b Suggest reasons for the large-scale industry in squares 4581 and 4681.
4 Describe the distribution of shopping centres, as shown on the map.
5 Contrast the pattern of roads in square 4280 with those in 4183 and 4083.
6 Give a four-figure grid reference for (a) Martindale Pond and (b) Port Weller Harbour.
7 Give a six-figure grid reference for Niagara College (near the centre of St Catherines).

Study Figure 16.

8 Give the square references for two (industrial) estates.
9 Compare the two sites that you have identified.
10 Suggest the reasons for the development at Tseung Kwan O industrial estate.
11 Approximately what size are Sheung Tak Estate and Tseung Kwan O?
12 How does the land in the two estates differ from the land on the rest of the island, as shown on the map?

● Pie charts and proportional pie charts

Pie charts

Pie charts are subdivided circles. These are frequently used on maps to show variations in the composition of a geographical feature, for example gross regional domestic product (Figure 17). The pie chart may also be drawn proportional in size, to show an extra dimension, in this case the size of GRDP.

Figure 15 1:50 000 map of St Catharines, Ontario, Canada

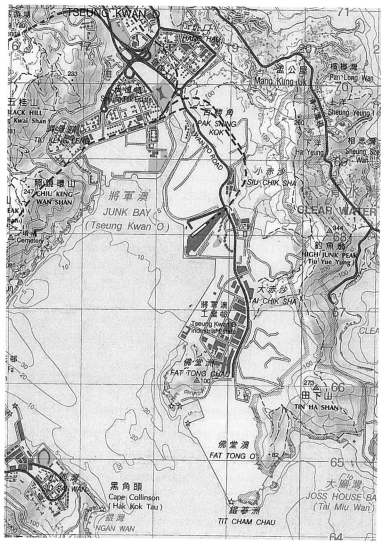

Figure 16 1:50 000 map of Hong Kong

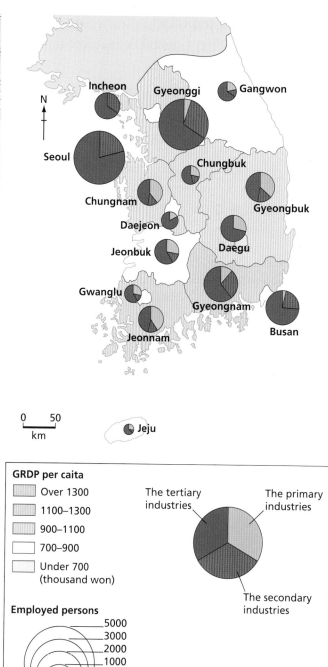

Figure 17 Example of the use of pie charts: employment and gross regional domestic product (GRDP) in South Korea, 2000

Plotting the pie chart

The following steps should be followed in the construction of a pie chart:

1 Convert the data into percentages.
2 Convert the percentages into degrees (by multiplying by 3.6 and rounding up or down to the nearest whole number).
3 Draw appropriately located circles on your base map.
4 Subdivide the circles into sectors using the figure obtained in step 2.
5 Differentiate the sectors by means of different shadings or colours.
6 Draw a key explaining the scheme of shading and/or colours.
7 Give your diagram a title.

Bar charts

In a bar chart, the length of the bar represents the quantity of the component being measured, for example places or time intervals. The vertical axis has a scale which measures the quantity. There are four main types of bar chart:

- **Simple bar charts** – each bar indicates a single factor.
- **Multiple or group bar chart** – features are grouped together on one graph to help comparison.
- **Compound bar chart** – various elements or factors are grouped together on one bar (the most stable element or factor is placed at the bottom of the bar to avoid confusion).
- **Percentage compound bar chart** – this is a variation on the compound bar chart. It is used to compare features by showing the percentage contribution. These graphs do not give a total in each category but compare relative changes in terms of percentages.

Activities

Table 1 Foreign investment into Korea (US$ million)

Year	Total	USA	Japan	UK
1985	532.2	108.0	364.3	12.3
1995	1947.2	644.9	418.3	86.7
2000	15216.7	2922.0	2448.0	84.0
2005	11563.5	2689.8	1878.8	2307.8

1 a Using the data in Table 1, draw a compound bar graph to show how foreign investment varied between 1985, 1995 and 2005. Draw one bar for 1985, one for 1995 and one for 2005.
  b Describe the changes as shown in your bar chart.

Table 2 Contribution of the main economic sectors to Korea's GDP

Agriculture	Industry	Services
2.6%	39.2%	58.2%

2 a Using the data in Table 2, draw a pie chart showing the contribution of each economic sector to Korean GDP.
  b Comment on the chart you have drawn.

● Scatter graphs

Scatter graphs show how two sets of data are related to each other, for example population size and number of services, or distance from the source of a river and average pebble size. To plot a scatter graph, decide which variable is independent (population size/distance from the source) and which is dependent (number of services/average pebble size). The independent is plotted on the horizontal or x axis and the dependent on the vertical or y axis. For each set of data, project a line from the corresponding x and y axis and where the two lines meet mark the point with a dot or an X (Figure 18).

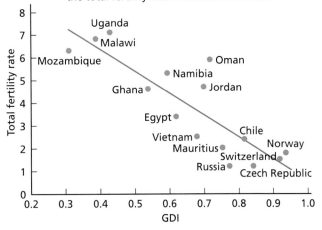

Figure 18 Scatter graph

Triangular graphs

Triangular graphs are used to show data that can be divided into three parts, for example soil (sand, silt and clay), employment (primary, secondary and tertiary), or population (young, adult and elderly) (Figure 19). They require that the data are in the form of a percentage and that the percentages total 100%. The main advantages of the triangular graph are:

- they allow a large number of data to be shown on one graph (think how many pie charts or bar charts would be needed to show all the data on Figure 19)
- groupings are easily recognisable, for example in the case of soils groups of soil texture can be identified
- dominant characteristics can be shown easily
- classifications can be drawn up.

Triangular graphs can be tricky and it is easy to get confused, especially if care is not taken, but they do provide a fast, reliable way of classifying large amounts of data which have three components.

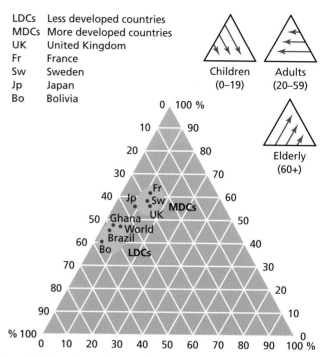

LDCs	Less developed countries
MDCs	More developed countries
UK	United Kingdom
Fr	France
Sw	Sweden
Jp	Japan
Bo	Bolivia

Figure 19 Triangular graph to show population composition in selected countries

Activities

1 Construct a scatter graph using the data in Table 3.

Table 3

Site	Discharge (m³/sec)	Suspended load (g/m³)
1	0.45	10.8
2	0.42	9.7
3	0.51	11.2
4	0.55	11.3
5	0.68	12.5
6	0.75	12.8
7	0.89	13.0
8	0.76	12.7
9	0.96	13.0
10	1.26	17.4

When all the data are plotted, a line of best fit is drawn. This does *not* have to pass through the origin. It is useful to label some of the points, for example the highest and smallest anomalies (exceptions), especially if these are referred to in any later description.

2 On a copy of Figure 20 and using the data in Table 4, show how the workforce of Korea has changed over time.

Table 4 Percentage of Korean workforce employed in primary, secondary and tertiary industries, 1970–2000

	Primary industries	Secondary industries	Tertiary industries
1970	50.4	14.3	35.3
1980	34.0	22.5	43.5
1990	17.9	27.6	54.5
2000	10.9	20.2	68.9

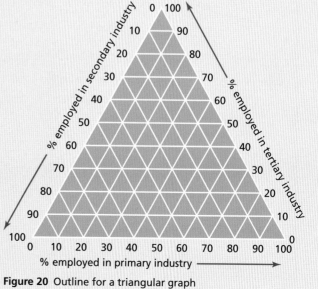

Figure 20 Outline for a triangular graph

● Sketch maps and annotated photographs

You can label a photograph or diagram to make it very informative. It is important that you label clearly all the important features.

Many photographs that are used in exams are aerial views which show industrial, residential, recreational and commercial land uses. In your projects, however, you are more likely to use much simpler photos. If you study these carefully you can find out a number of interesting features. Examine Figure 21 and then answer the questions in the following Activities.

Figure 21 Aerial view of Hyundai shipyard in Pusan, Korea

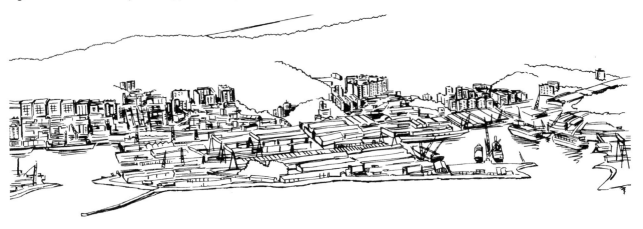

Figure 22 Sketch drawing of Figure 21

Activities

1 a Study Figure 21. Copy Figure 22 and add the following labels in the most appropriate locations.

   Harbour wall to reduce wave energy
   Flat land for large-scale industrial development
   Lack of development on steep ground
   Deep estuary allows development of port industries
   High-rise residential accommodation
   Large docks for ships to be repaired or built

   b Add a couple of other labels based on your own observations.

2 Study Figure 23. Make a sketch diagram of the resort and add the following labels:

   Purpose-built holiday resort
   Easy access to the beach
   Boat moorings
   Fine, white sandy beach
   Bay
   Lagoon

3 Make a sketch of Figure 24, the Caledonian Canal, and add the following labels:

   The main river
   The Caledonian Canal
   A lock
   Barges and river boats
   Large pastoral fields
   An arable field
   Areas of woodland
   Settlement – possibly tourist accommodation

Figure 24 Caledonian Canal

Figure 23 Jolly Harbour resort in Antigua

Geographical investigations: coursework and the alternative to coursework

Risk assessment is essential for all geographical investigations

For Paper 3, which is entitled 'Coursework', a school-based assignment set by teachers of up to 2000 words must be completed. The proposals for the coursework undertaken by candidates must be approved beforehand by the examination board, Cambridge International Examinations. An example of an outline submission to Cambridge International Examinations is provided in the syllabus document.

The coursework investigation can be based on human geography, physical geography, or on an interaction between the two. The assignment must be clearly related to one or more of the syllabus themes. The piece of coursework undertaken must cover the assessment criteria in the following proportions:

- Knowledge with understanding – 12 marks
- Skills and analysis (observation and collection of data, organisation and presentation of data, analysis and interpretation) – 12 marks each, giving a total of 36 marks
- Judgement and decision making (conclusion and evaluation) – 12 marks.

Examples of suitable coursework assignments can be found in the syllabus document. The wide-ranging scope of this syllabus offers many possibilities for coursework.

Paper 3 is not a compulsory paper, as Paper 4 'Alternative to coursework' can be taken instead. Here candidates answer two compulsory questions, completing a number of written tasks based on the three themes covered in Paper 1. The questions in Paper 4 involve an appreciation of techniques used

in fieldwork studies. The marks allocated to the assessment criteria for Paper 4 are exactly the same as for Paper 3. Guidance on the enquiry skills for Paper 4 is provided in the syllabus document.

This component of the syllabus (either Paper 3 or Paper 4) accounts for 27.5% of the total assessment, the same weighting as Paper 2. The information provided in this section of the book is applicable to both Papers 3 and 4.

The enquiry skills for Paper 3, known as the **route to geographical enquiry**, can be set out in the following stages:

1 Identification of an issue, question or problem
2 Defining the objectives of the study
3 Collection of data
4 Selection and collation of data
5 Presentation and recording of results
6 Analysis and interpretation
7 Conclusions, evaluation and suggestions for further work.

● Identification of an issue, question or problem

This section effectively forms your introduction to the investigation. Here you clearly identify a topic for investigation through observation, discussion, reading or previous study. Many geographical investigations begin by stating one or a number of **hypotheses** used to test the issue, question or problem. Hypotheses are the ideas you intend to test. Before you can set out your hypotheses with confidence you need to ensure that you have a good understanding of the topic (for example, sand dunes) under consideration. Studying the **geographical background** should ensure that you have a clear knowledge and understanding of the **theories** or **models** that are used to try to explain your enquiry. You will refer back to these theories and models in your conclusion.

Examples of hypotheses are:

- Pedestrian density is highest at the centre of the CBD, and declines with increasing distance from the centre.
- The sphere of influence of settlements increases with settlement size.
- The pH of sand dunes decreases with distance inland.

- Population density is higher in inner urban areas than in the suburbs.
- Average temperatures in urban areas are higher than in surrounding rural areas.

For each hypothesis you investigate you should describe what you expect to find and explain why. Within this section of the investigation you should also justify the **geographical location** of your inquiry. It should clearly be a good location to address the issue, question or problem you intend to investigate. Make sure that you include the area's site as well as its regional situation. Include clearly labelled location maps. Give each map or diagram used in this section a figure number. For example, a map showing the location of your study area within its wider region might be labelled 'Figure 1: The location of Studland in Dorset'. Follow this procedure for all illustrations used throughout your investigation. Also make sure that you refer to each 'Figure' in your text.

A geographical investigation of London Docklands

● Defining the objective of the study

You now move on to define the objectives of the study in specific terms, refining the content of the previous section. You will also make decisions concerning:

- what data are relevant to the study
- how the data can be collected.

It is useful in this section to briefly state the sequence of investigation you are going to follow. This should ensure that you are clear about the remaining stages of the investigation and that you tackle the route to geographical enquiry in a logical manner.

● Collection of data

Data can be collected on a group or individual basis which may include:

- fieldwork to collect primary data
- gathering data from secondary sources such as from census information and published maps.

Make sure that you clearly explain the difference between primary data and secondary data. Use as many different techniques as possible to gather information, for example interviews, observations, surveys, questionnaires, maps and looking at figures. Describe and justify each method. Describe the use of primary fieldwork methods and in particular the method and/or equipment used to collect each type of information. Equally, describe and explain the use of secondary sources, for example parish records.

Explain clearly how you decided to use your figures, maps, answers to questions etc. Some reasoning is necessary here – that is, justify why you used that method or source. Explain in detail, for example, how you questioned people, collected census figures, obtained maps etc. Write this up almost like the method for a scientific experiment. You can use a planning sheet here, stating when you collected data, where from, at what time, places you visited, observations you made, interviews you conducted etc. If you are using a questionnaire then you must justify the questions that you use, for example explain why you have recorded the age and gender of respondents in a shopping survey. You need a range of methods to obtain full marks.

To collect data in a sound and logical way so that valid conclusions can be drawn you should be aware of the characteristics and importance of:

- sampling
- pilot surveys
- questionnaires and interviews
- methods of observing, counting and measuring
- health and safety and other restrictions.

Sampling

The reasons for sampling

For many geographical investigations it is impossible to obtain 'complete' information. This is usually because it would just take too long in terms of both time and cost. For example, if you wanted to study the shopping habits of all 1000 households in a suburban area by using a doorstep questionnaire, it would be a huge task to visit every household.

However, it is valid to take a 'sample' or proportion of this total 'population' of 1000 households, providing you follow certain rules. The idea is that you are selecting a group that will be representative of the total population.

You might decide to take a 5% or 10% sample which would involve talking to 50 or 100 of the 1000 households in the area. But how do you decide which 50 or 100 households to sample? There are three recognised methods of sampling which are considered scientifically valid. All three methods avoid bias which would make results unreliable.

Sampling types

Before selecting the sampling method you need to consider how you are going to take a sample at each location. There are three alternatives:

- Point sampling – making an observation or measurement at an exact location, such as an individual house or at a precise six-figure grid reference.
- Line sampling – taking measurements along a carefully chosen line or lines, such as a transect across a sand dune ecosystem.
- Quadrat (or area) sampling – quadrats are mainly used for surveying vegetation and beach deposits. A quadrat is a gridded frame.

All three sampling types are shown in Figure 1. Here all of the sampling types are illustrated using the systematic method of sampling. When you have read the next section you might think how these diagrams would look using random and stratified sampling.

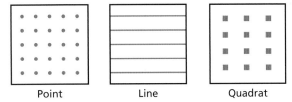

Figure 1 Point, line and quadrat sampling

Sampling methods

Random sampling

This method involves selecting sample points by using random numbers (Figure 2). Tables of random numbers can be used or the numbers can be generated by most calculators. The use of random numbers guarantees that there is no human bias in the selection process.

61	89	04	24	98	65	96	96
33	79	53	35	51	56	11	78
96	84	68	33	84	15	08	10
28	34	05	81	54	02	60	18
19	35	37	56	39	97	66	15
37	21	22	09	18	99	33	03
46	77	77	83	19	39	43	48
12	44	97	58	79	57	42	30
08	91	47	87	38	21	74	24
98	17	54	62	62	21	06	90
73	53	29	99	11	76	30	00
35	28	06	62	12	99	48	48
50	34	68	74	61	42	19	63
95	49	75	96	49	81	93	10
22	30	86	92	56	79	71	50
68	83	63	59	30	55	37	20
69	67	64	05	14	37	16	36
04	43	66	24	01	62	72	98
03	40	89	99	66	22	11	32
95	44	09	92	08	41	49	27

Figure 2 Section of a table of random numbers

Systematic sampling

With this method the sample is taken in a regular way. It might, for example, involve every tenth house or person. When using an Ordnance Survey map it might mean analysing grid squares at regular intervals.

Stratified sampling

Here the area under study divides into different natural areas. For example, rock type A may make up 60% of an area and rock type B the remaining 40%. If you were taking soil samples for each type, you should ensure that 60% of the samples were taken on rock type A and 40% on rock type B.

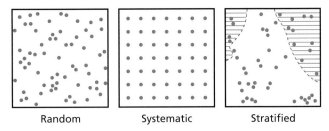

Figure 3 Random, systematic and stratified sampling

Deciding on the size of the sample

The larger the sample the more likely you are to obtain a true reflection of the total population. However, this could happen with a very small sample by chance. But equally a small sample could give a very misleading picture of the total population.

A good rule to follow at IGCSE with regard to sample size is to take as many samples as possible with regard to:

- the time available
- available resources
- the number of samples required for a particular statistical technique such as Spearman's Rank Correlation Coefficient
- your capacity to handle the data collected (there are many computer programs available to help with this).

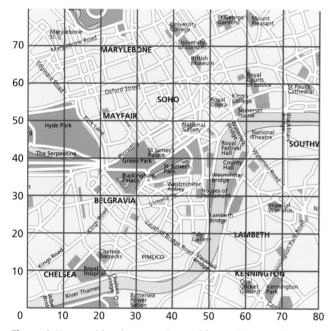

Figure 4 How a grid and map can be used for random sampling

Pilot surveys

A pilot study or trial run can play an important role in any geographical investigation. A pilot study involves spending a small amount of time testing your methods of data collection. For example:

- If you are using equipment, does all the equipment work and can everyone in the group use it correctly?
- If your data collection involves a questionnaire, can the people responding understand all the questions clearly?
- If a method of sampling is used, does everyone know how to select the sample points accurately?

A small-scale pilot study allows you to make vital adjustments to your investigation before you begin the main survey. This can save a great deal of time in the long run.

Activities

1 Work in groups to provide outlines of different geographical investigations that would involve (a) random sampling, (b) systematic sampling and (c) stratified sampling.
2 Why might it be beneficial to conduct a pilot study prior to beginning a geographical investigation?

Questionnaires and interviews

Questionnaire surveys involve both setting questions and obtaining answers. The questions are pre-planned and set out on a specially prepared form. This method of data collection is used to obtain opinions, ideas and information from people in general or from different groups of people. The questionnaire survey is probably the most widely used method to obtain primary data in human geography. In the wider world questionnaires are used for a variety of purposes, including market research by manufacturing and retail companies and to test public opinion prior to political elections.

One of the most important decisions you are going to have to make is how many questionnaires you are going to complete. The general rules to follow here are similar to those for sampling, set out in the previous section. Remember, if you have too few questionnaire results, you will not be able to draw valid conclusions. For most types of study, 25 questionnaires is probably the minimum you would need to draw reasonable conclusions. On the other hand it is unlikely you would have time for more than 100 unless you were collecting data as part of a group.

A good questionnaire:

- has a limited number of questions that take no more than a few minutes to answer
- is clearly set out so that the questioner can move quickly from one question to the next – people do not like to be kept waiting; the careful use of tick boxes can help this objective
- is carefully worded so that the respondents are clear about the meaning of each question
- follows a logical sequence so that respondents can see 'where the questionnaire is going' – if a questionnaire is too complicated and long-winded people may decide to stop halfway through
- avoids questions that are too personal
- begins with the quickest questions to answer and leaves the longer/more difficult questions to the end
- reminds the questioner to thank respondents for their cooperation.

The disadvantages of questionnaires are:

- Many people will not want to cooperate for a variety of reasons. Some people will simply be too busy, others may be uneasy about talking to strangers, while some people may be concerned about the possibility of identity theft.

● Research has indicated that people do not always provide accurate answers in surveys. Some people are tempted to give the answer that they think the questioner wants to hear or the answer they think shows them in the best light.

As with other forms of data collection, it is advisable to carry out a brief pilot survey first. It could be that some words or questions you find easy to understand cause problems for some people. Amending the questionnaire in the light of the pilot survey before you begin the survey proper will make things go much more smoothly.

Delivering the questionnaire

There are really three options here:

● Approach people in the street or in another public environment.
● Knock on people's doors.

● Post questionnaires to people. With this approach you could either collect the questionnaire later or enclose a stamped addressed envelope. This last method is costly and experience shows that response rates are rarely above 30 per cent. Another disadvantage is that you will be unable to ask for clarification if some responses are unclear.

If you are conducting a survey of shopping habits you may want to find out if there are significant differences between males and females and between different age groups. In this case you would use a stratified sample divided by gender and the percentage of population in each age group.

The time of day may also be important. In the example given above, very few people in some age groups may be around at a certain time of day. For example, most teenagers will be in school or college at mid-morning on a weekday.

A good questionnaire
Introduction: 'Excuse me, I am doing a school geography project. Could I ask you one or two quick questions about where you go shopping?'

1 How often do you come shopping in this town centre?

   More than once a week ☐

   Weekly ☐   Occasionally ☐

2 How do you travel here?

   Walk ☐ Car ☐ Bus ☐ Train/Tube ☐
   Other _____

3 Roughly where do you live? _____

4 Why do you come here rather than any other shopping centre?

   Near to home ☐   Near to work ☐

   More choice ☐   Pleasant environment ☐
   Other _____

5 What sort of things do you normally buy here?

   Groceries ☐   Clothes/shoes ☐

   Everything ☐
   Other _____

6 Do you shop anywhere else, and if so where?
   _____

7 Why do you go shopping there?

8 What do you buy there?

9 Sex: M ☐ F ☐   Age (estimate) under 20 ☐
                    20–30 ☐ 30–60 ☐ Over 60 ☐

'Thank you very much for you help'

A bad questionnaire
Introduction: 'Excuse me, but I wonder if I could ask you some questions?'

1 Where do you live?
   _____

2 How do you get here?
   _____

3 Do you come shopping here often?
   _____

4 Why do you come here?
   _____

5 Do you buy high-or low-order goods here?
   _____

6 Is this a good shopping centre and if so, why?
   _____

7 Where else do you go shopping?
   _____

8 Do you shop there because it is cheaper or nearer to your home?
   _____

9 How old are you? _____

'Right, that's it then.'

Figure 5 Two questionnaires, one good and one bad

Interviews

Interviews are more detailed interactions than questionnaires. They will generally involve talking to a relatively small number of people. A study of an industrial estate might involve interviews if you were trying to find out why companies chose to locate on the estate. An interview is much more of a discussion than a questionnaire, although you should still have a pre-planned question sheet. It can be a good idea to record interviews but you should ask the interviewee's permission first.

Health and safety and other restrictions

It may be sensible to work in pairs when conducting questionnaires as some people can act in an unfriendly manner when approached in the street. Working in pairs can also speed the process up considerably, with one person asking the questions and the other noting the answers. Also be aware that shopping malls, individual shops and other private premises may not allow you to conduct questionnaires without seeking permission beforehand.

Activities

1 Design a questionnaire that might be used as part of an investigation into tourism in a small resort.
2 Briefly outline a geographical investigation in your local area that could involve the use of interviews.

Observations, counts and measurements

Field sketches

Personal observations or perceptions may form an important element of a coursework investigation. A field sketch is a hand-drawn summary of an environment you are looking at. In both urban and rural environments field sketching is a very useful way of recording the most important aspects of a landscape and noting the relationships between elements of such landscapes. The action of stopping for a period of time to sketch the landscape in front of you will often reveal details that may not have been immediately apparent.

Figure 6 is an example of a good field sketch. This sketch highlights the important geographical features of the landscape. Key features should be clearly labelled but make sure that your sketch is not too cluttered. This will detract from the really important details. A good field sketch will be viewed as a higher-level technique by your coursework moderator.

Annotated photographs

Annotated photographs should be seen as complementing field sketches rather than just being an alternative to them. Like field sketches, good, fully annotated photographs are regarded as a higher-level skill. Always record the precise location and conditions of the photographs you take. This should include a grid reference, the direction the photograph was taken in, weather conditions and time of day.

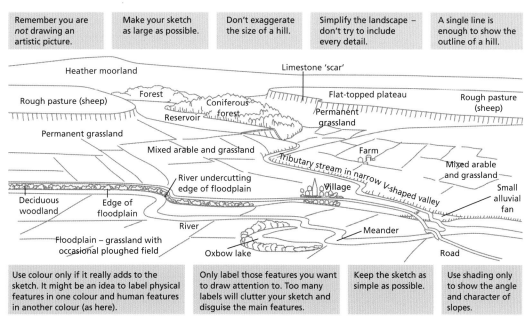

Figure 6 Example of a field sketch

Such information will make annotation quicker and easier in the long run.

An annotated photograph shows your key perceptions about a location you have visited on fieldwork. A series of such photographs might show:

- how the type and quality of housing varies in an inner city or suburban area
- how a river and its valley changes from source to mouth.

Annotations should be in the form of short, sharp sentences (Figure 7). Moderate abbreviation is fine providing the meaning of the comment remains clear. Some annotations will be just descriptive, but where the opportunity arises some explanation should also be included. Annotation can be most effective when the photograph is placed on the page in landscape format, which will allow more space for annotations on all four sides. As with field sketches, a series of annotated photographs could form a very effective part of your analysis. You should look to correlate annotated photographs with the tables and graphs showing your data analysis. Photographs are also useful to show how you carried out surveys and field measurements.

Fields inundated as the river has topped its banks

Flat, marshy floodplain

Road built on an embankment

Limited number of trees to intercept rainfall

Settlement built on slightly higher ground

Figure 7 Example of an annotated photograph

Activities

1 a Draw a field sketch of an urban or rural environment within easy reach of your school.
  b Suggest why this location has geographical interest.
2 Annotate a photograph of a location of interest you have visited.

Recording tables

The most straightforward method of observation is noting whether a physical or human feature exists in an area or not. Figure 8 is an example of a recording table showing park facilities. The objective here is to compare the facilities in four parks before attempting to explain the differences between them. Recording is done by placing a tick in the appropriate square. Notice that there is a final column to accommodate any unexpected findings.

Park	size (ha)	Woodland	Children's playground	Sports pitches	Tennis/basketball counts	Picnic site	Restaurant/café	Boating/fishing lake	Ornamental gardens	Pavillion/bandstand	Toilets	Car park	Info centre/gift shop	Other
High Lodge Forest Park	120	✔	✔			✔	✔				✔	✔	✔	Maze, jungle gym
Ditchingham Estate Park	25	✔						✔				✔		
Long Stratton Park	4	✔	✔	✔	✔					✔	✔	✔		Skate ramp
Castle Mall Gardens	2					✔	✔		✔					Viewpoint

Figure 8 Example of a recording table showing park facilities

Scoring systems

Scoring systems are used in quality of life and other types of survey. Figure 9 is an example of a scoring system used to study variations in environmental quality in different parts of a residential area. Figure 9a shows that in this example ten local environmental factors are being observed. Figure 9b shows how the scoring system works. Here a score of 5 is the maximum possible for the best environmental conditions. The minimum score is 1. For each location the individual environmental scores are added together to achieve a total environmental score. In this example, the lowest possible total score is 10, and the highest is 50. It can be useful to practise the system in class using photographs before going out to conduct fieldwork.

a

Ward name:		Location						
Factor	1	2	3	4	5	6	7
1	Condition of brickwork/paintwork							
2	Condition of pipes, guttering, windows							
3	Quality/state of repairs of pavements							
4	Quality/state of repair of roads							
5	Extent of litter							
6	Extent of graffiti							
7	Presence/condition of vegetation							
8	Availability of parking							
9	General condition of front of house							
10	Age/number of vehicles							
	Total:							
Ward total:								
Average Location Score:								

b

Explanation of the ranking system used for the environmental checklist			
	Factor	Explanations	
1	Condition of brickwork/paintwork	1 Worst condition	5 Best condition
2	Condition of pipes, guttering, windows	1 Worst condition	5 Best condition
3	Quality/state of repairs of pavements	1 Worst condition	5 Best condition
4	Quality/state of repair of roads	1 Worst condition	5 Best condition
5	Extent of litter	1 Most litter	5 No litter
6	Extent of graffiti	1 Large amount present	5 None present
7	Presence/condition of vegetation	1 Worst condition/none	5 Best condition
8	Availability of parking	1 None noticeable	5 Space for 1+ car
9	General condition of front of house	1 Worst condition	5 Best condition
10	Age/number of vehicles	1 Outdated/no vehicle	5 New models/+1 vehicle

Figure 9 Example of an environmental scoring system

Tally charts

Counts of various kinds are an element of many geographical investigations. Figure 10 is an example of a tally chart used to record visitor numbers at key locations in a park. The convention is to show counts in groups of five with the fifth count as a horizontal line drawn across the previous four counts.

	Children's playground	Sports ground
Male	‖‖ II	‖‖ ‖‖ III
Female	‖‖ III	

Figure 10 Example of a tally chart

You will notice that Figure 10 does not give a time when the count took place or state how long counting went on. In this example the number of visitors could vary significantly according to the time of day. It is therefore very important to plan carefully for your counts so that when you have collected and presented your data you can justify the conclusions you have drawn.

Pedestrian counts often form part of urban geography investigations. You could see how pedestrian counts decline with distance from the centre of the CBD. Pedestrian counts could be conducted every 50 or 100 m from this point.

Activities

1 Produce a recording table that could be used as part of a geographical investigation in your local area.
2 Look at the scoring system shown in Figure 9. Discuss the merits and limitations of this example.

● Presentation and recording of results

A wide variety of graphical techniques can be used to present geographical data. The skill is in choosing the best type of graph for the particular data set under consideration. Coursework marks can be lost by the incorrect use of graphical techniques. You should also consider the size of any graph or diagram you use. It is important that the labels of axes and all other information can be clearly read.

It is important to integrate all maps, graphs, photographs and diagrams with the text. The most elementary way of doing this is to use a sentence such as 'Figure 4 is a line graph showing temperature change in my garden'.

Line graphs

A line graph shows points plotted on a graph whereby the points are connected to form a line. This type of graph is used to show continuing data. It shows the relationship between two variables which are clearly labelled on both axes of the graph. Many line graphs show changes over time. However, time

does not have to be one of the variables of a line graph. Examples of the use of line graphs include:

● temperature changes during the course of a day
● pedestrian counts by time of day
● temperature change with altitude.

The axes of a line graph should begin at zero and the variable for each axis should be clearly labelled. Be careful with the choice of scale as this will determine the visual impression given by the graph. Figure 11 is an example of a line graph. Here, only one line has been drawn but it is valid to show a number of lines so long as the course of each line is absolutely clear from start to finish.

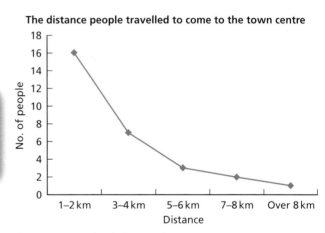

The distance people travelled to come to the town centre

Figure 11 Example of a line graph

Median-line bar graphs

Median-line bar graphs are useful when the objective is to show both positive and negative changes. The median line is set at zero with the positive scale above the median line and the negative scale below it (Figure 12). This type of bar graph can create a very good visual impression. You can see instantly whether changes are positive or negative and exactly what the extent of the individual changes are.

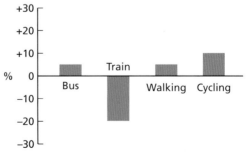

Percentage change in mode of travel to school, 2000–07

Figure 12 Example of a median-line bar graph

274

Histograms

A histogram is a special type of bar graph (Figure 13). It shows the frequency distribution of data. The x axis must be a continuous scale with the values marked on it representing the lower and upper limits of the classes within which the data have been grouped. The y axis shows the frequency within which values fall into each of the classes. A vertical rectangle or bar represents each class. The bars must be continuous without any gaps between them.

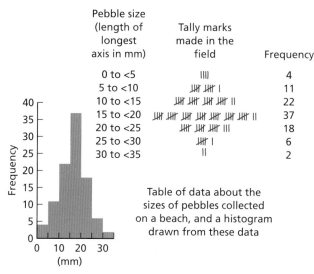

Pebble size (length of longest axis in mm)	Tally marks made in the field	Frequency
0 to <5	IIII	4
5 to <10	IIII IIII I	11
10 to <15	IIII IIII IIII IIII II	22
15 to <20	IIII IIII IIII IIII IIII IIII IIII II	37
20 to <25	IIII IIII IIII III	18
25 to <30	IIII I	6
30 to <35	II	2

Table of data about the sizes of pebbles collected on a beach, and a histogram drawn from these data

Figure 13 Example of a histogram and the data used

Choropleth maps

Choropleth maps can use variations in colour or different densities of black and white shading. The following steps should be followed in the construction of a choropleth map:

- Look at the range of data and divide it into classes. There should be no less than four classes and no more than eight.
- Allocate a colour to each class. The convention is that shading gets darker as values increase.
- Now apply each colour to the applicable areas of the map.
- Provide a key, scale and north point.

Proportional circles

Proportional circles are the next step up from pie charts. While pie charts are viewed as a basic graphical technique, proportional circles are a higher-level technique. Proportional circles are useful when illustrating the differences between two or more

amounts. They are particularly effective when placed on location maps. In Figure 14 the three circles shown are proportional in area to the total number of offences recorded in the three urban areas. The method used to decide the radius of each circle is as follows:

- Write out each of the total figures for which circles are to be drawn in the first column.
- Find the square root for each figure and write this down in a second column.
- Use the square root for the radius of each circle. By doing this the area of each of the circles will be mathematically proportional to the figures they are representing. For the radii you can use any units you want providing they are the same for each of the circles.

Table 1 shows a very simple example.

Table 1 To prepare proportional circles: an example

Totals	Square root	Radius of circle
4	2	2 cm
9	3	3 cm
16	4	4 cm

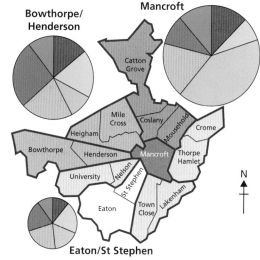

Number of offences committed		Types of crime	
☐	Under 750	▨	Violent/sexual assault
▨	750–999	☐	Burglary
▨	1000–1249	☐	Theft
▨	1250–1499	▨	Theft of/from motor vehicle (car crime)
▨	1500+	▨	Criminal damage
━	Boundary of beat areas	▨	Other (including robbery)

Figure 14 Example of a choropleth map with proportional circles

Flow-line diagrams

Flow-line diagrams and maps are used to illustrate movements or flows. One might be used to show the variation in volumes of traffic from different smaller settlements into a larger settlement. Straight lines are used but the width of the individual flow lines will be proportional to the amounts of traffic they represent. Thus, a line 10 mm wide may represent 500 vehicles an hour along a road. On the same scale a line 2 mm wide would represent 100 vehicles an hour. Flow lines could also be used to show the number of buses coming into a town on a particular day.

Ray diagrams

There are two main types of ray diagram: wind roses and desire lines. Ray diagrams are made up of straight lines (rays) which show a connection or movement between two places.

Wind rose diagrams (Figure 15) show the variations in wind direction for a certain time period. The direction of each ray to the centre is the direction from which the wind is blowing. Each ray is proportional in length to the number of days the wind blew from that direction.

Wind directions recorded for one year at a school weather station in Liverpool, UK

Direction of wind	N	NE	E	SE	S	SW	W	NW	Calm
Number of days per year	26	37	39	32	30	57	60	53	51

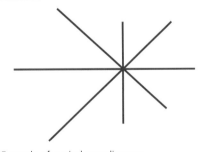

Figure 15 Example of a wind rose diagram

Desire-line diagrams show movement from one place to another. This type of diagram could be used to show where people live and the supermarket they use. If there are four supermarkets in an area then the rays would focus on four points rather than just one as in a wind rose diagram. Desire-line diagrams are therefore more complicated than wind rose diagrams.

Semantic differential profiles

Semantic differential profiles (SDPs) (Figure 16) are useful for recording perceptions of environmental quality. They consist of a series of pairs of words with opposite meanings. There should generally be a minimum of five gradations between each pair of words. The observer must decide where to place a cross or other mark to state the condition of the environment they are observing. When all the observations have been made the crosses are joined up with a ruler. If, for example, you were studying three different housing areas in a town you could show all three profiles on one SDP diagram by using different colours for each area.

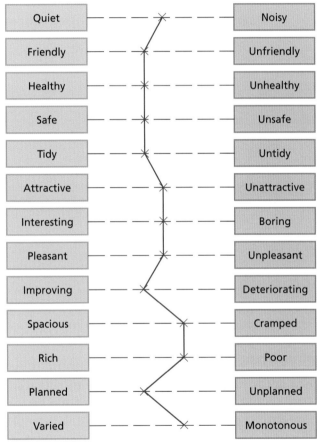

The average semantic differential profile for Graveny ward

Figure 16 A semantic differential profile

Radial (circular) graphs

Radial (or circular) graphs (Figure 17) can be used to plot:

- a variable that is continuous over time, such as temperature data over the course of a year
- data relating to direction using the points of the compass.

The two axes of a radial graph are the circumference of the circle and the radius. Values increase from the centre of the circle outwards.

Rate of weathering: Rahn's index

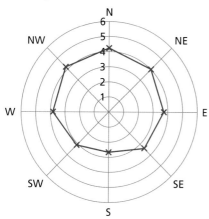

A radial graph to show the influence of aspect on gravestone weathering. The graph shows the mean of Rahn's index for each compass direction.

Rahn's index

Class	Description
1	Unweathered
2	Slightly weathered: faint rounding to corners of letters
3	Moderately weathered: gravestone rough, letters legible
4	Badly weathered: letters difficult to read
5	Very badly weathered: letters indistinguishable
6	Extremely weathered: no letters left, scaling

Figure 17 A radial graph

Isoline diagrams

Isolines join points of equal value on a map. They are similar to contours on an Ordnance Survey map. Isolines can only be drawn when the values under consideration change in a fairly gradual way over the area of the map. Data for quite a large number of locations are required in order to draw a good isoline map. Isoline maps are unsuitable for patchy data. Figure 18 is an isoline map showing pedestrian flow in and around a central business district.

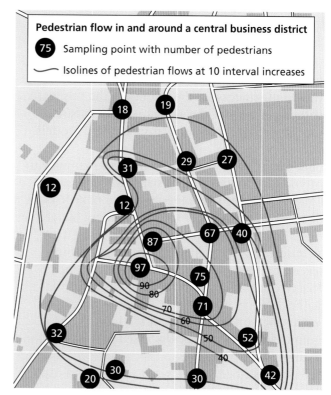

Figure 18 An isoline map

● Analysis and interpretation

Here you will analyse and interpret your findings on the issue/question/problem set out in the first section of your investigation. This will be done with reference to relevant geographical concepts.

You need to describe the patterns in data presented in your graphs and tables of results.

- After each graph or technique, describe fully the results or trend or association (using simple descriptive statistics). What do your results tell you? Describe your findings in detail by quoting the evidence from your methods of analysis.
- Do the graphs and other techniques used help to answer the question set?
- Make comments to link the data. For example, show how one diagram, graph or map relates to others.
- Where relevant, consider the values and attitudes of people involved.

Activities

1 Find two examples of line graphs in a geography textbook. What do the line graphs show in these examples?
2 What is the difference between a histogram and an 'ordinary' bar graph?
3 Discuss the merits and limitations of choropleth maps.
4 Draw a series of proportional circles using fieldwork data or data from a textbook.
5 Construct a semantic differential profile that could be used in your local area.
6 Look at Figure 18 which shows an isoline map. Suggest two more examples for which isoline maps could be drawn.

Making effective conclusions

Using the evidence from the data you should be able to make judgements on the validity of the original hypothesis or aims of the assignment. Compare the results of the data analysis against standard models and theories.

Link what you have discovered in your enquiry to what you have studied in the syllabus. For example, if you have looked at shopping, have you talked about high-order and low-order shops, shopping hierarchies and other relevant concepts? If your investigation is on leisure, have you linked this to the amount of leisure time people have, spheres of influence of leisure centres, how accessible places are, and other key aspects of the topic? These things should be mentioned briefly in the first section of your assignment and need to be discussed now in relation to your findings.

Having described your results, you need to explain and discuss them. Why have you arrived at such results? Do they confirm (accept) or refute (reject) your hypotheses?

Evaluation and suggestions for further work

Your investigation is likely to be slightly less than perfect, and you need to show the examiner that you are aware of this. It is important to realise that you will not be marked down for showing an awareness of the limitations of methods, results and conclusions.

There may be limits to where and when you could carry out a survey, or the number of people you could interview. There may be constraints in terms of expense, in terms of the equipment you are able to use, or how often you can visit the fieldwork location(s). The last item can also be due to a busy school timetable which can mean your time at a fieldwork location is limited. Investigations in physical geography may be hampered by unexpected conditions. For example, river measurements may be affected by floods or a drought.

You need to make an assessment and state whether or not such limitations have impaired your investigation. If they have, then your results

are likely to be compromised, and so will any conclusions that you have drawn. For example, a survey of tourists to Oxford on a Tuesday afternoon in April found that most people were attracted to the university buildings and the colleges, and listed the poor weather as their main criticism. Had the survey taken place in another kind of tourist destination, the attractions listed would have been very different. Similarly, had the survey taken place in July or August the weather might not have been mentioned as a problem. On the other hand, congestion due to too many tourists might have been mentioned. This example shows that the methods (including the date and time of any survey) produce results that can affect our conclusions and, therefore, our evaluation. Another survey at another time might give a different set of results and conclusions.

The final part of your evaluation can lead you to suggest future lines of enquiry from the insights you have developed by following the route to geographical enquiry in your investigation.

Case studies

Analysing sand dunes

Sand dunes at Studland, Dorset, UK

Sand dunes provide an interesting and manageable ecosystem for study at IGCSE. This is because significant changes can be identified over a relatively small area. In a sand dune system the most recently formed dunes are by the sea. The dunes become older with distance inland. Sand dunes form a series of ridges with intervening 'slacks' between them.

A useful starting point is to survey the morphology (size and shape) of the dunes. Figure 19a shows how measurements can be taken across a sand dune ecosystem using a tape measure and a clinometer. The transect line should be at right-angles to the coast. The first ranging pole is carefully placed where there is a distinct break in slope from the back of the beach, marking the beginning of the sand dunes. The second ranging pole is placed at the next break of slope. The angle of slope is read from the clinometer. This process is repeated for each break of slope. With about four people working as a team, all the measurements

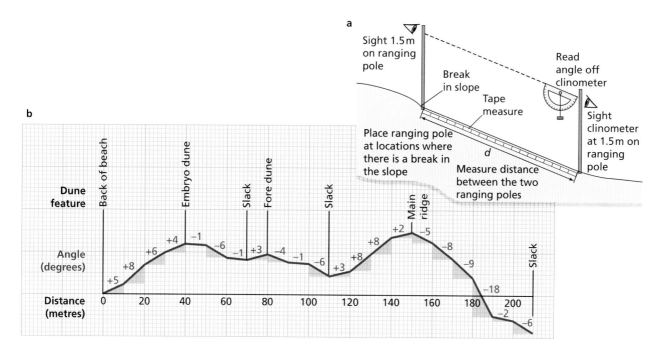

Sand dune transect recording sheet

LOCATION: _Nailsworth-upon-Sea_   DATE:_____

Site	Distance from sea (m)	Angle (°) (uphill)	Vegetation cover (%)	Number of species	Maximum vegetation height (m)	Dominant plant species	Wind speed (m/s)	Soil moisture content (%)	Soil organic content (%)	Soil pH
1	10	+5	50	2	0.51	Marram grass	7.3	8.7	0.2	6.9
2	10	+8	92	3	0.56	Sea couch grass	4.4	25.9	0.4	6.8
3	20	+6	86	3	0.48	Marram grass	3.4	11.2	1.8	6.8
4	30	+4	95	3	0.51	Sea couch grass	3.0	25.0	3.3	6.8
5	40	−1	96	2	0.69	Marram grass	2.4	54.4	3.6	6.8
6	50	−6	94	3	0.71	Marram grass	3.2	9.1	3.7	6.7
7	60	−1	89	5	0.75	Sea couch grass	3.0	24.1	4.6	6.9

Figure 19 Conducting a survey of a sand dune ecosystem

required to draw a cross-section of the dunes (Figure 19b) can be taken in a couple of hours. If a larger group of people is available a number of transects could be taken across the sand dunes. Transects could be compared and any differences discussed.

As the dune survey proceeds, other measurements can also be taken. At regular locations across the sand dune system the following can be measured:

- vegetation cover, with the dominant plant species noted
- maximum height of vegetation
- wind speed
- soil moisture content
- soil organic content
- soil pH.

Figure 19c provides an example of a recording sheet that could be used for such a survey. While the first few readings might take a little time, once you become familiar with what is required the process should speed up considerably.

All these measurements can be used to test the standard theories about sand dunes presented in textbooks, which can be set out in a series of hypotheses to be tested:

- Vegetation density increases with distance inland.
- The number of species increases with distance inland.
- The height of vegetation increases with distance inland.
- Soil organic content increases with distance inland.
- Soil pH decreases with distance inland.
- Wind speed decreases with distance inland.

Investigating rivers

Streams and small rivers are a popular focus for geographical investigation because most schools will not be too far from a suitable example. Figure 20 shows some of the measurements that can be taken at various locations along the course of a river. For safety reasons it is best to avoid working in streams above the height of your knees.

For most river studies you will want to produce a cross-section of the river channel. The method is as follows:

- Use a tape measure to assess the channel width. This should be done at right-angles to the course of the river. If you want to produce a cross-section of the river when discharge is at its highest you should look for evidence of the highest point the water reaches on each bank. This will give the bankfull width.
- Channel depth should be measured at regular intervals across the river using a metre stick or ranging pole. Every 20 or 30 cm should provide an adequate sampling interval.
- A cross-section can then be drawn using graph paper (Figure 21). As with all cross-sections, careful choice of scale is important.

Although various types of float can be used to measure river velocity, it is best to use a flow meter. The impeller (screw device) is pointed upstream at the same points across the river used to calculate the depth intervals. You will be able to see how velocity varies with distance from the banks and how velocity varies with depth. You could also calculate the mean flow rate for this stage of the river.

The discharge of the river can be calculated by multiplying the velocity by the cross-sectional area (Figure 21). The gradient of a river can be measured using ranging poles and a clinometer, in the same way that sand dune measurements were carried out in the previous example.

Bedload measurements can also be taken to assess the impact of attrition with increasing distance downstream. Ensure that the samples of bedload are selected randomly by a ranging pole or metre stick at intervals across the river. Collect the stones that are touching the pole or stick. Measure the long axis, shape and radius of curvature of each stone.

As in the sand dune case study, a series of hypotheses based on textbook theory can be set up to be tested.

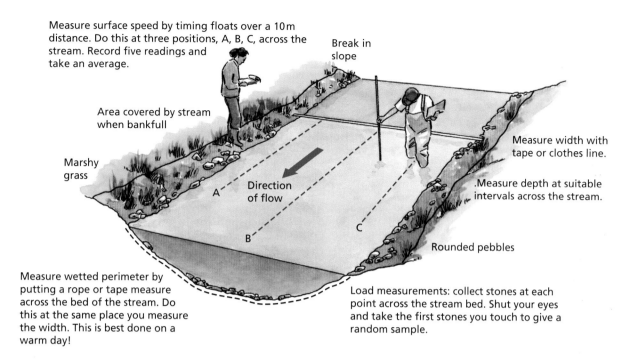

Measure surface speed by timing floats over a 10 m distance. Do this at three positions, A, B, C, across the stream. Record five readings and take an average.

Break in slope

Area covered by stream when bankfull

Measure width with tape or clothes line.

Marshy grass

Direction of flow

.Measure depth at suitable intervals across the stream.

A

B

C

Rounded pebbles

Measure wetted perimeter by putting a rope or tape measure across the bed of the stream. Do this at the same place you measure the width. This is best done on a warm day!

Load measurements: collect stones at each point across the stream bed. Shut your eyes and take the first stones you touch to give a random sample.

Figure 20 Taking river measurements

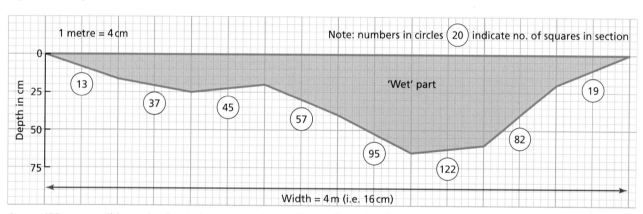

1 metre = 4 cm

Note: numbers in circles (20) indicate no. of squares in section

Depth in cm

'Wet' part

Width = 4 m (i.e. 16 cm)

Area = 470 squares. This number has to be converted, according to the scale. The scale 4 cm = 1 m means that each 2-millimetre square on the graph paper represents 0.0025 m² in reality ($^1/_{400}$). If the scale had been 1 cm = 1 m then the scaling factor would be $^1/_{100}$ = 0.01 m². In this example 470 × 0.0025 = 1.175 m².

Figure 21 A river cross-section

Glossary

Abrasion (or corrasion): a type of erosion in which rock fragments carried by waves or a river scrape and grind away a surface such as a cliff face.

Accessibility: the ease with which a place may be reached. An area with high accessibility will generally have a well-developed transport network and be centrally located.

Active volcano: a volcano currently showing signs of activity.

Afforestation: planting of trees in areas that have not previously held forests. Trees may be planted to increase interception by vegetation, to bind soil together and reduce overland runoff and thus soil erosion.

Age-specific mortality rates: mortality rates specific to a single year of age, for example the infant mortality rate, or an age range, for example the child mortality rate.

Agricultural technology: the application of techniques to control the growth and harvesting of animal and vegetable products.

Air pressure: the pressure at any point on the Earth's surface that is due to the weight of the air above it; it decreases as altitude increases. At sea level the average pressure is 1013 millibars (mb). Areas of relatively high pressure are called anticyclones; areas of low pressure are called depressions.

Altitude: measurement of height, usually given in metres above sea level. Temperature declines, on average 1 °C for every 100 m (and therefore rises 1 °C with every decrease in altitude of 100 m).

Anti-natalist policies: policies that aim to reduce population growth.

Aquifer: a body of rock through which large amounts of water can flow. The rock of an aquifer must be porous and permeable so that it can hold and transmit water.

Arable farms: involves cultivating crops; there is no involvement with livestock.

Arch: a natural bridge-like feature formed by erosion. Arches are formed from the erosion of a headland where two caves meet and break through the headland.

Arid: arid areas are usually defined as areas that receive less than 250 mm of rainfall each year.

Ash: very fine-grained volcanic material.

Aspect: the direction in which something faces.

Attrition: the process by which particles of rock being transported by a river or the sea are rounded and become smaller in size by being struck against one another. Particles become smaller and more rounded as the distance downstream in a river increases. Particles near the shoreline become smaller and more rounded due to more frequent attrition.

Backwash: the movement of water back down the beach due to the effect of gravity.

Bar: a depositional feature – a long ridge of sand or pebbles running parallel to a coastline that is submerged at high tide. Some bars develop as offshore bars when waves disturb sediments on the sea bed and form them into a submarine ridge or bar, while others form from the development of a spit across the whole of a small bay.

Bay: a wide, open curving indentation of the sea.

Beach: a feature of coastal deposition, consisting of pebbles on exposed coasts or sand on sheltered coasts. It is usually defined by the high and low water marks.

Bid rent: a model which states that land value and rent decrease as distance from the central business district increases.

Biodiversity: biological diversity, a measure of the variety of the Earth's plant and animal species, of genetic differences within species, and of the ecosystems that support those species.

Biofuels: any fuel produced from organic (once living) matter, either directly from plants or indirectly from industrial, commercial, domestic, or agricultural wastes. Fossil fuel substitutes that can be made from a range of crops including oilseeds, wheat and sugar.

Birth rate: the number of live births per thousand population in a year.

Braided channel: the subdivision of a river into several channels caused by deposition of sediment as small islands in the main channel. Braided channels are common in glacial meltwater streams.

Brand: a distinguished name and/or symbol intended to identify a product or producer.

Brown Agenda: the range of environmental problems associated with large cities.

By-product: something that is left over from the main production process which has some value and therefore can be sold.

Carbon credits: a permit that allows an organisation to emit a specified amount of greenhouse gases.

Carbon trading: a company that does not use up the level of emissions it is entitled to can sell the remainder of its entitlement to another company.

Carrying capacity: the largest population that the resources of a given environment can support.

Cave: coastal caves are large holes formed where relatively soft rock containing lines of weakness is exposed to severe wave action.

Central business district (CBD): the CBD of a town or city is where most of the commercial activity is found.

Chamber: the reservoir of magma located deep inside the volcano.

Cinders: small rocks and coarse volcanic materials.

Cliff: a rock-face along the coast, where coastal erosion, weathering and mass movements are active and the slope rises steeply (over 45°) and for some distance. The nature of the cliff depends on the nature of the rocks, their hardness and jointing pattern.

Climate: the combination of weather conditions at a particular place over a period of time – usually a minimum of 30 years. Climate thus includes the averages, extremes and frequencies of all meteorological elements such as temperature, atmospheric pressure, precipitation, wind, humidity and sunshine.

Cloud: water vapour condensed into millions of minute water particles that float in the atmosphere. Clouds are formed by the cooling of air containing water vapour, which generally condenses around tiny dust or ice particles.

Coastal management strategies: measures taken to prevent coastal erosion and/or flooding. To reduce erosion, several different forms of coastal protection are used. These can be divided into **hard engineering** and **soft engineering**.

Coastline: the area of contact between land and sea.

Collision boundary: a plate boundary where two plates are converging. These include (a) destructive boundaries where an oceanic plate meets a continental plate and (b) collision boundaries where the two plates are both continental plates.

Commercial farming: farming for profit, where food is produced for sale in the market.

Communications systems: the ways in which information is transmitted from place to place in the form of ideas, instructions and images.

Community energy: energy produced close to the point of consumption.

Community tourism: a form of tourism that aims to include and benefit local communities, particularly in developing countries.

Commuter: a person who travels into a large town or city for work but lives in a different settlement.

Concentric model (Burgess): a model of urban land where different activities occur at different distances from the urban centre. The result is a sequence of concentric circles or rings.

Cone volcano: steep volcano formed of sticky (viscous) acidic lava, ash and cinders.

Confluence: the point at which two rivers meet.

Conservation: allowing for developments that do not damage the character of a location.

Conservation of resources: the management of the human use of natural resources to provide the maximum benefit to current generations while maintaining capacity to meet the needs of future generations.

Conservative plate boundary (or transform plate boundary): where two plates slide past one another without loss of material.

Constructive plate boundary: a plate boundary where new material is being formed by the upwelling of magma from within the Earth's interior.

Constructive wave: a wave with a long wavelength and a low height. They help to build up beaches by deposition.

Consumer culture: the equation of personal happiness with consumption and the purchase of material possessions.

Convectional rainfall: rainfall associated with hot climates, resulting from the rising of convection currents of warm air. Air that has been warmed by the extreme heating of the ground surface rises to great heights and is cooled quickly. The water vapour carried by the air condenses and rain falls heavily. Convectional rainfall is usually associated with a thunderstorm.

Coral: living organisms that may form large reefs. Coral reefs provide a habitat for a diversity of living organisms.

Counterurbanisation: the process of population decentralisation as people move from large urban areas to smaller urban settlements and rural areas.

Crater: depression at the top of a volcano following a volcanic eruption. It may contain a lake.

Cross-profile: the cross-section of a river valley.

Cruciform settlement: occurs at an intersection of roads and usually consists of lines of buildings radiating out from the crossroads.

Cultural diffusion: the process of the spreading of cultural traits from one place to another.

Cumulative causation: the process whereby a significant increase in economic growth can lead to even more growth as more money circulates in the economy.

Dam: structure built to hold back water in order to prevent flooding, to provide water for irrigation and storage, and to provide hydro-electric power.

Death rate: the number of deaths per thousand population in a year.

Decentralisation: the movement of people or industry away from the centre of the city to the suburbs or the edge of town.

Deforestation: destruction of forest for timber, fuel, charcoal burning, and clearing for agriculture and extractive industries, such as mining. It causes fertile soil to be blown away or washed into rivers, leading to soil erosion, drought, flooding, and loss of wildlife.

Delta: a landform formed when a river, heavily laden with sediment, enters a body of standing water, such as a lake or a sea with negligible currents. The lack of velocity in the lake or sea causes the river to deposit its load.

Demographers: people who study human populations.

Demographic divide: the difference between countries where population growth remains high and those with very slow growing, stagnant or declining populations.

Demographic momentum: although the global population growth rate has been declining for decades, the number of people added each year remains very high because there are currently so many women in the child-bearing age range.

Demographic transition model: a model illustrating the historical shift of birth and death rates from high to low levels in a population.

Densely populated: having a high population density.

Dependency ratio: the ratio of the number of people under 15 and over 64 years to those in the 15–64 age group.

Depopulation: a decline in the number of people in a population.

Deposition: the laying down of material carried by rivers or the sea because of a reduction of velocity or discharge (both causing a loss of energy), often caused by increased friction with vegetation or coarse particles.

Deprivation: a condition in which a population group suffers from a poor quality of economic, social and environmental conditions.

Desert: a dry area with limited vegetation. Deserts can be either hot or cold. Characteristics common to all deserts include irregular rainfall of less than 250 mm per year.

Desertification: the gradual transformation of habitable land into desert.

Destination footprint: the environmental impact caused by an individual tourist on holiday in a particular destination.

Destructive plate boundary: plate boundary where an oceanic plate meets a continental plate. The ocean plate is more dense than the continental plate so it sinks below the continental crust.

Destructive wave: a wave with a high height and a short wavelength which helps erode beach materials and cliffs.

Development: the use of resources to improve the quality of life in a country.

Development gap: the differences in wealth, and other indicators, between the world's richest and poorest countries.

Diffusion: the spread of a phenomenon over time and space.

Discharge: the volume of water passing a certain point per unit of time. It is usually expressed in cubic metres per second (cumecs).

Dispersed settlement: a settlement pattern in which most of the houses are scattered in the countryside rather than being concentrated in towns and villages.

Dormant volcano: a volcano that has not erupted for a very long time but could erupt again.

Dormitory settlement: a settlement that has a high proportion of commuters in its population.

Drainage basin: the area of land drained by a river system (a river and its tributaries).

Drought: an extended period of dry weather leading to conditions of extreme dryness. Absolute drought is a period of at least 15 consecutive days with less than 0.2 mm of rainfall. Partial drought is a period of at least 29 consecutive days during which the average daily rainfall does not exceed 0.2 mm.

Dry point site: an area free from flooding in an otherwise wet region, for example a hilltop site surrounded by a marsh.

Dust storm: a severe windstorm that sweeps clouds of dust across an extensive area, especially in an arid region.

Earthquake: a sudden movement of the Earth's crust.

Economic leakage: the part of the money a tourist pays for a foreign holiday that does not benefit the destination country because it goes elsewhere.

Economic water scarcity: when a population does not have the necessary monetary means to utilise an adequate source of water.

Economies of scale: the reduction in unit cost as the scale of an operation increases.

Ecosystem: an integrated unit consisting of a community of living organisms (animals and plants) and the physical environment (air, soil, water and climate) that they inhabit. Individual organisms interact with each other and with their habitat.

Ecotourism: a specialised form of tourism where people experience relatively untouched natural environments such as coral reefs, tropical forests and remote mountain areas, and ensure that their presence does no further damage to these environments.

Emigration rate: the number of emigrants per thousand population leaving a country of origin in a year.

Energy mix: the relative contribution of different energy sources to a country's energy consumption.

Enhanced greenhouse effect: global warming caused by large-scale pollution of the atmosphere by economic activities.

Environmental impact statement: a document required by law detailing all the impacts on the environment of an energy or other project above a certain size.

Epicentre: the point on the Earth's surface directly above the focus of an earthquake. The strength of the shockwaves generally decreases away from the epicentre.

Erosion: wearing away of the Earth's surface by a moving agent, such as a river, glacier or the sea. In a river, there are several processes of erosion including hydraulic action, abrasion, attrition, and solution. In coastal areas, hydraulic action is the most potent form of erosion.

Evaporation: the process in which a liquid turns to a vapour.

Evapotranspiration: the combined water losses of **evaporation** and **transpiration**.

Extensive farming: where a relatively small amount of agricultural produce is obtained per hectare of land, so such farms tend to cover large areas of land. Inputs per unit of land are low.

Externalities: the side-effects, positive and negative, of an economic activity that are experienced beyond its site.

Extinct volcano: a volcano that has shown no signs of volcanic activity in historic times.

Fetch: the distance of open water over which wind can blow to create waves. The greater the fetch the more potential power waves have when they hit the coast.

Flood: a discharge great enough to cause a body of water to overflow its channel and submerge (flood) the surrounding area.

Floodplain: area of periodic flooding along the course of a river valley. When river discharge exceeds the capacity of the channel, water rises over the channel banks and floods the adjacent low-lying lands.

Focus: the position within the Earth where an earthquake occurs. Earthquakes may be divided into **shallow-focus** and **deep-focus** earthquakes depending on how far below the surface they occur.

Footloose industries: industries that are not tied to certain areas because of energy requirements or other factors.

Form: the shape of a settlement, mainly influenced by its physical geography and topography.

Formal sector: that part of an economy known to the government department responsible for taxation, and to other government offices.

Fossil fuels: fuels comprising hydrocarbons (coal, oil and natural gas), formed by the decomposition of prehistoric organisms in past geological periods.

Function: a classification of settlements based on their socio-economic functions, for example market towns, commuter towns and ports.

Gabion: a wire basket filled with rocks or stones used for stabilising slopes and protecting the base of cliffs in areas of coastal erosion.

Gentrification: the movement of higher social or economic groups into an area after it has been renovated and restored. This may result in the out-migration of the people who previously occupied the area. It most commonly occurs in the inner city.

Geothermal energy: the natural heat found in the Earth's crust in the form of steam, hot water and hot rock.

Gini coefficient: a technique used to show the extent of income inequality.

Global city: a city that is judged to be a significant nodal point in the global economic system. These are major financial and decision-making centres.

Globalisation: the increasing interconnectedness and interdependence of the world economically, culturally and politically.

Gorge: narrow, steep-sided valley that may or may not have a river at the bottom. A gorge may be formed as a waterfall retreats upstream.

Greenfield site: an area of agricultural land or some other undeveloped site that is a potential location for commercial development or industrial projects but has not yet been developed. Such sites are normally on the edge of town and have good transport links.

Green village: a village that consists of dwellings and other buildings, such as a church, clustered around a small village green or common, or other open space.

Gross National Product: the total value of goods and services produced by a country in a year, plus income earned by the country's residents from foreign investments and minus income earned within the domestic economy by overseas residents.

Gross National Product per capita: the total GNP of a country divided by the total population.

Groundwater: water stored underground in a permeable rock, e.g. chalk or sandstone.

Growth pole: a particular location where economic development is focused, setting off wider growth in the region as a whole.

Groyne: wooden or concrete barrier built at right angles to a beach in order to block the movement of material along the beach by longshore drift. Groynes are usually successful in protecting individual beaches. However, as they prevent beach material from being transported along the coast they can starve beaches further down the coast of sand and shingle; hence these beaches may be at increased risk of wave erosion.

Hamlet: a small rural settlement that is more than just an isolated dwelling but not large enough to be a village. Typically it has 11–100 people. In the UK, it may have a church and a pub but very little else.

Hard engineering: any coastal (or river) protection scheme that involves altering the natural environment with concrete, stone, steel, metal etc., for example the use of sea walls, gabions, groynes and revetments. Artificial structures are built in order to protect the natural environment from erosion.

Headland: a point of land projecting into the sea, also known as a cape or a promontory.

Hierarchy: the organisation and structure of settlement based on size and the number of functions that a settlement has. At the top of the hierarchy are cities and conurbations. At the base are individual farmsteads and hamlets.

High order goods/services/functions: expensive services and goods (comparison goods) such as electrical goods and furniture, that the shopper will buy only after making a comparison between various models and different shops.

Hotspot: a relatively small area where magma rises through a continental or oceanic plate. As the plate moves across the hotspot a chain of volcanoes may form, e.g. the Canary Islands and the Hawaiian Islands.

Human Development Index (HDI): the United Nations measure of the disparities between countries using life expectancy at birth, mean years of schooling for adults aged 25 years, expected years of schooling for children of school entering age, and GNI per capita (PPP$).

Humidity: the quantity of water vapour in a given volume of air (absolute humidity), or the ratio of the amount of water vapour in the atmosphere to the maximum amount the air can hold (relative humidity). At dew point the relative humidity is 100% and the air is said to be saturated. Condensation (the conversion of vapour to liquid) may then occur.

Hurricane (tropical cyclone): a region of very low atmospheric pressure in tropical regions. Hurricanes originate in latitudes between 5° and 20° north or south of the equator, when the surface temperature of the ocean is above 27 °C. A central calm area, called the eye, is surrounded by inwardly spiralling winds (anticlockwise in the northern hemisphere) of up to 320 km/hr.

Hydraulic action: the erosive force exerted by water. It is particularly effective on jointed rocks, especially during storm conditions.

Hydrological cycle: the water cycle, by which water is circulated between the Earth's surface and its atmosphere.

Hypermarket: a very large self-service store selling a wide range of household and other goods, usually on the outskirts of a town or city.

Immigration rate: the number of immigrants per thousand population entering a receiving country in a year.

Incidental pollution: one-off pollution incidents.

Industrial agglomeration: the clustering together of economic activities.

Industrial estate: an area zoned and planned for the purpose of industrial development.

Infant mortality rate: the number of deaths of children under one year of age per thousand live births per year.

Infiltration: the movement of water into the soil. The rate at which water enters the soil (the infiltration rate) depends on the intensity of rainfall, the permeability of the soil, and the extent to which it is already saturated with water.

Informal sector: that part of the economy operating outside official recognition.

Inner city: the area that surrounds the central business district of a town or city. In many cities this is one of the older industrial areas and may suffer from decay and neglect, leading to social problems. Inner cities are characterised by poor-quality terraced housing with old manufacturing industry nearby.

Inputs: (1) a farm requires a range of inputs such as labour and energy before anything else can happen; (2) the elements that are required for processes to take place. Inputs include raw materials, labour, energy and capital.

Intensive farming: characterised by high inputs per unit of land to achieve high yields per hectare.

Interception: the precipitation that is collected and stored by vegetation.

Internally displaced people: people forced to flee their homes due to human or environmental factors but who remain in the same country.

Internet: a group of protocols by which computers communicate.

Involuntary (forced) migration: when people are made to move against their will due to human or environmental factors.

Irrigation: supplying dry land with water by systems of ditches and also by more advanced means.

Knick-point: an indent or abrupt change in the smooth, concave long profile of a river. Knick-points usually mark the location of a waterfall.

Lagoon: a coastal body of shallow salt water, usually with limited access to the sea. The term is normally used to describe the shallow sea area cut off by a coral reef or a bar.

Landfill: a site at which refuse is buried under layers of earth.

Land tenure: the ways in which land is or can be owned.

Lava: molten magma that has reached the Earth's surface. It may be liquid or may have solidified.

Least Developed Countries (LDCs): the poorest of the developing countries. They have major economic, institutional and human resource problems.

Levée: raised bank found along the side of a river channel.

Life expectancy at birth: the average number of years a newborn infant can expect to live under current mortality levels.

Linear settlement: housing that has grown up along a route such as a road. Many settlements show this pattern, since roads offer improved access to employment centres.

Load: material transported by a river. It includes material carried on and in the water (suspended load), material carried in solution (soluble load) and material bounced or rolled along the river bed (bedload).

Long profile: the cross-section of a river from its source to its mouth.

Longshore drift: the movement of material along a beach. When a wave breaks obliquely (at an angle to the beach), pebbles are carried up the beach in the direction of the wave (swash). The wave returns to the sea (backwash) at right-angles to the beach (direction of steepest slope), carrying material with it. In this way, material moves along a beach. Longshore drift may erode beaches and form spits and bars. Attempts to halt longshore drift include the erection of barriers known as groynes.

Loss of sovereignty: this results from the ceding of national autonomy to other organisations.

Low-order goods/services/functions: items or services that are purchased/required frequently (convenience goods), such as milk or bread. People are not prepared to travel far to buy such items and there is no real saving in shopping around.

Magma: molten rock within the Earth. When magma reaches the surface it is called lava.

Managed retreat: the coastline is allowed to retreat (erode) in certain areas where the population density or the value of land is low, so that nature takes its course.

Mangroves: salt-tolerant forests of trees and shrubs that grow in the tidal estuaries and coastal zones of tropical areas.

Mass media: a section of the media specifically designed to reach a large audience. The term was coined in the 1920s with the advent of nationwide radio networks, and mass circulation newspapers and magazines.

Mass migration: a large-scale migration between a particular origin and a particular destination.

Maximum-minimum thermometer: a thermometer which shows both the maximum temperature in a given time period and the minimum temperature in the same time period.

Meander: shaped curve in a river flowing sinuously across relatively flat country.

Megacity: a city with more than 10 million inhabitants. In 1997 there were 18 megacities, 13 of them in developing countries.

Megalopolis: the term used to describe an area where many conurbations exist in relatively close proximity.

Mercalli Scale: a scale of earthquake intensity based on descriptive data.

Microgeneration: generators producing electricity with an output of less than 50 kW.

Migration: the movement of people across a specified boundary, national or international, to establish a new permanent place of residence.

Millionaire city: a city with more than 1 million inhabitants.

Mixed farming: involves cultivating crops and keeping livestock together on a farm.

Monsoon rain: the rainy phase of a seasonally changing pattern of rainfall.

Mouth: the point where a river enters the sea or a lake.

Multiplier effect: the idea that an initial amount of spending or investment causes money to circulate in the economy, bringing a series of economic benefits over time.

Natural hazard: a natural event that puts people, property and livelihoods at risk.

Natural vegetation: the vegetation type that would be found in an area if there was no human impact. For example, the natural vegetation of the British Isles is oak woodland, as that is the species best able to tolerate the temperate climate of that part of the world.

Newly industrialised countries (NICs): nations that have undergone rapid and successful industrialisation since the 1960s.

Nucleated settlement: a settlement in which houses and other buildings are tightly clustered around a central feature such as a church, village green or crossroads.

Optimum population: the best balance between a population and the resources available to it. This is usually viewed as the population giving the highest average living standards in a country.

Organic farming: does not use manufactured chemicals and thus occurs without chemical fertilisers, pesticides, insecticides and herbicides.

Out-of-town location: a location found on the edge of town (often a greenfield site) where land prices are lower, land is available for development, and accessibility to private cars is high.

Outputs: (1) what a farm produces, such as milk, eggs, meat and crops; (2) the finished product or products that are sold to customers.

Overpopulation: when there are too many people in an area relative to the resources and the level of technology available.

Overgrazing: the grazing of natural pastures at stocking intensities above the livestock carrying capacity.

Overland runoff: overland movement of water after rainfall.

Oxbow lake: curved lake found on the floodplain of a river. Oxbows are caused by the loops of meanders being cut off at times of flood and the river subsequently adopting a shorter course.

Package tour: the most popular form of foreign holiday where travel, accommodation and meals may all be included in the price and booked in advance.

Padi-fields: flooded parcels of land used for growing rice.

Pastoral farming: involves keeping livestock such as dairy cattle, beef cattle, sheep and pigs.

Physical water scarcity: when physical access to water is limited.

Plantation: a large farm or estate where one crop is produced commercially, such as palm oil in Malaysia or tea in Sri Lanka. Plantations are usually owned by large companies, often multinational corporations. Many plantations were established in countries under colonial rule, using slave labour.

Plunge pool: deep pool at the bottom of a waterfall. It is formed by the hydraulic action of the water and abrasion by the more resistant rock.

Pollution: contamination of the environment. It can take many forms – air, water, soil, noise, visual and others.

Population density: the average number of people per square kilometre in a country or region.

Population distribution: the way that the population is spread out over a given area, from a small region to the Earth as a whole.

Population explosion: the rapid population growth of the developing world in the post-1950 period.

Population policy: encompasses all of the measures taken by a government aimed at influencing population size, growth, distribution, or composition.

Population pyramid: a bar chart arranged vertically, that shows the distribution of a population by age and gender.

Population structure: the composition of a population, the most important elements of which are age and sex (gender).

Potable water: water that is free from impurities, pollution and bacteria, and is thus safe to drink.

Pothole: small hollow in the rock bed of a river. Potholes are formed by the erosive action of rocky material carried by the river (abrasion), and are commonly found along the river's upper course, where it tends to flow directly over solid bedrock.

Precipitation: water that falls to the Earth from the atmosphere. It is part of the hydrological cycle. Forms of precipitation include rain, snow, sleet, hail, dew and frost.

Preservation: maintaining a location exactly as it is and not allowing development.

Prevailing wind: the direction from which the wind most commonly blows in a region. In the British Isles, for example, the prevailing wind is south-westerly, blowing from the Atlantic Ocean and bringing moist and mild conditions.

Primary product dependent: countries that rely on one or a small number of primary products for all their export earnings.

Primary sector: industries that exploit raw materials from land, water and air.

Pro-natalist policies: policies that promote larger families.

Pro-poor tourism: tourism that results in increased net benefits for poor people.

Processes: (1) the operations that take place on a farm, such as ploughing and harvesting; (2) the industrial activities that take place in a factory to make the finished product.

Product chain: the full sequence of activities needed to turn raw materials into a finished product.

Product stewardship: an approach to environmental protection in which manufacturers, retailers and consumers are encouraged or required to assume responsibility for reducing a product's impact on the environment.

Programme food aid: food that is provided directly to the government of a country for sale in local markets. This usually comes with conditions from the donor country.

Project food aid: food that is targeted at specific groups of people as part of longer-term development work.

Purchasing power parity (PPP): income data that have been adjusted to take account of differences in the cost of living between countries.

Push and pull factors: push factors are negative conditions at the point of origin which encourage or force people to move. In contrast, pull factors are positive conditions at the point of destination which encourage people to migrate.

Pyroclastic flow: superhot (700 °C) flows of ash, pumice (volcanic rocks) and steam moving at speeds of over 500 km/hr.

Quaternary sector: industries using high technology to provide information and expertise.

Quota: agreement between countries to take only a predetermined amount of a resource.

Rainfall: a form of precipitation in which drops of water fall to the Earth's surface from clouds. The drops are formed by the accumulation of fine droplets that condense from water vapour in the air. The condensation is usually brought about by rising and subsequent cooling of air.

Rain gauge: an instrument used to measure precipitation, usually rain. It consists of an open-topped cylinder, inside which there is a close-fitting funnel that directs the rain to a collecting bottle inside a second, inner cylinder.

Ranching: a commercial form of pastoral farming which involves extensive use of large areas of land for grazing cattle or sheep. Ranches may be very large, especially where the soil quality is poor. In the Amazon basin some deforested areas are used for beef cattle ranching.

Range: the distance that people are prepared to travel to obtain a good or service.

Rate of natural change: the difference between the birth rate and the death rate. If it is positive it is termed **natural increase**. If it is negative it is known as **natural decrease**.

Rate of net migration: the difference between the rates of immigration and emigration.

Rationing: a last-resort management strategy when demand is massively out of proportion to supply. For example, individuals might only be allowed a very small amount of fuel and food per week.

Re-use: extending the life of a product beyond what was the norm in the past, or putting a product to a new use and extending its life in this way.

Recycling: the concentration of used or waste materials, their reprocessing, and their subsequent use in place of new materials.

Refugees: people forced to flee their homes due to human or environmental factors and who cross an international border into another country.

Relief food aid: food that is delivered directly to people in times of crisis.

Remittances: money sent back to their families in their home communities by migrants.

Renewable energy: sources of energy such as solar and wind power that are not depleted as they are used.

Resource management: the control of the exploitation and use of resources in relation to environmental and economic costs.

Revetment: a form of hard engineering in which the energy of the waves is absorbed by wooden planks or reflected by concrete structures.

Ria: a drowned V-shaped valley and its tributaries.

Richter Scale: an open-ended scale to record the magnitude of earthquakes: the higher the number on the scale the greater the strength of the earthquake.

River cliff: steep slope forming the outer bank of a meander. It is formed by the undercutting of the river current, which is at its fastest when it sweeps around the outside of the river.

Rural depopulation: population decline in a rural area. It is usually the most isolated rural areas that are affected.

Rural-to-urban migration: the movement of significant numbers of people from the countryside to towns and cities.

Rural-urban fringe: the boundary area of a town or city, where new building is changing land use from rural to urban. It is often a zone of planning conflict.

Saltation: the bouncing of rock particles along a river bed. It is the means by which bedload (material that is too heavy to be carried in suspension) is transported downstream.

Sand dune: a mound or ridge of wind-drifted sand common on coasts and in deserts. In coastal areas, sand is trapped by vegetation, notably sea couch grass and marram grass, to form stable dunes.

Sea level: an average level of the sea, between high water mark and low water mark.

Secondary sector: industries that manufacture primary materials into finished products.

Sector model (Hoyt): a model of urban land use in which the various land use zones are shaped like wedges radiating from the central business district.

Shanty town: unplanned, illegal shelters constructed from cheap or waste materials (such as cardboard, wood, and cloth). Shanty towns are commonly located on the outskirts of cities in developing countries, or within large cities on derelict land or near rubbish tips.

Shield volcano: a gentle, low-angled volcano formed of runny, basaltic lava, e.g. Mauna Loa in Hawaii. The lava is capable of flowing long distances before cooling.

Shifting cultivation: a farming system in which farmers move on from one place to another when the land becomes exhausted. The most common form is slash-and-burn agriculture: land is cleared by burning, so that crops can be grown. After a few years soil fertility is reduced and the land is abandoned. A new area is cleared while the old land recovers its fertility.

Site: the immediate location in which a settlement is located.

Situation: the relative location in which a settlement is located.

Slum: an area of poor-quality housing. Slums are typically found in parts of the inner city in developed countries and in older parts of cities in developing countries. Slum housing is usually densely populated, in a poor state of repair, and has inadequate services.

Social norms: the general attitudes of a population to important issues such as family size, contraception, religion, politics etc.

Soft engineering: any form of coastal (or river) protection that involves the use of natural means, e.g. sand dunes, saltmarshes, tree planting and/or beach replenishment.

Soil: the outermost layer of the Earth's solid surface consisting of weathered rock, air, water and decaying organic matter overlying the bedrock. Soil comprises minerals, organic matter (called humus) derived from decomposed plants and organisms, living organisms, air and water.

Soil erosion: the wearing away and redistribution of the Earth's soil. It is caused by the action of water, wind and ice, and also by unsustainable methods of agriculture.

Solution (or **corrosion**): the process by which the minerals in a rock, notably calcium ions, are dissolved in acid water. Solution is one of the processes of erosion.

Sparsely populated: areas with a low population density.

Sphere of influence: every settlement serves a specific area for a variety of functions such as education, healthcare, shopping and recreation.

Spit: a ridge of sand or shingle connected to the land at one end and in the open sea at the other end. It is formed by the interruption of longshore drift due to wave refraction, river currents, secondary winds and/or changes in the shape of the coastline.

Spring line settlement: a line of settlements at a site where water is available.

Stack: an isolated upstanding pillar of rock that has become separated from a headland by coastal erosion. It is usually formed by the collapse of an arch.

Stevenson screen: a box designed to house weather-measuring instruments such as thermometers. It is kept off the ground by legs, has louvred sides to encourage the free passage of air, and is painted white to reflect heat radiation.

Stump: an eroded stack that is exposed only at low tide.

Subduction zone: the area where one plate slides beneath another, a process known as subduction. One plate (usually a dense oceanic plate) plunges underneath a less dense continental plate. As it sinks, it melts and is destroyed, forming magma, which in turn may reach the surface through volcanic activity.

Subsidy: financial aid supplied by the government to an industry for reasons of public welfare.

Subsistence farming: the most basic form of agriculture where the produce is consumed entirely or mainly by the family who work the land or tend the livestock. If a small surplus is produced it may be sold or traded.

Suburb: outer part of an urban area. Suburbs generally consist of residential housing and shops of a low order (newsagent, small supermarket). Often, suburbs are the most recent growth of an urban area. Their growth may result in urban sprawl.

Supervolcano: a volcano with a Volcanic Explosive Index (VEI) of 8 or more.

Suspension: the movement of fine-grained material, such as clay and silt, in a river by turbulent flow.

Sustainable development: a carefully calculated system of resource management which ensures that the current level of exploitation does not compromise the ability of future generations to meet their own needs.

Sustainable tourism: tourism organised in such a way that its level can be sustained in the future without creating irreparable environmental, social and economic damage to the receiving area.

Sustained pollution: longer-term pollution.

Swash: the movement of material up the beach in the direction of the prevailing wind.

System: a situation in which there are recognisable inputs, processes and outputs.

Terrace: a levelled section of a hilly cultivated area.

Tertiary sector: the part of the economy that provides services to businesses and to people.

Thermal expansion: the increase in water volume due to temperature increase.

Threshold: the minimum number of people necessary before a particular good or service will be provided in an area.

Tombolo: a bar that links an island to the mainland.

Total fertility rate: the average number of children a women has during her lifetime.

Tourist generating countries: countries from which many people take holidays abroad.

Toxic: toxicity is a measure of the degree to which something is poisonous.

Traction: the movement of large-sized materials in a river bed by rolling.

Transnational corporation (TNC): a firm that owns or controls productive operations in more than one country through foreign direct investment (FDI).

Transpiration: the loss of water from a plant by evaporation.

Transport systems: the means by which materials, products and people are transferred from place to place.

Tributary: a stream or river that joins a larger river.

Tropical rainforest: dense forest usually found on or near the equator where the climate is hot and wet. The vegetation in tropical rainforests typically includes a canopy formed by high branches of tall trees providing shade for lower layers, an intermediate layer of shorter trees and tree roots, lianas, and a ground cover of mosses and ferns.

Tsunami: a wave in the water generated by an earthquake. Most tsunamis are generated by submarine earthquakes, volcanic eruptions or landslides.

Underpopulation: when there are too few people in an area to use the resources available effectively.

Underemployment: a situation where people are working less than they would like to and need to in order to earn a reasonable living.

Urbanisation: the process by which the proportion of a population living in or around towns and cities increases through migration and natural increase.

Urbanisation of poverty: the gradual shift of global poverty from rural to urban areas with increasing urbanisation.

Urban land use: a simplified model of the land use (such as industry, housing, and commercial activity) that may be found in towns and cities.

Urban renewal: an urban area where existing buildings are either demolished and rebuilt or renovated.

Urban sprawl: outward spread of built-up areas caused by their expansion. Unchecked urban sprawl may join cities into conurbations.

V-shaped valley: river valley with a V-shaped cross-section. These valleys are usually found near the source of a river, where the steeper gradient means that there is a great deal of abrasion along the stream bed and there is more vertical erosion than lateral (sideways) erosion.

Vent: the channel through which volcanic material is ejected.

Village: a small assemblage of houses; smaller than a town or city and larger than a hamlet.

Volcano: a cone-shaped mountain formed by material (magma, ash and cinders) erupted from below the Earth's surface.

Voluntary migration: when the individual has a free choice about whether to migrate or not.

Waste product: all manufacturing industries produce waste product which has no value and must be disposed of. Costs will be incurred in the disposal of waste product.

Waterfall: cascade of water in a river or stream. It occurs when a river flows over a bed of rock that resists erosion. Weaker rocks downstream are worn away, creating a steep, vertical drop and a plunge pool into which the water falls.

Water scarcity: when water supply falls below 1000 m³ per person a year, a country faces water scarcity for all or part of the year.

Watershed: a ridge or other line of separation between two river basins or drainage systems.

Water stress: a country is judged to experience water stress when water supply is below 1700 m³ per person per year.

Wave: circular or elliptical movement of water near the surface of the sea.

Wave-cut platform: a gently sloping rock surface found at the base of a coastal cliff. It is covered by water at high tide but is exposed at low tide. It is formed by the erosion (by waves) of a former cliff face.

Wave refraction: the way in which a wave changes shape and loses speed as it comes into contact with the sea bed. If refraction is complete, waves break parallel to the coastline. If refraction is not complete, longshore drift occurs.

Weather: day-to-day variation of atmospheric and climatic conditions at any one place over a short period of time. Such conditions include humidity, precipitation, temperature, cloud cover, visibility and wind, together with extreme phenomena such as storms and blizzards. Weather differs from climate in that the latter is a composite of the average weather conditions of a locality or region over a long period of time (at least 30 years).

Wet point site: a settlement with a reliable supply of water in an otherwise dry area.

Xerophyte: a plant adapted to live in dry conditions. Common adaptations include a reduction of leaf size, leaf hairs to trap a layer of moist air, water storage cells, sunken stomata, and permanently rolled leaves or leaves that roll up in dry weather (as in marram grass).

Index